T0244230

QUANTUM DRAMA

QUANTUM DRAMA

FROM THE BOHR–EINSTEIN DEBATE TO THE RIDDLE OF ENTANGLEMENT

JIM BAGGOTT

AND

JOHN L. HEILBRON

OXFORD
UNIVERSITY PRESS

OXFORD
UNIVERSITY PRESS

Great Clarendon Street, Oxford, OX2 6DP,
United Kingdom

Oxford University Press is a department of the University of Oxford.
It furthers the University's objective of excellence in research, scholarship,
and education by publishing worldwide. Oxford is a registered trade mark of
Oxford University Press in the UK and in certain other countries

Published in the United States of America by Oxford University Press
198 Madison Avenue, New York, NY 10016, United States of America

British Library Cataloguing in Publication Data
Data available

Library of Congress Control Number: 2023946262

ISBN 9780192846105

DOI: 10.1093/oso/9780192846105.001.0001

Printed and bound by
CPI Group (UK) Ltd, Croydon, CR0 4YY

For Latha

Contents

ACT IV. PRODUCTIVE INEQUALITIES

About the Authors

Jim Baggott is a freelance science writer. He gained a BSc (1978) in chemistry at the University of Manchester and completed a DPhil (1981) at Oxford University. He worked as a postgraduate research fellow at Oxford and at Stanford University in California, before returning to England to take up a lectureship in chemistry at the University of Reading. After five years of academic life he decided on a complete change of career and worked in the oil industry for 11 years before setting up his own independent business consultancy and training practice. He maintains a broad interest in science, philosophy, and history, and writes on these subjects in what spare time he can find. He was awarded the Marlow Medal by the Royal Society of Chemistry in 1989 and a Glaxo Science Writer's prize in 1992. His book *Mass: The quest to understand matter from Greek atoms to quantum fields*, won the 'Cosmos' Prize for Science Writing in 2020.

John L. Heilbron (1934–2023) was Professor of History and Vice-Chancellor Emeritus at the University of California, Berkeley, honorary fellow at Worcester College, Oxford, and sometime visiting professor at Cornell, Yale, and the California Institute of Technology. He received his AB (1955) and MA (1958) degrees in physics and his PhD (1964) in history from the University of California, Berkeley. His work has won prizes from the History of Science Society (Sarton Medal), the International Academy of the History of Science (Koyré Medal), the Royal Society of London (Wilkins Prize Lectureship), the University of Pisa (Premio internazionale Galileo Galilei), and the American Physical Society and the American Institute of Physics (Pais Prize) 'for his ground-breaking and broad historical studies, ranging from the use of renaissance churches for astronomy, through seventeenth and eighteenth-century electrical science, to modern quantum mechanics'.

'At the next meeting with Einstein ... our discussions took quite a dramatic turn'

—*Niels Bohr*

Prologue

Our drama is an episode in a long story. It has to do with the distinctly human compulsion to divide the world into domains defined by rules and expectations. Such domains, like religion and physics, can accommodate a wide range of human experience. They can also encourage fundamentalists to pursue the illusory goal of a final theory of everything belonging to the domain. Their aim is closure; to offer unassailable proof of finality and to declare that no plausible alternatives are possible. The American theorist Stephen Weinberg made clear what is at stake: 'A final theory will be final in only one sense—that it will bring an end to a certain sort of science, the ancient search for those principles that cannot be explained by deeper principles'.[1]

But the closure of a domain of experience ruled by a particular theory assumes the existence of criteria for unambiguously determining whether a given fact or question lies within the theory's competence. And that, as the long story of Euclidean geometry shows, is not always easy to decide.

The domain of Euclidean plane geometry includes all figures that can be constructed with an unmarked straight edge and compass, and its ruling theory is expressed in a set of axioms assumed to be true from which everything else can be deduced. Greek geometers deduced among much else that it is possible to construct only four fundamental regular polygons. These have 3 (equilateral triangle), 4 (square), 5 (pentagon), or 15 sides. An infinite number of others can be constructed by continual bisection of the sides of any of the fundamentals. On this understanding, no construction in keeping with the axioms and the rule of edge and compass can make any other regular polygon. And yet, only 2,000 years after Euclid, Carl Friedrich Gauss discovered that fundamental regular polygons with bizarre numbers of sides, such as 17, are theoretically possible.

During the same period, many Euclidean geometers believed incorrectly that the rules of their science permitted the construction of a square equal in area to a circle. They lacked the means to prove the task impossible. It took two millennia to show that π is a transcendental number—no algebra can be devised to compute it—and the squared circle is impossible. Mathematicians opened the long-closed domain of Euclidean plane geometry to admit bizarre regular polygons and closed it again to exclude the squared circle.

Our quantum drama centres on the question whether quantum phenomena are like squared circles or bizarre regular polygons when tested by the fundamental definitions and expectations of the 'classical' physics that our principal actors, Niels Bohr and Albert Einstein, learned at school. The analogy to Euclidean geometry is not perfect. Physics does not confine itself to arbitrarily limited ideal types and does not stop to prove the impossibility of one theory before adopting another. Still, we believe the analogy holds. Those who dissent; those who, almost as an obligation, resist the claims of closure advanced by fundamentalists; those who stubbornly search for bizarre regular polygons, can precipitate important discoveries.

Adherents of the classical physics of Isaac Newton declared that by reducing the properties of matter to size, shape, and motion, the rival physics of René Descartes could not give a satisfactory account of gravity. Cartesians replied by attributing to Newton the concept of a gravitational force acting at a distance and rejecting it as unphysical, incomprehensible, and 'occult'. Newton agreed with the objection but claimed that it did not apply. 'That one body may act on another at a distance through a vacuum without the mediation of any thing else & through which their action or force may be conveyed from one to another is to me so great an absurdity that I believe no man who has in philosophical matters any competent faculty of thinking can ever fall into it'.[2] However, he proffered no alternative explanation: 'I have not been able to discover the cause of those properties of gravity from phenomena, and I frame no hypotheses; for whatever is not deduced from the phenomena is to be called a hypothesis: and hypotheses, whether metaphysical or mechanical, have no place in experimental philosophy'.[3]

The discomfort attendant on the admission of non-mechanical, non-local forces into physics eventually led to the introduction of field theories that incorporated elements from both earlier systems. By 1900, the president of the French Physical Society, in opening an international congress of physics in Paris, could proudly proclaim that, once again, 'the spirit of

Descartes hover[ed] over modern physics'.[4] In our drama, Einstein plays the part of Gauss searching for bizarre regular polygons, Descartes rejecting action at a distance as unphysical and unintelligible, and Newton excluding certain sorts of 'hypotheses' by definition. Bohr plays the parts of the Euclidean declaring closure of his domain and the Newtonian willing to admit the unintelligible and non-intuitive in return for a consistent, quantitative account of the facts. As we will see, the instinct to reject action at a distance motivated many of the episodes in our drama.

Absurdities or, as Bohr preferred to call them, 'irrationalities', abounded in the forging of quantum mechanics and, for many physicists, including Bohr and Einstein, in the final product. Relative comfort with the absurd and the ambiguous measured the divide between them. Although never resolved, their disagreement prompted fruitful debates because they never lost their admiration for one another and shared the belief that physicists have a duty to discover and declare the foundations of natural philosophy. Their debates, regarded as 'philosophical' (and therefore irrelevant) by many physicists, did not result in experimental tests during their lifetimes, but their approaches to the problem eventually prompted a series of extraordinary experimental discoveries.

These discoveries, which support practical applications such as quantum encryption, teleportation, and computing, came a half-century rather than two millennia after the declaration, by Bohr's school, of the closure of the relevant theoretical domain: quantum mechanics. The pace of science has picked up since Euclid. Nonetheless, the speed with which Bohr and his school sought to close quantum mechanics, within two or three years of the discoveries on which they based their opinion, is breath-taking. Einstein and his followers objected to their overly hasty surrender of the possibility of alternatives more in keeping with an intuition grounded in classical physics. It was not a question whether light should be pictured as waves or particles, or atoms as solar systems, but of whether the microworld could be pictured at all. With enough time and talent, would it not be possible to show that the claimed failure of strict causality in the microworld was as mistaken as ruling a 17-sided regular polygon out of Euclidean geometry? Why was the move to closure on such fundamental questions so swift? Why was the opposition so ineffectual?

For answers, we must invoke not only the state of science but also the social and cultural commitments and psychological motives of the participants. World War I was a watershed for physics as for other high cultural pursuits of Western Europe. It divided the creators of quantum physics, who would stay close to classical ideas, from the creators of quantum mechanics who, with some easily explainable exceptions, began their careers with their minds full of quantum ideas. The first generation grew up between the Franco–Prussian War of 1870 and World War I, an era of relative peace, stability, confidence, security, and the complacency associated with the Victorian era. The second generation began their careers after experiencing the anxieties and scarcities of the war and a troubled peace, and while confronting threats from the growing political appeal of fascism and communism.[5]

Classical physics, with its emphasis on mechanical or picturable models, the all-encompassing principles of thermodynamics, and everyday notions of space, time, and causality mirrored the political and social stability of its heyday. Quantum mechanics, with its rejection of norms associated with the regime that had brought on the war and its emphasis on the uncontrollable, acausal, uncertain behaviour of the microworld, agreed well with the uncertainties of its development in the 1920s and 1930s.

Already by 1900 classical physicists had reasons to worry about foundations. They had the new experimental discoveries of X-rays, radioactivity, and the electron to absorb and their greatest theoretical achievements, electrodynamics and statistical mechanics, had encountered serious problems. Lord Kelvin, one of the great architects of classical physics, identified these problems as 'clouds' over the future of the discipline. The same analogy was invoked by commentators on the political scene to refer to rising socialism and nationalism, and to colonial and commercial competition, within and between the great powers. But by and large the mood of physicists and the larger society that supported them during Victorian times was optimistic and cosmopolitan. We would be hard pressed to account for a physics of the late nineteenth century distinguished by doubt and a physics of the 1920s marked by certainty.

Many of the worriers entertained the idea that physics aimed not at the fundamental truth of things, but only at an accurate description of phenomena. No doubt 'descriptionism', as we may call the collection of fin-de-siècle instrumentalist philosophies, was no more useful in the everyday work of scientists then than it is now. But it figured promiscuously in general talks and essays, such as the famous address that the mathematician and physicist Henri Poincaré gave at the opening of the International Congress of Physics of 1900. You should not think yourselves above librarians, Poincaré told his surprisingly receptive audience, for your theories are no more fundamental than the principles of a library catalogue and, like it, should be judged by how efficiently it accommodates new acquisitions.[6] The most influential descriptivist for our purposes was the Austrian physicist Ernst Mach, who for a time had Einstein's attention.

Mach taught that science built on and predicted sensory experience and that theory should attempt no more than to connect the input with the output. In its extreme version, Mach's physics denigrated models of the world, like atoms or ethers, and gave no priority to mechanics. All reductionist schemes, all systems for referring sensory experience to the activities of hypothetical sub-microscopic particles, were inherently misleading and inconsistent with sound science. Among those who shared Mach's criticism of atomism was the professor of theoretical physics at the University of Berlin, Max Planck. That was one reason that at the turn of the century he was working at a problem expected to yield to the combined equations of thermo- and electrodynamics without reliance on atomic theories.

The problem concerned the properties of radiation maintained at a constant temperature inside a cylindrical porcelain vessel closed apart from a small hole. This construction made it a good approximation to an ideal 'black body', which can absorb and emit radiation of all frequencies. The setup may seem arcane in theory and far from useful, but it turned out to touch on fundamental questions and to have practical application in the then new industry of electric lighting. When the vessel is heated, its interior glows like a furnace, red, orange, bright yellow, brilliant white, depending on the temperature. Planck's problem was to find a 'radiation law' to describe the colour distribution (the intensity of radiant energy at each frequency) as a function of temperature and frequency. He failed to find one that agreed with experiment until he made what amounted to the assumptions that radiation in the container could exchange energy with its walls only through discrete amounts or *quanta* of energy, and that these permissible amounts had to be proportional to the frequency of the radiation participating in the exchange.

Being a theorist, Planck sought a theory that would allow for the quanta. In what he later described as an act of desperation, he turned to atomistic ideas that gave him what he needed. They not only appeared to ground his theory but also enabled him to deduce reliable information about the size of atoms. With this achievement, Planck contributed to the accumulating evidence supportive of atomism, for example, radioactivity and Brownian motion, and to the vigorous contemporary discussion of its status. Should physicists regard the atomic theory as a mere tool for describing an increasing range of phenomena or a discovery of the real, true, objective state of matter? Theories of radioactive decay and the initial successes of the nuclear model favoured atomism. Paradoxes in the behaviour of X-rays and gas molecules favoured a physics of description, even of contrary descriptions. The confusion of fundamental issues that would confound quantum mechanics were joined long before its invention.

The first two acts of our drama centre on issues brought up in arguments between Einstein and Bohr and their primary followers. Act I covers the generation born before 1900 and runs to the invention of quantum mechanics in 1925/6. The invention was the accomplishment primarily of young men who embarked on their careers after the catastrophic war that made a watershed in physics as in most other sectors of high European culture. Bohr and Einstein belonged to the previous generation. They established the positions from which they began their debate while making the contributions for which they received Nobel prizes in the early 1920s.

Act II opens with the debate prompted by the invention of quantum mechanics, beginning at the fifth Solvay conference held in Brussels in 1927. The Act continues with the consolidation of the 'Copenhagen spirit' (the approach of Bohr's school) and ends with the effort to spread its insights into psychology, biology, and thinking in general. World war then again intervened to change the situation fundamentally. With the mobilization and Americanization of physics, interest in fundamental problems plummeted. Questions that were centrally important to Bohr and Einstein vanished from the agenda.

Discussion revived through opposition to the Copenhagen spirit among mavericks centred at Princeton, where Einstein lived and worked until his death in 1955. By then a few of his leftist followers took on the challenge of reopening the supposedly closed matter of interpretation. Eventually they devised crucial experiments that, as some of them anticipated, would decide against Bohr. This is the matter of Act III, during which Bohr, as an international senior statesman of science, was busy with other things. He died in 1962 before techniques had developed to enable the crucial experiments. These and the surprising developments to which they gave rise occupy Act IV.

The entry into our drama of experimentalists precipitated a shift in emphasis. Although overtly in pursuit of an answer to the eternal question of closure, their efforts exposed a bizarre regular quantum polygon that resided, previously only glimpsed, within the domain of quantum mechanics. It is called entanglement. The practical demonstration of entanglement over large distances sponsored new technological developments that may yet transform our world.

The questions at the heart of our quantum debate remain open. Followers of Bohr continue to hold that quantum theory in its present form is final, and the quantum-mechanical domain closed. Followers of Einstein continue to search for evidence that will undermine the claim to finality or support a Cartesian challenge to the occult action at a distance that entanglement appears to imply. Their opposition, which rests on conflicting ideas about the ultimate purpose of science, has been fruitful and inspirational; full of 'deep human interest' even to Bohr's bulldog Léon Rosenfeld, who had little sympathy for Einstein.[7] It is a mistake, made in many accounts of the debate, academic, popular, past, and recent, to credit one side or the other with following the only true path. The motion was, and continues to be, necessarily zigzag, dialectical, reciprocal.

Galileo informed the world that '[Physics] is written in the language of mathematics, and its characters are triangles, circles, and other geometrical figures, without which it is humanly impossible to understand a single word of it; without these, one is wandering about in a dark labyrinth'.[8] And even with them. The physicist's calculations begin by translating a problem from ordinary language into symbols, continue by manipulating the symbols according to the applicable algebra, and end by translating the results into ordinary language for discussion or experimentation. During the manipulation the symbols may come to relate to ordinary language in new ways unperceived by the algebraist. One way of describing the consequences is confusion. A more fruitful elucidation calls attention to the multivalent content of a symbol, not all of which may be present to the algebraist's mind.

A simple case is 'frequency'. Its root meaning of number of recurrences in a given time is familiar, in both concept and quantity, from the 440 air pulses per second by which an orchestra tunes itself. The same concept is applied to visible light, although its frequency is ten billion times that of sound and the 'ether' that was supposed to carry its vibrations does not exist. Representing both by the same symbol, ν, has obvious perils as well as the potential fruitfulness of ambiguity. Here are two examples. The discovery and clarification of the ambiguity of ν as the frequency with which an electron traverses its orbit and ν as the frequency of the light so generated, were fundamental to Bohr's quantized atom. And the unresolved contradiction between ν as light frequency and ν as measure of light energy was fundamental to posing and elucidating the 'wave-particle duality'.

We give these examples to emphasize the power and advertise the dangers of symbolic representations that, in the end, must be translated into ordinary language. Bohr expressed the situation more perfectly than he realized when, unintentionally substituting Danish metaphor for English idiom, he declared that we are 'hanging in' (for 'dependent on') language. In this uncomfortable posture quantum physicists generated many apparent paradoxes, which prompted Bohr to invent his far-reaching and often-contested doctrine of complementarity. Its chief opponent was Einstein.

Galileo was only half right in saying that no one can understand the book of nature who cannot read its mathematical symbols. We believe, as did Bohr and Einstein, that the foundational concepts in physics can be understood without mastering its argot. All that is required is the attention necessary to acquire unfamiliar ideas. We have used a few mathematical symbols, however, to avoid ambiguity and circumlocution, and occasionally history and brevity have demanded a short equation, like $E = mc^2$. These belong to the rhetoric of our subject just as 'ex nihilo nihil fit' (nothing comes from nothing) does to philosophy, cosmology, and the kitchen. The symbol h (Planck's constant) designates a quantum quantity; phrases like $E = h\nu$ and $\Delta q \Delta p \sim h$ express physical constructs, mathematical relations, and slogans that drive our drama.

In the hope that it might be useful to an understanding of our approach to the drama, we end this prologue with a few words about what each of us brings to our collaboration. Jim Baggott's fascination with quantum mechanics, awakened while an undergraduate in 1975, spilled over to obsession some twelve years later with the discovery that using quantum mechanics is by no means the same as understanding it. He has written several books about it for a general audience, the first published in 1992. An encounter with John led him to suggest a collaboration on *Quantum drama*. We decided to try to write a book that captures the deeper riddles of the subject and humanizes the people who created and struggled with them.

John Heilbron was lucky enough to participate as a graduate student in a project to interview physicists active in the early days of quantum physics and to collect and microfilm their correspondence. He lived in Copenhagen in 1962/3 and met Bohr and other physicists associated with the mainstream interpretations of quantum physics, among them Werner Heisenberg, Pascual Jordan, Oskar Klein, and Léon Rosenfeld. After immersion in the science of early modern times he returned to the twentieth century to write a brief biography of Bohr and, at Jim's suggestion, to collaborate in exploring the sequel to the story of quantum physics that he began to study in Copenhagen 60 years ago.

ACT I

Correspondence
to Complementarity

Mutual Admiration

The Royal Swedish Academy of Sciences, which awards the Nobel prizes in physics and chemistry, advisedly acted neutrally during World War I. When it could not balance a winner from the Allied Powers against one from the Central Powers, it withheld judgement. According to its rules, if it does not find a recipient for a withheld prize during the following year, the prize disappears and the Academy retains the money. In the first post-war year the Academy rewarded two Germans: the prize for 1919 went to Johannes Stark for an experimental discovery made in 1913, and the reserved prize of 1918 to Max Planck for the introduction of the quantum into physics in 1900. They were an ill-matched pair. Planck was a strong supporter of the liberal Jew Einstein; Stark would become a Nazi and a fierce opponent of relativity and quantum theory, the 'Jewish Physics' of Einstein and Bohr.

The Academy would have been oddly insensitive if it had not picked someone from the Allied side for the physics prize of 1920. Its cognizant committee chose the embodiment of the science it admired. He was Charles Edouard Guillaume, the French director of the International Bureau of Weights and Measures, 'for his precision measurements'. And for 1921? Astonishingly, the Academy could not find a worthy recipient. By then Einstein had been nominated by former belligerents on both sides some 45 times in all, and Bohr, a citizen of neutral Denmark, 9 times.

In 1922 the pressure became too great to ignore: Einstein received 17 nominations, Bohr 11. The Academy gave Bohr the prize for 1922 and Einstein the reserved prize for 1921. In their brief explanation of their choice of Einstein, the judges in Stockholm revealed the resistance many physicists then still felt about the direction of their science. Einstein deserved his award, they said, 'for his services to theoretical physics, and especially for his discovery of the law of the photo-effect'. This involves the liberation of electrons from the surface of a metal by the action of light. Not a word about his invention of relativity or his fundamental contributions to quantum theory; his special merit was merely an equation, although one with the simplicity and power of his famous declaration, $E = mc^2$.

Einstein's prize-winning equation, $T = h\nu$, specifies that the energy of motion (the kinetic energy, T) of the liberated electron is proportional to the light's frequency ν. The proportionality factor, h, is Planck's constant. It is a small number, 6.6×10^{-34}, in the system of units used in present-day atomic physics. Its presence made Einstein's equation much more than a simple empirical law of the photo-effect. It had the same value as the 'quantum of action' that Planck had been obliged to include in his formulation of the law of black body radiation in December 1900. Like Planck's formula, Einstein's law was implicated in the deep mysteries of light.

Einstein had derived his law from the assumption that the energy causing the photo-effect came in particle-like bundles ('light quanta'), the energy of each bundle determined by the light frequency according to $h\nu$. That was to reduce the photo-effect to a game of marbles, in which the shooter (the light) has an energy mysteriously proportional to its frequency. It would have been much more reasonable to assume that the effective energy of the light quanta depends on their number rather than their frequency: many quanta should imply high intensity, and thus a large energy irrespective of their frequency. But only light quanta with a frequency above a threshold characteristic of a metal will dislodge its electrons. No wonder the Nobel authorities, and most other physicists, rejected Einstein's theory!

At most the marble game could represent light only in its interaction with the electron, for everyone knew that light behaves like a wave as it travels through space or squeezes through small holes or past sharp edges or reflects from oil slicks. In these cases, its defining mark was 'superposition', the property of overlapping waves to combine their motions. The standard demonstration of this property, which physicists invoked so often in the development of quantum theory that it can be taken as an actor in our drama, dates from around 1700, when the wave theory became competitive with Newton's particulate 'rays of light'. This demonstration, or, better, icon, is the so-called 'double-slit experiment' first described by the English polymath Thomas Young in lectures published in 1807.[1]

In Young's form of the experiment, light from a point source to the left of the screen (not shown) slips through two tiny holes or narrow slits A and B that serve as secondary light sources, as shown in Fig. 1(a), which is adapted from Young's *Course of lectures on natural philosophy and the mechanical arts*. The two beams of light they produce superpose or interfere at every point to the right of the screen. The solid curves indicate loci of maximum upwards swing of the carrying medium which define the wavefront and whose motion propagates the waves. The dotted curves indicate the maximum downward swing. Where two maxima of the same kind overlap,

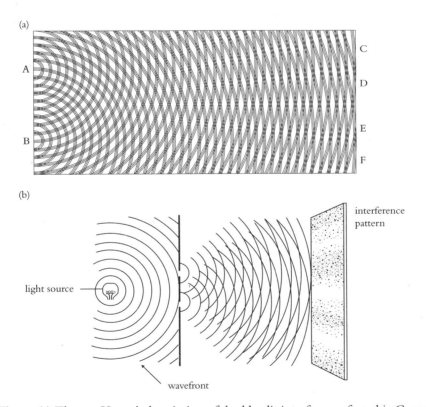

Fig. 1 (a) Thomas Young's description of double-slit interference from his *Course of lectures on natural philosophy and the mechanical arts* [1807], Kelland, ed. (1845). Coherent light passing through two holes or narrow slits at points A and B interferes to produce fringes at points C to F. This is shown more clearly in (b).

the swing is doubled and the light is bright; where two opposite maxima overlap, the swings annul one another and the light goes out. Fig. 1(b) shows a modern version of the experiment with an interference pattern consisting of alternating light and dark fringes produced by the superposed beams on a screen or photographic plate parallel to the plane of the double slits.

Young observed that both slits must be open to produce the interference phenomena just described. If one is closed, light coming through the other produces a 'diffraction' pattern on the screen caused by superposition of beams propagating from the slit's edges. It is not possible, however, to build up the two-slit interference pattern from light alternately diffracted from one slit and then the other, as would be expected if light were a stream of particles. A similar situation arises when light strikes a tiny obstacle. The interference that occurs beyond it can be eliminated by cutting the light from only one side or edge of it, 'a crucial experiment', as one of Young's defenders rightly remarked.[2] With his principle of superposition, Young could explain all this and more complicated phenomena like the colour of thin plates, which, as he put it, presented the particulate theory with 'difficulties absolutely insuperable'.[3]

Although by the end of the nineteenth century the wave theory could explain all sorts of optical and electromagnetic phenomena, it was running into difficulties that in its turn would prove insuperable. No one had managed to find a plausible account of the entity supposed to carry the waves. Since light propagates in 'empty' space, this light-bearing entity (the 'luminiferous ether') could not be a material substance like the air that spreads sound or the water that carries swells. Young regarded the discovery of the omnipresent ether as more important than that of the wave theory that led to it, for it opened 'an ample extension of our views of the operations of nature, by means of our acquaintance with a medium, so powerful and so universal, as that to which the propagation of light must be attributed'.[4]

The medium could not sustain its extensive responsibilities after light became an electromagnetic phenomenon, and Einstein felt obliged to eliminate it from his theory of relativity. That was in 1905, in a paper published three months after he applied the light quantum to the photo-effect. He thereby coupled relativity and radiation theory by eliminating the ether from both. It was much easier to dry up the ether than its waves, however, and Einstein did not insist. His light quantum was merely a 'heuristic hypothesis', useful where he used it, but powerless against the many phenomena adequately treated by Young's theory of light.

When Einstein's supporters put him forward for election to the Berlin Academy of Sciences in 1913, they felt obliged to apologize for his crazy rejection of light waves. 'It should not be held too much against him that occasionally his speculations overshoot the mark, as in his hypothesis of light quanta'.[5] Among the strongest and most persistent opponents of his hypothesis was Bohr. He condemned it not only because it could not account for the effects arising from 'superposition' but also from elementary logic. The light quantum contradicted itself. Its definition required a frequency, Bohr said, and frequency is a concept associated with waves.

Bohr's Nobel citation reads: 'for his investigations of the structure of atoms and of the radiation emanating from them'. The citation might suggest that Bohr's investigations overlapped Einstein's, since both involved interactions between light and matter. Not at all. Bohr won his spurs by ignoring the physical nature of radiation. He was concerned to calculate the frequencies or colours of light absorbed and emitted by various atoms, their 'spectral lines'. In the simplest case, the hydrogen atom, the spectrum of a set of its discrete emission frequencies spanning the visible region could be expressed by a simple empirical formula deduced in 1885 by a numerological mathematician named Johann Balmer. His formula, written for frequencies (he used wavelengths) is almost as simple as $T = h\nu$. It is $\nu_n = R\left(\frac{1}{4} - \frac{1}{n^2}\right)$, where the integer n takes values greater than 2. This says that the frequencies of the lines in the Balmer series increase in a simple progression to a maximum of $\frac{1}{4}R$, where R is a constant. This very same R appeared in Balmer-like expressions for the spectral series of other elements. Evidently, it was an important quantity, hinting, as did Planck's constant h, at something much deeper than a number turned up by experiment.

The basic structure of the Bohr atom was a swarm of electrons arranged in concentric circular orbits around a central nucleus, like the particles in Saturn's rings or, in its simplest form, the Earth–Moon system. Bohr took this 'planetary' system as a model for hydrogen, the simplest of all chemical atoms, consisting of a positively charged nucleus and a single orbiting negatively charged electron. In such systems the distance r of the Moon from its planet or of the planet from its central star bears a simple relation to its orbital frequency ω, Greek 'omega', the number of times the planet completes its orbit in a unit of time. This relation, $\omega^2 \propto 1/r^3$, is Kepler's third law of planetary motion ($\propto$ signifies 'is proportional to'). It is two centuries older than Young's theory of light.

Why should anyone have been surprised that the planetary model did not apply without restriction to the atomic domain? Gravity regulated the solar system, some thousand billion billion times the size of an atom, and it permitted planets to circulate at any distance from the Sun, whereas all the atoms of a given element in their normal state have a fixed size, supposed to be the same wherever they are in the universe. (This supposition also has an ancient pedigree: it is an application of Newton's third rule of philosophizing.[6]) What fixes this size, supposing that the partial analogy to a solar system holds? Bohr suggested a quantum relation like Einstein's and Planck's: $T = \frac{1}{2}h\omega$, where T again signifies kinetic energy, but now of the orbiting electron rather than an electron dislodged by its interaction with light.

A relation of this type, which involves Planck's constant h, can be called a 'quantum rule' and one without it, like Kepler's third law, a 'classical rule'. In an orbit characterized by $T = \frac{1}{2}h\omega$ and Kepler's third law—a discomforting mix of quantum and classical rules—the atom is fixed in size but, in Bohr's theory, cannot communicate with the outside world. To get it to 'speak'—to absorb and emit light—Bohr had to endow it with excited states in which the electron revolved in orbits larger than its lowest-energy state, called the 'ground state'. Bohr characterized these larger orbits by an easy though arbitrary generalization of the ground-state relation: $T_n = \frac{1}{2}nh\omega_n$, where n is a whole number and T_n and ω_n are the kinetic energy and orbital frequency of the nth orbit. Geometrically, n measures the diameter of the circular orbit or the major axis of an elliptical one. Eventually other 'quantum numbers' would come into play related to other features of the electron orbit such as its eccentricity.

An electron describing an orbit defined by the generalization $T_n = \frac{1}{2}nh\omega_n$ remains speechless as long as it stays in the nth orbit. Its state, in Bohr's terminology, is 'stationary'. Bohr allowed his atom to break its silence, to absorb or emit a light wave of sharp frequency ν, only through an abrupt, unanalysable transition of its electron from one orbit to another. Such a *quantum jump* would become the icon of Bohr's model, as puzzling and disagreeable as Einstein's light quantum.

Bohr's process of light emission anticipated the rationale of Einstein's Nobel prize, for, just like the judges in Stockholm, Bohr invoked the formula, while rejecting the theory, of light quanta. He assumed that the energy emitted in the jump was $h\nu$, ν being the frequency or colour of the emitted light. Together with the postulated relation between the kinetic energy T and orbital frequency ω, the specification that the emitted energy is $h\nu$ had a great and unpleasant consequence. It implied that the frequency of the emitted light ν does *not* equal the frequency ω of the electron motion that produces it. That violated a basic assumption of the wave theory of electromagnetic radiation.

It is useful to see how this catastrophe occurs to appreciate Bohr's basic strategy in developing physical principles. Every quantum jump involves two orbits and consequently two different ω's, but gives rise to a single ν. We can easily demonstrate the catastrophe from the observation that forces like gravitation and electrodynamics, which diminish in strength as the square of the distance between the interacting bodies, have the agreeable property that the total energy E of the revolving planet or electron is the negative of its kinetic energy, $E = -T$. (This relation depends on the convention that a free electron, one removed to 'infinity' from its nucleus, has zero 'potential' energy; since it takes energy to remove it from bondage, its energy when bound is negative relative to this arbitrary zero.)

Let the electron jump between state n and a lower-energy state m. It becomes more tightly bound and so loses energy in the amount $T_m - T_n$. The energy lost is emitted as $h\nu$. For good bookkeeping, we must label this frequency—that of the emitted light—by the quantum numbers of both the orbits involved in the electron jump. And so, Bohr arrived at $h\nu_{mn} = E_n - E_m = T_m - T_n = \frac{1}{2}h\left(m\omega_m - n\omega_n\right)$, a most peculiar, unenlightening, and off-putting formula. Fig. 2 illustrates the relationship between Bohr's electron orbits and three prominent visible emission lines in the Balmer series, denoted H_α (at the red end of the spectrum), H_β (a blue line), and H_γ (violet), corresponding to quantum jumps to $m = 2$ from $n = 3$, $n = 4$, and $n = 5$, respectively.

Experiments done in 1913 and 1914 showed that the frequencies ν_{mn} for spectral lines emitted by hydrogen and ionized helium came out exactly as anticipated by Bohr's odd calculus. Only the bold could be pleased. Einstein immediately judged the theory to be 'an enormous achievement'.[7] It indirectly supported the light quantum by undermining the classical rule that oscillating electrons produce electromagnetic radiation at precisely the frequency (and multiples of the frequency) of their motion. No doubt Einstein also appreciated that Bohr's method revealed the universal

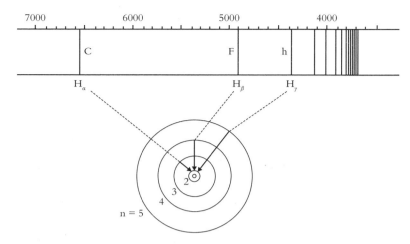

Fig. 2 The spectrum of atomic hydrogen in the upper part of the figure appears in *Lærebog i physik* (1910) by Christian Christiansen, Bohr's physics teacher at the University of Copenhagen. Wavelengths are recorded in Ångströms (tenths of a billionth of a metre) along the top. The lower part of the figure shows how Bohr related the characteristic spectral lines H_α, H_β, and H_γ to quantum jumps between electron orbits.

significance of the constant R. Multiplying both sides of the Balmer formula by h and reading the result as an energy equation, Bohr could infer that $T_n = hR/n^2$.

That made possible a demonstration that ranks high in the poetry of physics: R, like h, is a number of universal significance. Using the postulate $T_n = \frac{1}{2}nh\omega_n$ and Kepler's law applied to an electron of charge $-e$ and mass m_e orbiting a hydrogen nucleus with the balancing charge $+e$, Bohr calculated the kinetic energy T_n in terms of the electron charge e, its mass m_e, Planck's constant h, and the quantum number n. These are all fundamental physical constants apart from n; but the n's cancelled, and Bohr ended with the magnificent result $R = 2\pi^2 m_e e^4/h^3$. With the values of these physical constants available to him, Bohr calculated R for the hydrogen atom that came within 0.06% of the experimental measurement. By relating R accurately to this strange string of constants, Bohr convinced many physicists that there was something to his miscegenated atomic model.

We mentioned that Bohr turned the off-putting discrepancy between orbital and radiation frequencies to his advantage without worrying about its potentially destructive implications for the wave theory of light. But the wave theory worked very well within its domain. Bohr consequently imposed a condition that would help and haunt quantum physicists for decades. At some limit atomic quantities calculated by quantum rules must become equal to similar quantities calculated by classical rules. At the outermost fringes of an atom, where n is very large, the frequencies ω of adjacent orbits are almost the same. Bohr found that by requiring this common ω to equal the frequency ν of the radiation emitted in a jump between the orbits, he recovered his formula for R. He elevated this coincidence into a fuzzy demand that he called the 'correspondence principle'. At appropriate, high-energy limits the *numerical* value of an observable quantity should be the same whether calculated using classical or quantum rules.

Einstein's special theory of relativity reduces mathematically to ordinary physics for situations in which all the speeds occurring are much less than the speed of light. It is alluring to think that something similar occurs between quantum and classical mechanics in the limit of large quantum numbers, where the effects of the small value of h can be neglected. But a major difference obtrudes. The speed of light, the figure of merit in the relation between relativity and classical mechanics, is a concept of classical physics, whereas the quantum is entirely foreign to it. Whether classical mechanics can be considered a limiting case of quantum physics is still contested. In Bohr's view, it cannot.

That Bohr could make essential progress with his fundamentally ambiguous correspondence principle exemplified his peculiar genius. Most quantum physicists could not make it work. The professor of theoretical physics at the University of Munich, Arnold Sommerfeld, who developed much of the early machinery of the quantum theory of electron orbits, expressed perplexity as well as admiration at what Bohr could do with what Sommerfeld called the 'magic wand' of correspondence.[8]

In the middle of World War I, Einstein had taken time off from preaching pacifism and perfecting relativity theory to make a decisive contribution to the quantum theory of the atom. His intervention had the unintended consequences of showing Bohr where to apply the correspondence principle most effectively and of introducing an idea that Einstein himself would later regret. This hostage to fortune was *probability*.

When an assemblage of atoms bathes in radiation of sufficiently high frequencies, electrons in these atoms jump from one state to another. At equilibrium, upward jumps from a state m to a higher state n must equal, on average, downward jumps from n to m. Einstein assumed that the rate of depopulation of the lower state must be proportional jointly to the intensity of the radiation and the probability of the upwards jump. He took the corresponding rate of depopulation of the upper state to have two sources: a spontaneous downward jump and one 'stimulated' by the radiation. This 'stimulated emission', a purely theoretical concept that Einstein invented, has had practical applications more revolutionary than the mass-energy equivalence of relativity, for light amplification by the stimulated emission of radiation is the principle of the laser. Einstein's purpose in introducing probabilities was to enable a new, more satisfactory derivation of Planck's radiation law, which indeed does follow from equating the numbers of upward and downward jumps. From the point of view of the correspondence principle, Einstein's treatment brought in a new set of atomic quantities: probabilities for jumps (or, less picturesquely, transitions) between two states which of necessity are labelled by two numbers.

What Einstein had in mind by transition probabilities is not easy to say. He likened the probability of spontaneous emission to that of radioactive decay, but he considered it a weakness that 'the time and direction of the elementary processes are given over to "chance"'.[9] Most physicists had accepted that radioactive decay occurs spontaneously entirely at hazard, the signature of chance and randomness. Although Einstein had introduced the concept of probability in his analysis, he later explicitly rejected the assimilation of light emission to radioactivity. His distaste for the idea prompted his famous quip that he would rather work in a gambling den than at physics if he had to accept the notion that electrons can jump whenever and wherever they please.[10]

With all this we have omitted what Einstein regarded as the most important result of his analysis: the endowment of the light quantum with an essential property of a particle. In previous considerations he and others had attributed only energy to it. Now, to secure the equilibrium between radiation and molecules that he needed to obtain Planck's formula, he had to allow a light quantum to carry directed momentum: it must move (at the speed of light) in a specific direction in space. The molecules must come to heat equilibrium under the combined processes of mechanical collisions, resistance to their motion presented by the radiation, and emission and absorption. The collisions involved exchanges of momentum; so, therefore, must the radiative processes. Einstein took considerable pains to prove this plausible assertion. Henceforth, a light quantum of energy $h\nu$, whether it existed or not, had a directed momentum $h\nu/c$.[11]

By temperament and research trajectory Bohr was a reformer of quite a different stamp from Einstein. Bohr had no problem granting electrons full freedom to jump as they pleased and accepted transition probabilities in the radical sense they had in radioactive decay. With the help of the correspondence principle and a student who became an essential collaborator, Hendrik Kramers, he was able to associate Einstein's transition probabilities with the intensities of spectral lines emitted in jumps between elliptical orbits. In order not to make a mystery of it, we mention that the association employed a standard technique for decomposing the electron's elliptical motion into a sum of oscillations at frequencies equal to integral multiples of the orbital frequency (that is, Fourier analysis). But when this and other sophisticated versions of classical mechanics failed to support correspondence in complicated cases, Bohr cheerfully entertained violent reworkings or replacements of classical ideas. By 1920 he was contemplating slaughtering so sacred a cow as the conservation of energy.

Behind this Viking boldness stood a line of Danish thinkers whose work Bohr had studied with the professor of philosophy at the University of Copenhagen, Harald Høffding. In his pertinent teaching, Høffding held that in time every physical theory would hit a wall that could be surmounted only by a step that appeared irrational from within the theory. It was the business of the true scientist to seek out, aggrandize, and overcome these impasses. Bohr welcomed the planetary nuclear atom as a research site because it accounted well for certain phenomena under some classical rules while failing utterly to remain stable under others. He leapt over the impasse by introducing 'stationary' states, quantum jumps, and an irrational relation between the colour of the light observed and the motions of the electrons responsible for it.

He fully recognized the inconsistencies in his creation but held fast to them all with the help of vague wordplay. Using a terminology that would dimly echo through physics for more than a century, he characterized the classical quantities defining the orbits as merely 'symbolic'. They served as stand-ins for the quantum quantities that he expected to replace his

classical-quantum hodgepodge. To identify these quantities and their relations he analysed the orbital motions of the electrons in ever greater detail
in the hope that the correspondence principle would point the way to
translate classical concepts into a quantum language that could then develop
independently.

The arena for his struggle was the Institute for Theoretical Physics that
he established in Copenhagen in the early 1920s, with public and private
financing. In it he installed Kramers, a graduate of the celebrated Dutch
school of classical physics at the University of Leiden, and a set of brilliant ambitious young men eager to make their reputations by solving 'the
quantum riddle'. Circumstances were ripe for incubating a new natural
philosophy. The war had laid waste to many traditional assumptions and
unrest in Europe continued the destruction. Bohr directed the hunt for a
consistent atomic physics with incandescent energy and made his institute
a hothouse for fast-growing hybrids of quantum and classical ideas. And he
had the persuasive power of the purse, the offer of stipends in stable Danish
kroner when runaway inflation in Germany was making life more insecure
than usual for young academics.[12]

He recruited two of his jejune geniuses, Wolfgang Pauli and Werner
Heisenberg, during lectures he gave at the University of Göttingen in the
summer of 1922. The pair were students of Sommerfeld, who had introduced new quantum numbers to extend the scope of the Bohr atom to
elliptical orbits and spectra more complicated than that of hydrogen. Sommerfeld had written the first textbook of quantum physics, *Atombau und
Spektrallinien* (*Atomic structure and spectral lines*), in which he sang of the
'atomic music of the spheres' and transcribed it into formulas featuring his
several quantum numbers. The first edition of *Atombau* appeared in 1919,
a year after Pauli matriculated. Pauli quickly mastered both it and the rapidly
expanding new material that he helped Sommerfeld stuff into subsequent
editions. As Sommerfeld's assistant, Pauli had the job of helping to instruct
Heisenberg, who matriculated in 1920 and immediately began the quest
for quantum harmonies. Heisenberg was an expert musician and was soon
manipulating atomic music as well as he could play the piano.[13]

The brilliant youngsters were exotic sprigs of academic families. Pauli's
father taught biochemistry at the University of Vienna, Heisenberg's Byzan-
tine Greek at the University of Munich. The elder Pauli numbered Mach,
the physicist-philosopher whose positivist, anti-metaphysical approach Einstein had once admired, among his close friends. Mach reciprocated the

feeling strongly enough to stand as godfather to the younger Pauli at a baptismal ceremony he regarded as nonsense. Mach's magic proved stronger than the priest's; Pauli made a poor Catholic but a good positivist. He was an even better mathematician and physicist. By the time he left the gymnasium he was able to write an encyclopaedia article on relativity, taking on an assignment Sommerfeld had originally accepted. Einstein was astonished that a man not yet 20 had mastered a theory that then still stumped most professional physicists.[14] And Bohr was astonished when Heisenberg, at 21, had the knowledge and confidence to question him in public before a hundred scientists who had come to hear him lecture in Göttingen.

An unlikely source of Heisenberg's and Pauli's self-confidence was the chaotic situation in Munich when they entered the university. The end of the war brought anarchy to the city, socialist and communist governments, political assassination, successive red and white terror, and food shortages. Heisenberg and his fellow boy scouts (the 'Pathfinders') slipped through blockades to procure food for their families and guided partisans of the side they favoured into and out of the city. Nonetheless, life in Sommerfeld's institute went on as usual, or, rather, more feverishly than before. Thrown in on themselves, the able and dedicated students did nothing but physics, and some almost nothing but quantum physics. Heisenberg got a mediocre grade on his doctoral examination because he could not answer straightforward questions about ordinary physical phenomena.[15] After graduating at the research front armed with Sommerfeld's techniques, he and Pauli each spent time at Göttingen as assistants to the professor of physics, Max Born, to learn the more sophisticated mathematics practised there.

The professor and his two assistants, all future architects of quantum mechanics, were peculiarly ill-assorted. Born, fastidious, Jewish, as much mathematician as physicist, often suffered from self-doubt. Pauli, cocky, Jewish by descent, unobservant Catholic through his father's conversion, was known as the 'Scourge of God' for his pitiless criticism of himself and others. Heisenberg, Lutheran, boyish, sportive, musical, and competitive, harboured no doubts at all. Neither assistant liked Göttingen, though for opposite reasons. Heisenberg condemned it as a 'desolate hole' because it lacked wholesome pathfinders. Pauli regarded it as provincial because it lacked unwholesome nightlife. Even so he found enough to do to keep him in bed until late in the day. He often missed Born's morning lectures at which he was supposed to assist but made a special effort to attend Bohr's presentation of quantum physics.[16]

Bohr's Göttingen lectures were an astonishing performance, the high point in the career of the planetary atom. To general amazement, he described the individual orbits (assumed elliptical) of all the electrons in all the atoms in the periodic table of elements and claimed that his assignments had the approbation of the correspondence principle. The description called on two quantum numbers, n, which measures the energy and major axis of the orbits, and k, which measures angular momentum and orbit eccentricity. A fold-out plate that accompanied the printed version of the lectures depicts these structures, the largest, of radium, including every one of its 88 orbits (see Fig. 3).

This masterpiece of the symbolic method derived only rhetorically from the correspondence principle. Bohr made the assignments of n and k largely from chemical and spectroscopic evidence and, above all, from symmetry considerations bordering on numerology. The symmetry required that an unoccupied space in the periodic table, which chemists had reserved for a rare-earth element, had to be chemically like zirconium. Shortly before the Nobel ceremonies of 1922, workers in Bohr's new institute in Copen-hagen confirmed that element 72 had the properties Bohr had predicted. It was named *hafnium* after the Latin name for Copenhagen. It could have been found any time anyone chose to look. It is no rarer than gold. Bohr

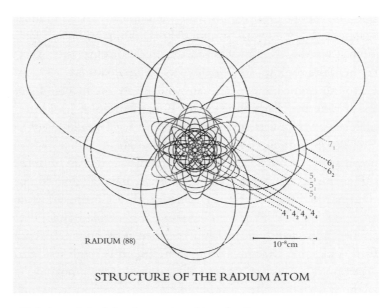

STRUCTURE OF THE RADIUM ATOM

Fig. 3 Bohr's model of the electronic orbits of radium.

announced the detection in his Nobel lecture, at once confirming his concept of the atom and the judiciousness of the members of the Nobel prize committee.

Einstein was unable to join Bohr at the altar in Stockholm for the good reason that he was lecturing in Japan. The trip was protective as well as eye-opening. It allowed Einstein to escape the rampant anti-Semitism directed against him and other prominent Jews in Berlin that climaxed in June 1922 in the assassination of his friend Walther Rathenau, who as foreign minister was trying to find an equitable solution to the reparation payments imposed on Germany at the end of World War I. The far right killed him for being Jewish, wealthy, internationalist, cultured, and decent. Einstein suffered from the same defects, apart from wealth. The confirmation of a romantic prediction of his relativity theory—the bending of starlight observed during a total solar eclipse—had gained him a notoriety that he feared could threaten his life.

A month before Rathenau was gunned down, Einstein had accepted a call to join the new International Commission for Intellectual Cooperation of the League of Nations. He would be the only German member of an entity intended to help revive pre-war relations among the nations. He resigned almost immediately (he would rejoin later) on recognizing that the additional publicity, especially the selection of a pacifist left-leaning Jew with Swiss citizenship as Germany's representative, would be very dangerous for him.[17]

Bohr too was Jewish though, like Einstein, he had no use for religious belief. He was thoroughly cosmopolitan and well-read in German and English, as well as in Danish literature. His institute in Copenhagen was an international retreat. In addition to culture and science, Bohr and Einstein had in common a fierce commitment, even a feeling of responsibility, to the world of ideas. At their first meeting in Berlin in 1920, they immediately kindled (the word is not too strong) an enduring friendship fuelled by admiration for one another's human qualities and intellectual power.

Einstein certainly was smitten. How wonderful that humanity could boast a Bohr! 'He is just brilliant'.[18] And wonderfully sensitive. Bohr had worried that he might receive the Nobel prize before Einstein. The decision to award Einstein the unassigned prize of 1921 cancelled Bohr's concern. It is 'my greatest honour and joy', he wrote Einstein, that they would receive their prizes simultaneously. Einstein: 'Dear Bohr, or rather dearest

Bohr', that you should have worried about precedence! 'I can say without exaggeration that your letter delighted me as much as the Nobel prize'. Another bit of reading that delighted Einstein during his travels in Japan was Bohr's Göttingen lectures. The naming of all the orbits of all the electrons in all the atoms in the universe recognized by Earth-bound chemists 'increased my admiration for your genius'.[19]

An Honourable Funeral

Precisely a week before and approximately 4,000 miles away from the Nobel ceremonies of 1922, Arthur Holly Compton, Professor of Physics at the Washington University in St Louis, announced results to the annual meeting of the American Physical Society that in five years would bring him too to Stockholm. For years he had been trying to understand the patterns created by scattering X-rays—a high-frequency form of light—from the surface of graphite. He had at last concluded that an X-ray collides with an electron in graphite much as one billiard ball bounces off another. He had tried different explanations, but only by assuming that the X-ray behaves as a particle could he account quantitatively for all the facts. Although the Compton effect differs essentially from the photo-effect in that the X-ray does not lose all its energy in the interaction, Einstein's notion of the light quantum applies to both.

Einstein was delighted. Here was a confirmation of the light quantum worthy, as he proposed, of a Nobel prize.[1] He had been desultorily devising experiments to decide between the wave and particle descriptions of radiation and a model for endowing his light quanta with the wave-like properties they would need to account for all the old experiments that buttressed the wave theory. He had tried out these ideas privately in 1921 before a favourite critical audience consisting of two professors of physics at the University of Leiden, Hendrik Antoon Lorentz and Paul Ehrenfest. Their opinions counted.

Lorentz was the greatest living architect of classical physics as well as a living classic himself. 'Everything that emanated from his supremely great mind was as clear and beautiful as a good work of art . . . For me personally [it is Einstein writing] he meant more than all the others I have met on my life's journey. Just as he mastered physics and mathematical structures, so he mastered himself, with ease and perfect serenity'.[2] We are told that whenever Einstein had some new theory to hand, he wanted to know 'what Lorentz would say about it'.[3] The admiration went both ways. When Lorentz decided to retire from his professorship in 1912, he wanted Einstein as his successor. But by then Einstein had committed himself to the Polytechnic in Zurich at which he had been a student a decade earlier. Lorentz cast around for a substitute and fished up Ehrenfest, who would become an intimate friend of both Einstein and Bohr.

Ehrenfest had an unusually perceptive and clear mind (when it came to physics) and exceptional human qualities. But he was not the perfectly balanced human being he replaced. On the contrary, he was a great rarity in Leiden, an excitable, restless, deracinated, bipolar, atheistic, vegetarian Viennese Jew. When Lorentz's brilliant choice fell on him, he was living with his wife Tatyana Afanasyeva, a Russian mathematician, in St Petersburg and vainly seeking an academic job. Because he had not bothered to qualify ('habilitate') as a university lecturer, he could not expect to secure a professorship in Germany. But in one bound he succeeded to the most distinguished chair in theoretical physics in the world. Einstein thought the gigantic jump deserved. Ehrenfest was fiercely honest, independent, critical, penetrating, and, above all, a real physicist, 'one of the few theoreticians whom the prevailing scourge of mathematics has not robbed of reason'.[4]

Einstein's new ideas about the light quantum did not pass the scrutiny of the hypercritics in Leiden. Lorentz objected that Einstein wanted to reduce light to a 'ghost field' that carries no discernible energy but merely measures by the intensity of its impotent ripples the likelihood of the presence of a light quantum. He and Ehrenfest found flaws in the concept of ghost waves and in the theory of the experiment Einstein had proposed to rule out real ones.[5] They did not persuade him to give up the ghost although, he admitted, his efforts to make quanta 'comprehensible' had so far failed.[6] Soon Ehrenfest proved to him that the proposed experiment could not be decisive.[7] Einstein emerged from their criticism 'a little more prudent and one hope poorer', but no less fertile in concocting experiments.[8] He continued to work on his knock-out blow with the help of a cooperative experimentalist who did indeed find the desired result and could have found any result desired, for he faked his experiments.[9]

Bohr and Einstein would meet from time to time in Leiden at Ehrenfest's invitation, notably in December 1925, at a grand celebration of the jubilee of Lorentz's doctorate. By then the problems of quantum theory had reached a

Bohr and Einstein met from time to time in Leiden at Ehrenfest's invitation. This photograph was probably taken at Ehrenfest's home at Witte Rozenstraat 57, ca. 1925–1930. Restoration of original negative and print by William R. Whipple.

new high and Ehrenfest set aside time for wide-ranging discussions between the only two men in the universe who, in his opinion, knew that 'completely radical new concepts' were needed to save physics.[10] Having set forth the terms of engagement (the paladins were to stay with him and allow him to control their schedules), Ehrenfest confided that despite his relative nothingness, he always felt that they encouraged him to pursue his own projects, 'whereas contact with other theorists totally discourages me'.[11]

Einstein responded as Galileo did when comparing himself and Kepler to eagles and the run of philosophers to starlings: 'I am not surprised by what you say about the contrast between Bohr and me and the other theorists in our appreciation for your work. There are people who worry about foundations (*Prinzipienfuchser*) and virtuosi. All three of us belong to the first type and have (at least the two of us) little gift for virtuosity'. As an example of an accomplished virtuoso Einstein mentioned Born.[12] He might also have picked Sommerfeld, who preferred definite calculations to hazy exploration, a dependable baton to conduct atomic music rather than the magic wand of the correspondence principle.

By the time of Lorentz's jubilee, confirmation of Compton's billiard-ball collisions between X-rays and atomic electrons had forced Bohr to abandon a last bold manoeuvre against the light quantum. This had grown from ideas supplied by Compton's compatriot, John Slater, then a recent Harvard PhD, who had devised a way to reconcile the wave with the particle. He supposed that notwithstanding its planetary structure, an atom can also be represented as a collection of vibrating strings. The piano thus associated with a given stationary state or stable orbit can sound only notes with frequencies and intensities corresponding to the transitions available to the atom in that state. In another state the piano would sound another complicated chord. The radiation absorbed or emitted by a collection of atoms would be analogous to the sound produced by an immense orchestra of pianos. This concert was no more real, however, than the instruments that played it. Or, to speak almost plainly, the radiation was only 'virtual' and could do no more than induce a certain probability of a quantum jump in the atoms attuned to it. The appropriate exchange of light quanta followed or accompanied the stimulated transitions.

When Slater and his ideas reached Copenhagen around Christmas 1923, the quantized planetary atom, which Bohr had presented so effectively 18 months earlier in Göttingen, was trembling under the scrutiny of the many physicists drawn to it by his lectures and Sommerfeld's *Atombau*. The challenges came primarily from the spectra of atoms more complicated than hydrogen and from the splitting of spectral lines emitted by atoms exposed to electric and magnetic fields.

The Bohr–Sommerfeld model had not fulfilled its promise. Its astonishing quantitative success for a system of one electron orbiting a single nucleus could not be duplicated for more complicated structures. Pauli tried his strength on the hydrogen-molecule ion (two hydrogen nuclei bound together by one electron, H_2^+) for his doctoral dissertation and failed to calculate the energy levels nature had chosen. He felt his failure deeply, in contrast to Heisenberg, who cheerfully wrote to him that work in Göttingen proved that no reasonable model of the helium atom (in Bohr's theory a doubly charged nucleus with two orbiting electrons) could be made to conform with experiment. 'All models of helium so far proposed are as false as the entire atomic physics'.[13]

The Bohr–Born group tried all sorts of expedients—unmechanical forces, electronic double-dealing ('two-valued') electrons, half-integer quantum numbers—Unsinn, Schwindel, Schimmel, nonsense, trickery, blight. 'Schimmel' became a term of art expressing the 'mouldiness' of the outdated orbital model. By June 1923, Pauli, then in Copenhagen learning to do physics Bohr's way ('of an entirely different order of magnitude from the rest of physics'), confided to Sommerfeld his worry that failure to extend the mouldy model to three bodies—two nuclei and an electron or a nucleus and two electrons—undermined all atomic theory. 'Of course, Bohr knows that very well'.[14] So did Born. In the summer of 1923, in a celebration of the tenth anniversary of the Bohr atom, he offered the inopportune opinion that atomic physics had to be rebuilt, 'from the ground up'.[15]

Having learned something of Copenhagen physics, Pauli offered Bohr an assessment of Heisenberg. He is very 'unphilosophical', Pauli observed, 'he does not bother to work out basic assumptions and their connections with existing theory', though personally he is a very nice fellow and even a genius. 'I hope that he will return [from a scheduled visit to Copenhagen] with a philosophical attitude toward his ideas'. Heisenberg knew he needed the visit and the lesson. 'I realize ever more that Bohr is the only person who, in the philosophical sense, understands something of physics'.[16] And the only physicist he thought his mother could like; 'he is not only a physicist, but much more, always especially kind to me'. We talk and talk, 'and with the help of one or many glasses of port wine physics is enriched with new discoveries'. Port alone did not solve problems. Luck also was needed. 'I have had a lot of luck', Heisenberg wrote his parents, but that might not continue. 'Which can hold out longer, physics or I?'[17]

Meanwhile Pauli was finding it very difficult to do physics Bohr's way. He moped about the streets of Copenhagen distressed over his inability to devise, by a wave of the magic wand of the correspondence principle, a satisfactory formula for the complicated splitting of spectral lines in a magnetic field. In the simplest, 'normal' form of this effect, first detected by Pieter Zeeman in 1896, a single spectral line becomes three, a phenomenon that Lorentz explained entirely on classical grounds. A new phenomenon explicable by established theory was the sort of science admired in Stockholm. Zeeman and Lorentz shared the Nobel prize for 1902.

The judges soon had their general reservation towards theory uncomfortably reinforced. Zeeman splitting into triplets showed itself to be a relatively uncommon form of the effect and Lorentz could not find a satisfactory theory for the many other patterns, or 'anomalous' multiplets, into which magnetic fields shattered spectral lines. Sommerfeld succeeded without much trouble in translating the normal form into quantum theory. But try as he might, Pauli failed to conquer the 'anomalous' effect; 'it would not come out'.[18] His frustration drove him out of quantum physics, temporarily, not to film comedy, as he had fancied after watching Chaplin movies, but to a good hard problem in classical physics. There he proposed to stay, 'hoping that Bohr will save us with a new idea'.[19]

While relaxing classically, he stumbled over a new physical idea, trickier than Heisenberg's trickiest. Unphilosophical, opportunistic Heisenberg had boasted the invention of 'swindles of a higher order' in suggesting half-quantum numbers to explain certain spectra. Pauli's new swindle, which belatedly won him the Nobel prize of 1945, was to ignore the correspondence principle and to treat the orbits merely as sets of quantum numbers. In 1923, three quantum numbers were used to describe the atom: n, k, and a newcomer j, needed to classify the sublevels in the anomalous Zeeman effect. Both k and j were subordinate to n; for a given n, k could take on only integral values from 1 to n; for a given k, j was restricted to a total of $2k - 1$ values. How many different sorts of electrons with principal quantum number n are there in an atom?

It is just arithmetic. The answer is n^2. For $n = 1$, $k = 1$, and there is only one possible value for j (so $n^2 = 1$). For $n = 2$, k may take values 1 and 2; j has one value for $k = 1$, three different values for $k = 2$, and the total—4—is n^2 once again. The pattern 1, 4, 9, 16 is just half the lengths characteristic of the seven periods in the periodic table of elements, 2, 8, 8, 18, 18, 32, and 32. Pauli therefore had to double the theoretical result, to $2n^2$, which he did by making the duplicity of the electron official. He assigned to every electron a fourth quantum number that could take only two values. Each of the $2n^2$ sublevels would then correspond to an entry in the periodic table: a chemical element. The quantum number $n = 1$ implies just two entries—hydrogen and helium—and accounts for the first period. The eight possibilities for $n = 2$ fill the second period, lithium through to neon.

This simple pattern breaks down for the 18 sublevels of $n = 3$. The electrons in the first eight elements fall in like their counterparts at $n = 2$, accounting for the third period from sodium to argon. But then, because they have lower energy than the remaining electrons with $n = 3$, two $n = 4$ electrons open the fourth period, followed by the 10 remaining electrons with $n = 3$ and another 6 from $n = 4$, making 18 elements from potassium to krypton. The fifth period, rubidium to xenon (18 entries) develops like the fourth; the sixth period, caesium to radon (32 entries) and the seventh, francium to oganesson (32 entries, most of which do not exist in nature and none of which beyond uranium were known in 1922), though more complicated than the fourth and fifth periods, follow the same principles.

With the doubling of the electron's role, the lengths of the periods came right if no more than one electron with a given set of the four quantum numbers could be present in the same atom. Otherwise stated, each possible orbit can accommodate up to two electrons, provided they have different values of the fourth quantum number. Assigning this quantum number (symbol s) a fixed value of ½ immediately resolved the splittings of the anomalous Zeeman effect. The interpretation and exploitation of s will be a frequent theme in our drama.

Pauli sent Heisenberg a draft of his understanding of the periodicity of the elements with the remark that the usual concept of the electron orbit must go. Heisenberg liked it all, the rejection of the orbits, the prohibition of identical twins (Pauli's 'exclusion principle'), and the formal character of the treatment. It looked more like a social than a mechanical rule. 'You have raised the swindle to hitherto undreamt-of swindling heights, you have broken all records!'[20] Bohr wrote in the same vein: we are all enthusiastic about your *Wahnsinn* (madness), it shows the extent of the whole *Schwindel* (trickery) and gives us hope that we are near a turning point. But let us not give up on the correspondence principle!

Pauli responded that he could not believe in electron orbits or share the hope for a quick resolution of the problems of atomic physics.[21] The path to the exclusion principle had been rough. It ran through the thicket of the anomalous Zeeman effect and required jumping several conceptual hurdles. In awarding the Lorentz Medal to Pauli in 1931, Ehrenfest expressed 'wonder at the beauty and mystery of this principle and at its fertility . . . [and] at the way you struggled through to this discovery'.[22] Later, after psychoanalysis by Carl Jung, Pauli found a deep meaning in increasing the number of electronic quantum numbers from three to four.[23]

An obvious way to grant the electron the extra freedom it needed to justify a fourth quantum number was to allow it to spin, just as the Earth rotates on its axis while also revolving around the Sun. Pauli spurned this suggestion when first made to him as a counter-revolutionary return to literal orbits and persuaded its proposer—Ralph de Laer Kronig—to give it up. But the idea resurfaced ten months later in the minds of two of Ehrenfest's students,

Samuel Goudsmit and George Uhlenbeck. They had already submitted a paper for publication when Lorentz told them that their idea was impossible in classical electron theory since it required velocities greater than light. Fearing they had made a significant error, they sought to withdraw their paper before it could be published. Luckily, they were too late.[24]

Many years later Bohr recalled his first reactions to Goudsmit and Uhlenbeck's proposal.[25] He had arranged to meet Pauli in Hamburg, where Pauli had obtained a teaching position, during a break in his journey to Leiden to attend Lorentz's doctoral jubilee. Pauli and his fun-loving Hamburg colleague Otto Stern wanted to know what Bohr thought about the Dutch electron. 'Very very interesting', Bohr replied, which, knowing Bohr-speak, they understood as 'probably wrong'. In Leiden, Einstein gave strong reasons to believe in a spinning electron.[26] Bohr then passed through Göttingen, where he was met at the station by Heisenberg and Heisenberg's collaborator Pascual Jordan. What do you think of the spinning electron? Bohr: Probably right. He continued to Copenhagen via Berlin. Pauli had come from Hamburg to meet him at the station and interrogate him once again. Answer: electron spin represented a great advance. Alas for Kronig! *'Der Kronig hätt' den Spin entdeckt, hätt' Pauli ihn nicht abgeschreckt'*. Kronig would have discovered spin if Pauli had not scared him off.[27]

Pauli now could rate spin 'a new Copenhagen heresy' as well as a reactionary *putsch*. 'I am convinced that . . . I am absolutely right', he wrote to Kramers, signing himself with the title Ehrenfest had given him, 'the scourge of God'. 'There is only one in physics, thanks be to God!!!!'[28] When technical objections to spin were resolved (they rested on a relativistic effect that the relativity expert Pauli had missed) he capitulated. Later he explained that he wrote at a time of 'spiritual and human confusion, caused by the requirement of literal orbits that could be visualized. I was so stupid when I was young!'[29] He was right, however, in warning that electron 'spin' is a quantum effect unlike the revolution of a spinning planet or a top. It should not be taken literally.

While Pauli distressed himself over Zeeman effects and edged toward the exclusion principle, Bohr employed his own brand of madness by turning to Slater's ideas about virtual radiation. He collaborated with Kramers in persuading Slater to forego the detestable light quantum, which, in Slater's model, followed the guidance of virtual radiation. There was a very high price to pay for this indulgence. The conservation of energy and momentum became a matter of statistics, of averages, not necessarily applicable in *individual* processes like a Compton collision. In the Bohr–Kramers–Slater (BKS) theory of early 1924, the virtual radiation induces probabilities of transition, but the emission of light is not causally connected with its absorption.

A Compton process occurs when the virtual wave associated with the incoming X-ray persuades a carbon atom in graphite to release an electron. A pre-established statistical harmony makes the loss of energy of all the colliding X-rays equal to the gain in kinetic energy of all the released electrons. Several physicists in addition to Bohr's school inclined toward the BKS approach, notably Erwin Schrödinger, a philandering Viennese then Professor of Physics at the University of Zurich. A previous serious flirtation with downgrading conservation laws to statistical regularities had prepared Schrödinger to go further than BKS. Why treat radiation as virtual rather than the electron orbits and their transitions?[30] Perhaps a piano is a better model of an atom than a solar system. Why should the great conservation principles apply to the nuclear atom and its supposed electronic structure? Those principles were the glory of a classical physics unacquainted with the microscopic quantum world. Perhaps strict conservation had to be added to the losses recently suffered by Western civilization in its catastrophic civil war.

Einstein did not like the BKS theory although he recognized in the virtual radiation something akin to the ghost field he had tried to develop. Chief among his objections was the surrender of strict causality implied by the failure of conservation in individual events; 'a definitive abandonment would be unbearable for me'.[31] Heisenberg, delighted with every *Schwindel*, inclined toward BKS. So too did Pauli, in so far as Bohr's loud advocacy could silence his 'scientific conscience'. But he soon recovered his critical voice to protest against BKS as another attempted coup against sound physics.[32]

The question could be settled experimentally, and soon was, by detection of both the scattered X-ray and the electron recoiling from the same event. Energy and momentum are indeed conserved for individual atomic processes. Bohr bowed to the result; '[we have] nothing else to do than to give our revolutionary efforts as honourable a funeral as possible'.[33] Nonetheless, the episode had significant enduring consequences. It forced Bohr to admit the light quantum and the obligation to attend to the particle as well as the wave aspects of radiation. Einstein came away more convinced than ever that physics must aim at explanations that respect causality and allow descriptions of events taking place in space and time.

These premises of classical physics did not belong to physicists alone. They were a common heritage of the disappearing civilization of the West, consequences of a way of thought that supported the foundations of society as well as of science. Einstein felt obliged to alert the educated public to the seriousness of the issues his colleagues were agitating. The alert came to readers of the *Berliner Tageblatt* under the title, 'Is science done for its sake alone?' Einstein's answer: yes and no. Yes, because science must be cultivated without any concern for practical application. Otherwise, it would waste away and neglect its great educational task, 'which is to awaken and keep awake a striving for causal knowledge of everything . . . one of the most valuable ideas of mankind'. And, therefore, no. Science should not be cultivated for itself alone. Researchers must always give society the opportunity to understand the great problems of science at first hand. 'Only when science also respects this important obligation will it earn the right to exist from a social point of view'.[34]

Einstein's continuing objections to quantum theory were obligations he thought he owed to science and society: to challenge his colleagues, inform the general educated public, and maintain the old principles in probing the deeper problems posed by the existence of the quantum.

New Ways to Calculate

Kramers did not surrender BKS easily. This was fortunate since his application of its virtual oscillators to the problems of dispersion paved the way to Heisenberg's invention of his singular version of quantum mechanics. Dispersion is familiar to anyone who has enjoyed the sight of a rainbow or observed the spread of light's colours by a prism. The classical formula for dispersion depends on the ratio of the frequencies of the incoming light to the frequencies occurring naturally in the dispersing atoms. To agree with the experiment, the natural frequencies must be the observed spectral ν's, not the suppositious electronic ω's: the notes of the BKS piano rather than Sommerfeld's atomic music of the spheres. It would be better, therefore, Kramers wrote, to purge the electron orbits and admit 'only such quantities as allow of a direct physical interpretation on the basis of the fundamental postulates of the quantum theory'.[1]

To proceed, he invented new ways to translate the continuous quantities occurring in the classical theory into the discontinuities characteristic of a physics obeying quantum rules, hoping to advance the programme to achieve what Born was calling 'atomic mechanics' in anticipatory lectures. Together with Heisenberg, Kramers extended the range of translations from classical into quantum argot; and with the help of this lexicon Heisenberg derived an expression for Planck's constant h in terms of observable frequencies and intensities. That gave him enough to recast the existing theory into a form free from the mechanical conditions (but not from the symbolic quantities) that Bohr had used to define the stationary states. Born and Bohr immediately recognized the recasting as the breakthrough to the quantum mechanics they had been pursuing.

As Heisenberg later told the story, the breakthrough came on Helgoland, an island in the North Atlantic to which he had fled to assuage the hay fever that was choking him in Göttingen. The place though remote was pleasant, a well-developed resort for sufferers who could afford it. His material requirements catered for, Heisenberg devoted himself to the problem he had been working on with Kramers, finding a calculus involving symbols with double indexes, much like the spectroscopists' T_{mn} and ν_{mn}, that gave the frequencies and probabilities of quantum jumps. No quantities referring to electron orbits were allowed. Guided by the correspondence principle, he learned how to multiply together infinite arrays of these doubly indexed numbers and compute them for some simple systems. He recalled that in his excitement he made so many mistakes that he was not sure of his success until dawn.

He then climbed a hill to watch the Sun rise over the birthplace of quantum mechanics. This romantic hill cannot bear witness to the tale, for the British blew it up, not from opposition to quantum mechanics but because it capped a fortress the Germans had used during World War II. Heisenberg wrote up his insights and showed them to Born, who was moping about, 'very unsure of myself' as he wrote Einstein, incapable of keeping up with Heisenberg and Pauli. Or with the equally brilliant Jordan, with whom he was stuck in the old mire, 'examining every imaginable correspondence relationship between classical, multiply-periodic [orbital] systems and quantum atoms'.[2]

Reading Heisenberg's new paper gave Born's self-esteem a needed tonic. He was able to recognize in the arrays of numbers and the rules for their multiplication something Heisenberg had not known. The juggling on Helgoland had produced a primitive instance of an established variety of algebra. Quantities A and B subject to this algebra are free from the requirement imposed on ordinary numbers, $A \times B = B \times A$, more usually written $AB = BA$. Mathematicians say that A and B *commute* if the order of a mathematical operation on them (in this case multiplication) does not affect the result. This can be expressed in a 'commutation relation', written $AB - BA$. If A and B commute then $AB - BA = 0$.

Born was able to evaluate the commutation of the arrays of numbers (or 'matrices', symbolized in what follows by bold-face type) that Heisenberg had turned up. He found that the matrices representing the 'position' q and 'momentum' p of—well, it is not easy to say of what—do not commute. Instead, they satisfy the commutation relation, $qp - pq = ih\mathbf{1}/2\pi$, which may be regarded as a remote descendent of the Bohr–Sommerfeld quantum condition on the orbits. In this expression i is the 'imaginary' square root of -1, and $\mathbf{1}$ is the 'identity matrix', a square array of numbers with all entries along the leading diagonal equal to 1 and all the remaining entries zero.

What can be said securely about Born's commutator is that the right-hand side stands for a square matrix in which every diagonal entry is the imaginary quantum $ih/2\pi$ and every other entry is nothing at all. No wonder some quantum physicists claim not to understand quantum mechanics.

The pre-existence of a non-commutative matrix algebra proved to be a blessing and a curse. A blessing because it enabled Born, Heisenberg, and Jordan to generalize Heisenberg's insight into formal 'matrix mechanics' and a curse because, as Pauli told it, the formality obscured the physical content of Heisenberg's insight and sophisticated the treatment of microscopic physics beyond the reach of most physicists. To solve a problem, the practitioner had to find numbers that simultaneously 'diagonalized' the matrix representing energy, satisfied Born's commutation rule, and obeyed the applicable classical equation of motion, now to be understood as a relation among matrices. By retaining the equation of motion, matrix mechanics shifted and deepened the conundrum of the connection between classical and quantum physics.

For all their mathematical agility, the Göttingen triumvirs could not calculate the energies of the stationary states of the hydrogen atom, and hence the pattern of lines in the Balmer series. Pauli succeeded in November 1925 after weeks of effort. Physicists who rotated around the Bohr–Born axis hailed his *tour de force*. Most others shuddered at the mathematical prowess and exotic concepts apparently needed to solve the simplest of useful problems using matrix methods. This criticism touched a sore spot with Born. To compensate for what he regarded as his want of physical intuition, he had prided himself on his ability to clarify the mathematics that physicists used. In elaborating the matrix formulation, he was exploiting his competitive advantage.[3]

Although the physical meaning of matrix mechanics was obscure, Bohr welcomed it as a realization of his goal of a quantum mechanics unpolluted by the mix of classical and quantum considerations that had given birth to it. He had plumped for a theory based entirely on its own quantum rules and saw in Heisenberg's breakthrough the completion of the programme of the correspondence principle or even (as Born did) as the final expression of it, 'a brilliant realization of . . . the guidelines in the development of atomic theory'.[4] The connection of classical physics to the quantum world did not disappear in the fulfilment of the correspondence principle; classical concepts would re-emerge as the essential connection between the new quantum physics and the old world of experience.

That the elements of the new theory, the matrix entries, were not things but probabilities gave physicists who worried about fundamentals a twinge of regret.[5] An honest scientist must bear it, however. Had not Bohr himself tried with all his strength to escape the paradoxes of the planetary atom? And in the extremity into which demonstrations of the light quantum had placed him, had he not been willing to sacrifice the laws of conservation of energy and momentum for individual events? With BKS as well as matrix mechanics in mind, Bohr concluded that no visualizable description of radiative interactions between distant atoms was possible. We must admit 'an essential failure of the pictures in space and time on which the description of natural phenomena has hitherto been based'.[6]

The only part of Bohr's evaluation of matrix mechanics with which Einstein agreed was that it incorporated the correspondence principle. To stop with probabilities at the outset of a new physics was to him abject and premature surrender. Why the hurry? 'Heisenberg's theory does not claim to be complete'. Why rush to conclusions about the dualism of waves and quanta? Perhaps this will all be resolved 'on a simple basis that we can see some necessity in'. Matrix mechanics might deliver some desirable numerical control for those who can use it. But do not worship mere mathematics. '[It] alone does not bring salvation'.[7] When he wrote these lines in the late spring of 1926, Einstein had strong reasons to believe that Heisenberg's system was not only incomplete, but also superseded. An alternative quantum mechanics had come into being for which Einstein had twice been the midwife.

Its first father was a French aristocrat, Louis de Broglie who, like his elder brother Maurice, had become a physicist rather than follow family tradition and cap his career as an admiral or minister of state. Owing to the war, Louis did not finish his doctorate until 1924, when he was 32, a decade older than Heisenberg, and stubborn enough to stick to his idiosyncratic ideas despite the misgivings of his thesis committee. Since these ideas arose from relativity theory, the committee referred the thesis to Einstein. The master could not fault its fundamental principle, a wave to guide a particle. Not a light quantum this time, but an electron.

Having pondered the association of a light quantum with a wave, de Broglie saw good reason to reverse the traffic, and associate a wave with an electron. 'After long reflection in solitude and meditation', he wrote in 1963, 'I suddenly had the idea, during the year 1923, that the discovery made by Einstein in 1905 should be generalized by extending it to all material particles and notably to electrons'.[8] This was the sort of symmetry Einstein liked. 'He has lifted a corner of the great curtain', Einstein wrote to a French colleague, 'he has thrown a first flicker of light on this worst of our physical riddles'.[9] De Broglie's electron resembled the white rabbit in *Alice in wonderland*, running around clutching its watch, and his invention of the associated wave looked like pulling a rabbit from a hat.

To get the electronic watch to tick, de Broglie equated Einstein's expression for the energy of a *particle*, mc^2, with the expression for the energy of the light quantum derived from some suppositious *wave* frequency, $h\nu$. Voilà! Let there be an 'internal' frequency, $\nu = mc^2/h$. Call the velocity of the ν-wave V and that of the particulate electron v and further let the speed of light be a mean proportion between them: $vV = c^2$. Since electrons possess mass and are obliged to travel more slowly than light, this equation implies that the ν-wave has a velocity faster than light. De Broglie waved away concerns that it violated relativity. Being imaginary and propagating only phase—the position of the wave in its crest-to-trough cycle—it could not carry any energy or useful information, and so posed no threat to the founding principles of Einstein's relativity. It is an imaginary '*phase wave*', a nothing.

Well, not quite nothing. The phase wave has a connection with the momentum of the particle that it represents. We have all the equations to hand to derive this connection, another icon of quantum physics. The frequency of a wave equals its velocity divided by its wavelength, thus $\nu = V/\lambda$. And so, from $\nu = mc^2/h$ and $c^2 = vV$, we have $\nu = mvV/h$. Replacing ν by V/λ, the elementary algebraist discovers that the electron's momentum p (mass $\times$ velocity) $= mv = h/\lambda$. The resulting relation reversed, $\lambda = h/p$, connecting the white rabbit's ν-wave to his external momentum is known as the de Broglie wavelength. We place $\lambda = h/p$ with our other iconic expressions, $E = mc^2$, $E = h\nu$, and Born's commutator among the inspired symbolic rhetoric of physics.

De Broglie's icon had two immediate appealing consequences: a natural representation of the quantum numbers of the stationary states and the prospect of a splendid experiment. First, the representation. Bohr's definition for the kinetic energy of the nth stationary state, $T_n = {}^1\!/_2 nh\omega_n$, easily transforms by elementary concepts of ordinary physics and trivial algebra into $p_n r_n = nh/2\pi$, an expression for the angular momentum $p_n r_n$ of the electron in its nth orbit of radius r_n.[10] Substituting the electron's wavelength for p_n, de Broglie could write $n\lambda_n = 2\pi r_n$, which states that the circumference of a stationary orbit contains an integral number of de Broglie wavelengths.[11]

Do not take this literally! The phase wave does not relate to the orbit as the standing vibration of a string does to a guitar. The phase wave is a formal device, another symbol. Supposing it to have a material basis or carrier ends in grief!

Although the phase wave does not exist materially, in the strange world of the quantum it predicted a decisive material effect. A beam of electrons of momentum p should give the same diffraction effects as a light wave of wavelength λ, which, for convenient p, lies in the region of X-rays. De Broglie suggested the experiment to an expert in such things. The expert declined; he had exact measurements to make.[12] Sloppier experimentalists in the US and Britain, Clinton Davisson and George Thomson, confirmed the effect in 1927 and, a decade later, received their Nobel prizes. The Thomson family did not complain that the son won the prize for detecting the electron's wave properties whereas the father, J.J. Thomson, had won his for developing its theory as a particle.

The family of matrix mechanicians was not so tolerant. By 1927 they faced a powerful rival in the sweepstakes for the description of the microworld. This rival came from an invention stimulated, once again, by a contribution from outside the narrow circle of concerned European theorists. And once again Einstein was the midwife.

In 1924, he received a paper from an Indian physicist unknown to him, Satyendra Nath Bose, who had succeeded in removing the primary remaining ambiguity from Planck's radiation formula by treating the radiation as a gas consisting of light quanta. Being full of de Broglie, Einstein applied Bose's analysis to material particles, idealized as a perfect gas, and inferred that at low temperatures the properties of the gas would deviate from classical expectations, much as classical radiation formulas diverged from Planck's. Einstein's considerations had an immediate spectacular payoff—not in observing the deviations, which did not occur for 70 years—but in drawing Schrödinger's attention to de Broglie's ideas.[13] Einstein praised de Broglie's 'very notable contribution' and Schrödinger promptly acquired a copy of the controversial thesis.

Physicists at the University of Zurich, where Schrödinger held his professorship, had established bi-weekly colloquia jointly with colleagues at the neighbouring Eidgenössische Technische Hochschule (ETH), the 'Polytechnikum' that Einstein had attended. On 23 November 1925, at the request of the ETH's Peter Debye, Schrödinger presented a colloquium on de Broglie's thesis. In the audience was a Swiss student, Felix Bloch, who recalled the event more than 50 years later:[14]

> Schrödinger gave a beautifully clear account of how de Broglie associated a wave with a particle and how he could obtain the quantization rules of Niels Bohr and Sommerfeld by demanding that an integer number of waves should be fitted along a stationary orbit. When he had finished, Debye casually remarked that this way of talking was rather childish. As a student of Sommerfeld he had learned that, to deal properly with waves, one had to have a wave equation.

Schrödinger now sought an equation for de Broglie waves that would relate to the trajectories of material particles as classical waves related to light particles. An obvious route existed for the standing waves presumed to characterize the stationary states of a one-electron atom: introduce the electron via its momentum into a standard wave equation by de Broglie's formula for the wavelength, $\lambda = h/p$, and re-express the momentum in terms of the kinetic energy ($\frac{1}{2}mv^2$, which, with $p = mv$, becomes $p^2/2m$). Schrödinger chose more difficult and obscure ways, but eventually lighted on such an equation, whose solution recovered the energy levels of the Bohr model of hydrogen with the great improvement that the quantum numbers appeared naturally, as the order of the vibration modes of the standing waves that were the solutions to his equation. Thus, a quantum number introduced ad hoc by Bohr and Sommerfeld emerged 'in the same natural way as it does in the case of [the integers specifying the number of nodes] of a vibrating string'.[15]

In Schrödinger's wave equation, these standing waves are described by their 'wavefunctions', denoted by the Greek letter ψ (psi), and their quantum numbers fix the energy states and thence the frequencies of the transitions between them. Arguments that suggest it is always perilous and often wrong to suppose that the ψ wave is real will shape much of our unfolding drama. Schrödinger was content to interpret its amplitude at any place as a measure of the amount of electronic charge present there. Or rather, since ψ can be an imaginary quantity featuring i (a good hint that it is not real!), by its modulus-square, $|\psi|^2$, since $|i|^2 = -i \times +i = 1$, a real number. A transition between stationary states could then be pictured as the simultaneous excitation of two different vibratory modes, like beats in sound waves. 'It is hardly necessary [Schrödinger wrote] to emphasise how much more congenial it would be to imagine that at a quantum transition the energy changes over from one form of vibration to another, than to think in terms of a jumping electron'.[16]

Schrödinger's wave mechanics offered a more digestible explanation of non-commutivity than the infernal matrix mechanics. In his mathematical manipulations, classical quantities such as linear momentum had undergone a remarkable transformation. They were no longer just 'numbers' with associated units, such as kilogram-metres per second. They had become mathematical 'operators', constructed in ways that would allow the numbers to be notionally extracted from the expression for ψ. The algebra of wave mechanics is sensitive to the order in which mathematical operations are applied, just like matrices. Applying the linear momentum operator to ψ followed by the position operator does not yield the same result as applying the position operator followed by the linear momentum operator.

To demystify non-commutation a little for readers acquainted with differential calculus, we note that in wave mechanics multiplication of ψ by q or p is the same as multiplying by q or differentiating with respect to q, respectively. To distinguish q or p as 'operators' from their use as ordinary numbers, they are often written $\hat{p}$ and $\hat{q}$. In this format, $\hat{q}\hat{p}\psi$ is not equal to $\hat{p}\hat{q}\psi$ since if $\hat{p}$ comes first, it generates a term proportional to ψ, $ih\psi/2\pi$ to be exact, to which $(\hat{q}\hat{p} - \hat{p}\hat{q})\,\psi$ is equal. Usually, ψ is omitted and the commutation relation exhibited as $\hat{q}\hat{p} - \hat{p}\hat{q} = ih/2\pi$, identical in form to the relation discovered by Born. Despite their different conceptual bases, wave and matrix mechanics describe the same physics. Schrödinger himself showed that the two theories are indeed equivalent.[17]

Planck and Einstein rode happily on Schrödinger's wave. Planck: it is 'the solution to the puzzle', an 'epoch-making work'; 'what a crossfire of ... acclamations await you!'[18] Einstein: 'a most ingenious theory of the quantum states', far better than the 'great quantum egg' Heisenberg had laid ('They believe it in Göttingen, I don't'.). With Schrödinger's ψ, physicists had no such 'infernal machine [as matrices] but a clear thought—"obligatory" for calculation'.[19] As calculations went forward, however, the clarity diminished. Just as with electron orbits, serious difficulties arose with ψ waves for atoms more complicated than hydrogen.

The problem lay in the necessity of assigning coordinates of position and momentum to every electron; ψ accordingly depended on many position coordinates and, in general, could not be expressed or pictured as a wave in ordinary space and time.

Schrödinger's realistic vibrating, pulsating electron clouds had to be taken with a shaker of salt. Einstein: 'they are indigestible'.[20] Schrödinger's colleagues at Zurich: 'Gar Manches rechnet Erwin schon/Mit seiner Wellenfunktion/Nur wissen mocht' man gerne wohl/Was man sich dabei vorstell'n soll'. 'Erwin collects without compunction/Many results from his wavefunction/To do the like we would be keen/If we knew what his ψ might mean'.[21]

Schrödinger suggested that his waves could form 'packets' that might represent particles. The construction he had in mind, based on classical optics, superposes travelling ψ waves of slightly different frequencies that destroy one another by interference everywhere but over a small space where they combine in phase (Fig. 4).

That would preserve a description cast in space and time, a necessity of human thought, for 'what we cannot comprehend within it, we cannot comprehend at all'.[22] A ψ packet might contain or constitute an electron in the same sense that a packet of light waves might a light quantum. And it was an added virtue that the velocity of the packet v was related to the velocity of the component waves V as required by classical physics. Unfortunately,

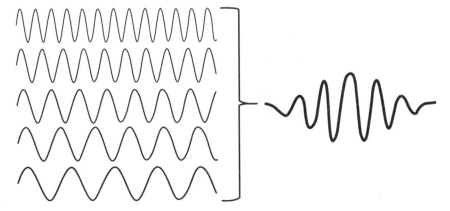

Fig. 4 A wave packet is created by the superposition of many waves with slightly different frequencies or wavelengths, such that they destroy one another by interference everywhere but over a small space where they combine in phase.

classical physics makes a ψ packet rapidly dissipate in its travels. It cannot represent an enduring particle.

In the spring of 1926, Born found a role for ψ that, unlike Schrödinger's wave packets, did prove to be enduring. He had to move out of the atom—the area of competence of matrix theory—to do it. The phenomenon of interest was a collision between a fast electron and an atom. Born represented the incoming electron as a plane de Broglie wave, the scattered electron far from the atom as a collection of possible plane de Broglie waves, and the atom as a system able to change its state during the collision. In his analysis of the outgoing wave, Born arrived at the interpretation of the ψ function belatedly judged worthy of a Nobel prize. In a preliminary discussion of his innovation, published in June 1926, he distanced himself from the very concept that he would follow to his probabilistic interpretation. 'I intentionally avoid the term Übergangswahrscheinlichkeiten' ('transition probabilities'). Indeed, he avoided the matrix approach altogether as it could not deal with collision phenomena. 'And precisely for this reason I regard [Schrödinger's method] as the deepest foundation of quantum laws'.[23]

Despite its power, Schrödinger's method could give only the probability of an outcome of the collision. Should we hope someday to discover ways to determine the internal motions and the state of the scattered particle? 'I am inclined to give up determinateness [*Determinierheit*] in the atomic world. But that is a philosophical problem, for which physical arguments alone are not decisive'.[24] This first presentation of the probabilistic interpretation of the wavefunction does not suggest strong conviction, although its laboured form and weak conclusion may have owed something to Born's habitual caution and self-doubt. Heisenberg had trouble untangling its meaning. Pauli had to inform him that it interpreted the construction $\psi^2 (q)\, dq$ as the 'likelihood' of finding a particle described by a wavefunction which varies with position, $\psi (q)$, within a very small volume dq centred around a specific position, q.[25]

In July 1926, a month after declaring his 'inclin[ation] to give up determinateness', Born explained that he pursued a middle way, between the statistical approach built into matrix mechanics and the realistic interpretation Schrödinger sought to give his ψ wave; and he designated Einstein's concept of electromagnetic waves as a 'ghost field' guiding photons as the inspiration for his compromise. Like the soundless symphonies of the BKS theory, which Born had enjoyed, Einstein's ghost field determined a probability, in this case of finding a light quantum at a given place and time. Born's idea was to interpret ψ as the ghost field for de Broglie's electrons.[26]

'Advisor' would be a better word than 'guide' for the relation Born saw between the ghost field and its particles. For although ψ develops determinatively in space and time like a classical quantity, it could not force, but only advise, the electron. '[T]he motion of the particle follows probability laws while the probability itself $[\psi(t)]$ spreads in agreement with the law of causality'. Once again, however, Born did not insist. He observed that anyone dissatisfied with his probabilistic treatment could assume the existence of unspecified parameters governing the electron that, if known and employed, would determine individual atomic events. He deemed a non-probabilistic account of quantum mechanics possible but improbable.[27] In this permissiveness, some commentators have discovered the origin or licence for 'hidden variable' theories.[28]

Born's main result became a rallying cry. 'There is no answer to the question, "what is the state after the collision", but only to the question, "how probable is a given result of the collision"'. Quantum mechanics, 'our quantum mechanics', gives no basis for supposing the existence of 'internal features of the atom [like the suppositious electron orbits] that determine a definitive outcome'. Perhaps such features (hidden variables) exist. Should we hope to discover them eventually? If theory agrees with experiment without them, why bother? Born offered the problem to philosophers along with a valediction that would rile up later actors in our drama.[29] He remarked that the question whether hidden variables existed had no practical import, since 'they must lead to the same results as the [probabilistic] theory'.[30] This was to ignore the possibility that the introduction of such variables might not only retrieve the old results, but also produce new ones.

Schrödinger conceded that Born's middle way solved the collision problem. And he appreciated Born's concession that, by facilitating difficult computations against which matrix mechanics had proved powerless, wave mechanics was 'the deepest formulation of the quantum laws'.[31] In the summer of 1926, while Born was making these friendly nods to ψ waves, Heisenberg was trashing them as yet another counter-revolutionary move: 'revolting', he wrote Pauli, just 'crap' (*Mist*).[32] By November Born had been educated by his aggressive student. Schrödinger's ψ's were now mathematical crutches; 'his physics is really pitiful'.[33] Schrödinger countered that abandoning a space-time description in principle did not reach the level of pitiful physics, it was not physics at all. 'The purpose of atomic research is to incorporate our experiments on the subject into our usual way of thinking; and this entire way of thinking moves in space and time as far as the external world goes. If this incorporation in space and time does not succeed, the entire enterprise fails'.[34]

Schrödinger brought his fight for a physical world picture to Bohr's Institute on 4 October 1926. Bohr's new assistant Heisenberg was present, Kramers having taken a professorship at Utrecht. In his later spiced-up account,[35] Heisenberg has Bohr hectoring his guest to abandon hope for a return to a space-time description of the atomic world. Schrödinger resisted: 'Surely you realise that the whole idea of quantum jumps is bound to end in nonsense'. The relentless badgering continued until Schrödinger succumbed to a feverish cold, which, however, did not save him, since Bohr pursued him into the Institute's guest suite and might have killed him with physics had Bohr's wife Margrethe not intervened.

No doubt the interaction was strong and warm, the stakes high, the opponents earnest. But they were not enemies, fellow travellers rather, intent on forcing their way through a thicket to an unknown beyond. Bohr described their common efforts to an English colleague. 'The discussions centred themselves gradually on the problem of the physical reality of the postulates of the atomic theory'. Schrödinger stuck to his conviction that stationary states with sudden transitions could be avoided. 'But I think we succeeded in convincing him that for the fulfilment of this hope he must be prepared to pay a cost . . . formidable in comparison with that hitherto contemplated by the supporters of the ideas of a continuity theory of atomic phenomena'. Bohr had the BKS theory in mind and the surrender of energy conservation as only an instalment of the costs that Schrödinger would be obliged to incur.[36]

The discussions disoriented Schrödinger. He wrote a sympathetic German colleague that Bohr held firm to the 'remarkable' belief that we cannot understand the microworld in the usual sense of the word. 'Therefore conversation [with him] is almost immediately driven into philosophical questions, and soon you no longer know whether you really take the position he is attacking, or whether you really must attack the position he is defending'.[37] Disorienting, to be sure, but also often inspiring as well as irritating to those Bohr thought it worth labouring to convert. We can glimpse this response in Schrödinger's letter to 'the great Niels Bohr' expressing his thanks for the privilege of 'talking with him for hours about things so close to my heart and hearing from him about the positions he now takes toward the many attempts to build a bit more on the sound foundation he has given to modern physics. That was for a physicist, who is one most earnestly, a truly everlasting experience!'[38]

And a warning that there would be a struggle for the soul of physics. The founder of quantum physics had wrestled with its first agonizing discontinuities, 'wrestled . . . with your whole soul', so Schrödinger addressed Planck nine months after his visit to Bohr. 'I believe that one is bound to take up this struggle anew with the same seriousness . . . [against] those who today already announce categorically: the discontinuous exchange of energy must be adhered to'.[39]

Schrödinger qualified as a *Prinzipienfuchser* in Einstein's classification of theorists and could not bring himself to accept, even as a temporary 'resting place', Bohr's view that visualizable pictures are only symbolic. That would be a 'painful limitation of our right to truth and clarity'.[40] Nor could he accept that the ψ wave gave probabilities for the outcome of experiments on a great many systems and no information about the behaviour of an individual one. Are there no single systems we can describe? Even if the concepts we have previously used do not enable us to do so, we do not have the right to solve our problems with contradictory invocations of waves and particles. 'Certainly we can weaken [the contradictions] by saying, e.g., that the whole atom behaves itself "in certain circumstances, so, as if . . . and in certain [other] circumstance, so, as if . . . ", but that is so to say only a logical quibble, which cannot be transformed into clear thought'.[41] Schrödinger had something much simpler in mind, 'just this one proposition: even if a hundred trials fail, we must not give up the hope of reaching the goal—not, do I insist, through classical pictures, but through logical concepts free from contradictions'.[42]

This last remark struck at Bohr's core. He prided himself, rightly, on his ability to detect 'ambiguities' that even an Einstein could overlook. He immediately threw himself into saving his self-image by purging the new mechanics and its interpretation of contradictions.

New Ways to Think

B ohr brought to the work the conviction, which he had held since his university days, that no single truth could express the entire content of any domain of experience. He therefore did not privilege waves or particles in the microworld but sought a way to have them both. It was the same technique he had used in forging his model of the hydrogen atom. Put contradictory concepts together, show the fertility of the miscegenation, and strive to reconcile the ingredients. As usual, he struggled with his problem in conversation with an accomplished assistant, who in this case was Heisenberg. The ambitious inventor of matrix mechanics was not so fond as his boss of the starting assumption that put Schrödinger's waves on a par with Göttingen's particles.

Their conversations grew warm. Bohr would not let go. Heisenberg came close to tears. The professor went skiing to cool down and his assistant seized the opportunity to pre-empt him. Relieved of Bohr's pressure, Heisenberg developed consequences of the quantum formalism that quickly became as emblematic as quantum jumps. These consequences are irreducible uncertainties in the simultaneous measurement of certain pairs of mechanical quantities, like the momentum p and position q of a particle, or the time t of its passing a point with an energy E ('conjugate quantities', as physicists say). Classical physics has no constraint in theory on the precision to which measurements of the conjugate quantities of material particles can be made. In Heisenberg's physics, the quantum-mechanical uncertainties cannot be eliminated. They are related reciprocally through Heisenberg's uncertainty relations $\Delta q \Delta p \approx h$, $\Delta E \Delta t \approx h$, where h is Planck's constant and Δ indicates 'uncertainty'.

These relations do not imply an insurmountable obstacle to the precision with which any single quantity can be measured. For example, $\Delta q \Delta p \approx h$ does not mean that quantum theory always limits the exactness of a measurement of position or momentum. It means that higher precision in the measurement of the position of a particle implies greater uncertainty in the *simultaneous* measurement of its momentum; the smaller Δq the larger Δp and *vice versa*. An exact measurement of position, with an infinitely small uncertainty ($\Delta q \approx 0$), implies an infinite uncertainty in the simultaneous measurement of momentum. We can in principle know exactly where a quantum object such as an electron is at any moment, but then we can have no way to determine its speed or heading.

Heisenberg's decision to submit an account of his new doctrine for publication during Bohr's absence left no room for uncertainty about Bohr's attitude on his return. Heisenberg may have suspected that everything was not in order in his exposition. But he knew that Bohr would scrape over every syllable with his relentless verbal scalpel. He was right. Bohr found fault with the argument and made Heisenberg correct the worst error in page proof. Bohr faulted Heisenberg's presentation for its one-sided description of the phenomena, for supposing that both the quantum object under investigation and the experimental instrument used to study it were particles (electrons, light quanta, atoms). Bohr insisted on democracy, on taking wave-like properties into account.

Heisenberg began his paper on uncertainty with a thought experiment that almost completes our collection of quantum icons. Let there be a microscope sufficiently powerful to reveal the position of an electron. Classical theory makes the resolution of a microscope (the smallest distance it can distinguish) proportional to the wavelength of the light it employs. To locate a particle within a distance Δq equal to a wavelength λ requires illumination by light with wavelength no greater than λ: $\Delta q \approx \lambda$. Given the smallness of the electron, light of very short wavelength, such as high-energy gamma rays (of even higher energy than X-rays), would be required for an exact measurement. Observing the gamma rays bounced off the electron would be sufficient to reveal its position.

But, Heisenberg reasoned, the smaller λ and the better the resolution, the greater the momentum of the light quanta ($p = h/\lambda$) and the more uncertain the uncontrollable exchange of momentum between it and the observed electron. Heisenberg took as the uncertainty in the particle's momentum Δp the recoil of the gamma-ray quantum as calculated from the Compton effect, $\Delta p = h\nu/c$, and so obtained $\Delta q \Delta p \approx \lambda h\nu/c$. Since for a wave travelling with velocity c, $c = \nu\lambda$, this relation retrieved the reciprocal uncertainty, $\Delta q \Delta p \approx h$. The position of the electron is uncertain because nature determines that the act of observing it disturbs it uncontrollably.[1]

No, no, Bohr must have ranted when he read Heisenberg's argument: if $\Delta p = h\nu/c\,(= h/\lambda)$ then it would be known as exactly as λ, and there would be no uncertainty. That was not the greatest paralogism, however. Heisenberg had not followed through on the classical functioning of the microscope (the very problem he could not answer in his doctoral examination). According to classical optical theory, the minimum distance resolvable by light of wavelength λ is $\lambda/2\sin\theta$, where the angle θ has the significance shown in Fig. 5. Bohr accepted this minimum distance as the uncertainty in position Δq. For Δp he took the maximum horizontal component of momentum that the illuminating light (assumed incident with momentum p along the microscope's axis) can acquire from the object and yet be scattered into the lens, both to the left and to the right: $\Delta p \approx 2p\sin\theta$. The product of uncertainties, $\Delta q \Delta p$, is now $\lambda/2\sin\theta \times 2p\sin\theta \approx \lambda p$. So far Bohr reasoned from the wave theory of light. Then, hauling in the quantum via de Broglie's rule, $\lambda p = h$, he had the uncertainty relation $\Delta q \Delta p \approx h$. Heisenberg acknowledged the 'depth and refinement' of Bohr's corrections in a note added to the printer's proof of his paper.[2] The note, which Heisenberg did not willingly add, healed their painful disagreement. 'Thank God!', Heisenberg wrote to his parents. 'My friendship with Bohr is of course more important than physics'.[3]

Heisenberg used a version of Bohr's interpretation of the gamma-ray microscope experiment in lectures on quantum mechanics he delivered in Chicago in 1929.[4] But he remained stubbornly unreconciled to the argument that the origin of uncertainty has its roots in a wave–particle duality of quantum objects.[5] He had introduced his microscopical analysis to provide operational definitions of 'position' and 'velocity', but it became instead a tool to establish uncertainty 'as an essential characteristic of the electron'.[6] The characteristic does not depend on an ambiguous duality; it manifests itself whether the analysis is purely wave-like or purely particle-like.

The physicist-philosopher Henry Margenau did not think much of Heisenberg's efforts to illustrate the uncertainty principle by thought experiments. 'Legendary apparatus' used to prove trivialities, obscure the issue, and achieve the impossible! Heisenberg purported to show that the 'physical interference of the measuring apparatus with the state of a system *conceived classically* produces an uncertainty in other variables'. So what? Similar interferences occur in classical physics; yanking in h by gratuitous appeal to

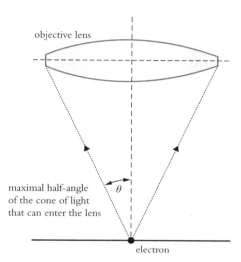

objective lens

maximal half-angle
of the cone of light
that can enter the lens

θ

electron

Fig. 5 Bohr's analysis of Heisenberg's infamous gamma-ray microscope experiment employs the classical wave approach to deduce a minimum resolvable distance, Δq, and a quantum particle approach to determine a classical uncertainty in momentum, Δp. Applying de Broglie's relationship to the result reproduces Heisenberg's uncertainty principle, $\Delta q \Delta p \approx h$.

a quantum postulate adds nothing. And emphasizing interference misleads in suggesting that the measuring process is the only cause of uncertainty.[7]

To bring his microscope into sharper focus, Heisenberg charged his young student Carl Friedrich von Weizsäcker to redo the argument using the fledgling quantum form of Maxwell's classical electrodynamics that he and Pauli had devised to describe the light-quantum. The purpose of this analysis was not to derive the uncertainty relations, which were assumed to hold; it was rather to demonstrate that uncertainty would still prevail in the new representation. Von Weizsäcker published his demonstration, which formed the core of his doctoral dissertation, in 1931.[8]

Von Weizsäcker made the incoming gamma-ray a plane wave striking an electron confined to the microscope stage. He laboured to prove that the outgoing ψ-wave scattered from the electron would move through the microscope much as a classical light wave and produce much the same effects. If the outgoing wave is represented as spherical, it converges to form an image in the *image plane* of the microscope, as shown in Fig. 6(a). A photographic plate in this plane will capture the image, from whose position the initial position of the electron can be inferred. In modern quantum terminology, we have adopted a 'position basis' for the electron.

(a)   (b)

image plane

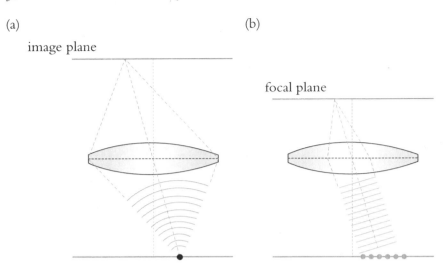

focal plane

Fig. 6 In von Weizsäcker's analysis of Heisenberg's gamma-ray microscope experiment, the outgoing photon scattered from the electron is described as a spherical wave if the photographic plate is placed in the image plane of the microscope (a), or a plane wave if it is placed in the focal plane, (b). The arrangement in (a) allows determination of the position of the electron, but not its momentum. The arrangement in (b) allows determination of the momentum of the electron but not its position. The arrangements are clearly mutually exclusive.

But if instead we place the photographic plate in the *focal plane* of the lens, we are required to assume a plane wave description for the outgoing wave from which we can determine only the direction from which it entered the microscope, Fig. 6(b). The analysis now demands a 'momentum basis' for the electron. We can use the direction of the plane wave and the location of the image it forms in the focal plane to deduce the momentum of the outgoing wave and hence the momentum imparted to the electron, but the assumption of planarity implies that we can no longer say anything about the electron's position. Positioning the photographic plate in the image plane or the focal plane allows us to determine the position or the momentum of the electron. But these measurements are evidently mutually exclusive. They cannot be performed simultaneously, and uncertainty prevails.[9]

Is the outgoing ψ-wave spherical or is it planar? The question is natural and insistent. Alas, in this analysis it must be rejected as unanswerable and irrelevant. The ψ-wave is 'purely symbolic' and can be whatever it needs to be in the context of the analysis of a specific experimental arrangement. Lurking in von Weizsäcker's analysis is a further consequence of the quantum treatment that, like Banquo's ghost, will very soon return to haunt our drama.

During his refreshing skiing trip, Bohr framed the quantum philosophy to which, when fully developed, he gave the awkward name 'complementarity'. The development included another fiery engagement with Heisenberg and more temperate help from Pauli and a Swedish theorist, Oskar Klein, who became Bohr's assistant after Heisenberg acquired a full professorship at Leipzig in the summer of 1927. Although a collaborative endeavour, its driving force was Bohr's compulsion to get to the bottom of things by constant revision of his ideas. Sometimes the reformulations seemed mere word play to his collaborators who recorded his mumblings as he gazed into depths deeper and murkier to him than to them. He talked and explored as if he would live forever, probing simultaneously for the concepts peculiar to the quantum world and the boundaries that separated it from classical physics and whatever might lie beyond the reach of quantum concepts. He could anticipate from his own experience and Høffding's teaching that quantum mechanics would fail eventually, and he had in Einstein's relativity a demonstration that deep difficulties could be surmounted by questioning accepted concepts of space and time.

In April 1927 Bohr wrote out the kernel of complementarity for Einstein. Heisenberg, then on vacation, had asked Bohr to send Einstein a proof copy of his paper on uncertainty, and Bohr, unwilling to be entirely upstaged by his student, included indications of his distinctive alternative. He began with the unexpected lesson that the unavoidable uncertainties that Heisenberg had detected gave just the leeway needed to enable physicists to apply classical concepts to the microworld. It takes some adroitness, however, to escape having to choose between the Charybdis of a wave and the Scylla of a particle description. We do it by recognizing that we cannot sail into both at once. Proper analysis of thought experiments like the gamma-ray microscope shows that the two possible aspects of the phenomena, undulatory and particulate, cannot show themselves fully at the same time and place.

Even so, Heisenberg's picturable thought experiments are profoundly misleading. They are altogether at odds with the rejection of electron orbits, the supposed inspiration for the breakthrough in Helgoland. The microworld cannot be visualized! There is no such thing as a classical particle there. Electrons dissolve into de Broglie waves just as light quanta into light waves. We can, indeed must, use both wave and particle descriptions, but we must also bear constantly in mind that in the quantum domain they are merely symbolic, applicable only when limited by the consequences of acts of observation. Bohr ended his letter with his usual optimism, as usual unrealized. He would have a fuller account ready 'soon', although, he hedged, the concepts needed were slippery and the situation fluid. 'Advances come at such a pace that everything new quickly becomes banal'.[10]

Three months later Klein noted, in Danish, 'All information about atoms expressed in classical concepts / All classical concepts defined in space and time/ ... /Complementary aspects of experiment that cannot be united into a space-time picture based on classical theories/ ... /Modern ... methods symbolic, and the treatment fundamentally statistical'.

These are headings for a featured lecture Bohr was to give in Como in September at a grand international meeting to celebrate the centenary of the death of Alessandro Volta and the rise of fascist Italy to its rightful place among the nations.[11] Draft after draft of the Como lecture was written during the unhurried summer until someone noticed that the world had slipped into September. Bohr had managed to compose only a short note that for a time he thought final and proposed to send to the British journal *Nature*. In his haste he made the distressing mistake of signing the article 'Niels Bohr' instead of his customary 'N. Bohr'. And so, as he prepared to leave for Italy without a text for his featured lecture (none has been found) he botched the paternity of his *Nature* note (he withdrew it permanently); and when he presented himself at the railway station in Copenhagen, he botched that too (he had forgotten his passport).[12]

Bohr's lecture at Como was a flop even for him. The stenographic report is incomprehensible. He talked for an hour (all other speakers had 15 minutes), inaudibly, in lengthy 'Bohrish' sentences hard to unscramble when read and impossible when only heard. It is said of his last recorded lecture that it took a team of linguists a week to discover the language he was speaking. Eugene Wigner, an important future actor in our drama, heard Bohr at Como. Wigner had spent time among the mathematicians of Göttingen and judged that he could speak for all Bohr's would-be auditors: 'This lecture will not induce any one of us to change his own meaning [opinion] about quantum mechanics'.[13]

Members of the Göttingen–Copenhagen physics circle also present at Como tried to help their bewildered colleagues in the discussion that followed Bohr's lecture. Born began it with two false statements. One was that Bohr had expressed so well the views 'we' hold about the foundations of quantum theory that little more needed to be said. The other statement was that matrix and wave mechanics constituted a 'unified structure of thought'. Born then asserted that quantum theory requires acceptance of the proposition that nature itself prohibits a complete specification of the state of a 'closed' system (one unsullied by contact with an outside world) and so forces physicists to resort to statistics. Kramers reiterated that Bohr had omitted nothing fundamental and showed how his elucidation saved most of the BKS theory. Heisenberg then declared that if his uncertainty relation could be beaten, quantum mechanics could not stand. The only Italian recorded in the discussion, the rising star Enrico Fermi, ignored the Göttingen–Copenhagen puzzles altogether.[14]

Bohr stayed in congenial Como for a week after the conference to work with Pauli on a manuscript for its proceedings. A version Bohr thought fit to print went to *Nature*'s German equivalent, *Die Naturwissenschaften*, on 11 October and was published, after revisions so wearisome that Pauli, who helped with them, refused to visit Copenhagen until Bohr could certify that the final proofs had been returned to the journal.[15] The article does not differ significantly from the paper Bohr eventually published in the conference proceedings, where few colleagues consulted it.

The definitive initial version of complementarity that *Prinzipienfuchser* would ponder appeared in *Nature* as well as in *Die Naturwissenschaften* in mid-April 1928. In the *Nature* variant, which we take as definitive, the persevering reader will find shrewd elucidations of thought experiments qualified by long sentences heavy with inevitability ('it is just this very circumstance that demands . . .'), obligation (we must 'renounce' the classical goal of a causal space–time description), and mystery ('symbolical utilization' of classical ideas). The difficulties of the presentation, apart from its language, lie in its two (or more) intertwined discourses. One dwelt on the peculiar behaviour of the microworld when the physicist tries to describe it using classical concepts. The other associated the physicist's difficulty with deep problems of cognition, the nature of science, and the human predicament.[16]

The quantum postulate, symbolized by h, records the brute fact that the atomic world suffers from 'an essential discontinuity, or rather individuality'. Consequently, any measurement made on an atomic system disturbs it in an uncontrollable way. Unlike the classical case, where the disturbance is either negligible or calculable, the quantum case involves an exchange between a 'measuring apparatus' and an 'observed system' that cannot be determined with an accuracy greater than one quantum of action. The quotation marks in the preceding sentence indicate that since it is arbitrary how the experimenter allocates h between the apparatus and the system, to neither can an 'independent reality' be ascribed. To infer the state of the system when untouched by measurement would not be safe. Free particles and radiation in empty space are abstractions useful symbolically, perhaps, but disqualified in principle as subjects of investigation. Bohr raised these extrapolations from the quantum postulate into a resounding principle. We must 'regard the space–time coordination and the claim of causality, the union of which characterizes the classical theories, as complementary but exclusive features of the description, symbolizing the idealization of observation and description respectively'.

Bohr accepted the uncertainty relations and offered a neat derivation of them that did not rely on Heisenberg's problematic microscope. The derivation represented a single electron by a packet of de Broglie–Schrödinger waves constructed from a superposition of waves with a frequency between ν and $\nu \pm \Delta\nu$, or wavelengths between λ and $\lambda \pm \Delta\lambda$, where $\Delta\nu$ is small compared with ν, and $\Delta\lambda$ is small compared with λ. The length of the resulting packet is our Δq (see Fig. 4 on p. 47) and the time taken for the bulk of the packet to pass some fixed point our Δt. In classical optics, these variables satisfy the approximations $\Delta\nu\Delta t \approx 1$ and $\Delta q\Delta(1/\lambda) \approx 1$.

The more precise the location of the classical wave packet (the smaller Δt and Δq), the greater the spread of wave frequencies (and reciprocal wavelengths) required to produce it. Convert now to the unpicturable quantum picture using the Planck–Einstein relation in the form $\Delta E = h\Delta\nu$ and de Broglie's rule in the form $\Delta(1/\lambda) = \Delta p/h$, and you recover the uncertainty relations $\Delta E\Delta t \approx h$ and $\Delta q\Delta p \approx h$. The spread of frequencies transforms into a spread in energy; the spread of reciprocal wavelengths into a spread in momenta.

The symbolic character of quantum theory appears immediately from Einstein's light *particle* with its energy $h\nu$, for its definition requires appeal to a quantity taken from the classical theory of *waves*. The electron wave packet consists of waves that do not exist in a normal physical sense. Unlike particulate electrons, classical wave packets do not persist; they disperse. Unlike sound waves, the de Broglie–Schrödinger waves have no medium to wave in and, unlike electromagnetic waves after Einstein destroyed the ether, they do not carry energy. They are 'purely symbolic', supporting derivations that are little more than rhetorical crutches, appeals to a physics that cannot be visualized or taken too literally.

Symbols, emblems, hieroglyphs everywhere! Matrix mechanics, a symbolic calculus, developed from the concepts of stationary states and quantum transitions by 'symbolic application' of classical theories. Wave mechanics too is but 'a symbolic transcription of the problem of motion of classical mechanics'. In these depths there are few places to drop anchor. 'The very idea of individual particles' is no more real than the superseded concept of stationary states. Still, with due prophylaxis, physicists can make observations and describe them to one another in classical terms.

Perhaps most striking in this elucidation was Bohr's hazardous handling of classical concepts in describing the microworld. Why not invent entirely new concepts, Schrödinger objected, concepts made for the job, rather than use patched-up inadequate old ones? Bohr dismissed the suggestion. 'We have not only ... no basis so far for any such modification, but the "old" empirical concepts appear to me inseparably tied to the foundation of human means of visualization'. We cannot do otherwise. New, purposefully forged quantum concepts cannot be conceived by ordinary, classical mortals. 'The quantum theory is approaching a stable provisional completion. Indeed, I believe that already we can say that any application of classical concepts that admits an unambiguous definition may also be given a physical interpretation'.[17] In this fashion the correspondence principle lived on in Bohr's physical philosophy.

The point is fundamental. Although Bohr, being human, could not avoid inconsistencies in pronouncements made over 40 years, his teaching about the place of classical concepts in the application of quantum theory remained remarkably stable if not always readily decipherable from his first formulations of complementarity. The requirement that physicists speak classically about the functioning of their instruments and the correlated aspects of the microworld the instruments bring out enables them to make and report measurements. But this requirement by no means makes a classical world: aspects of quantities conjugate to those measured cannot be determined together with them and remain out of reach.[18] Classical concepts define what we can learn about the microworld in any well-defined experimental situation. The difficulty about measurement does not arise because we disturb a world already defined, but because whatever undisturbed world might possibly exist is altogether inaccessible to us.

The necessity of classical concepts is plain from the fundamental formulas of matrix and wave mechanics, which take over the ordinary energy equation and its symbols unaltered. The bottle remained; the wine changed. In Heisenberg's form of quantum theory, the symbols stood for matrices; in Schrödinger's, for mathematical operations on ψ waves. The quantum condition could not be imposed via the old correspondence principle because classical physics has no h. But in both forms of quantum mechanics, the quantum condition can be set up to resemble classical formalisms closely. Does it signify that h enters the equations multiplied by the imaginary square root of -1, in wave mechanics in the wave equation, in matrix mechanics in the non-commutative algebra? Since observable results, which are the same in both systems, must be expressed as real numbers, quantum mechanics would seem to conjure a real world from imaginary symbols.

Bohr ended his account of the 'stable provisional completion' of quantum foundations with a warning. Relativity forced us to give up classical concepts of space and time. When honestly wedded to quantum theory, it will require more sacrifices than causality. 'We must be prepared to meet with a renunciation as to visualization in the ordinary sense going still further than in the formulation of the quantum laws'. We are not perfectly prepared to meet new problems because the words at our disposal refer only to ordinary perceptions. With the help of complementarity, we have found ways to talk rationally about 'the feature of irrationality characterizing the quantum postulate'. We can confidently expect more irrationality as we plunge deeper into the microworld. 'The situation . . . bears a deep-going analogy to the general difficulty in the formation of human ideas, inherent in the distinction between subject and object'.

With this hint at a question that had occupied him since his time with Høffding, Bohr expressed the hope, which in fact was a conviction, that the 'idea of complementarity' could deal with the irrationalities that would continue to arise as the human mind took on the universe.

ACT II

Uncertainty to Orthodoxy

Incompatible Conceptions

Einstein checked into the Grand Hotel Britannique in the early afternoon of Sunday 23 October 1927. He did not find the place inviting. 'The dining room is no bigger than our living room', he wrote his wife, 'and this is the leading hotel in Brussels'. He went to his room: 'it is like a haunted castle'. Was that why, in freshening up, he brushed his teeth with shaving soap? 'It was enough to make a normal man retch'.[1]

The non-normal man had arrived the day before the opening of a special conference convened, with his help, under the auspices of the Solvay Institute for Physics. This was a private, public, and international organization associated with the University of Brussels and patronized by the royal family and descendants of its creator, Ernest Solvay, a wealthy industrial chemist whose unusual ideas included confiscatory taxes on wealth. The first of the small, invitation-only Solvay conferences on physics had assembled in 1911. It took up urgent problems of radiation and the quantum. The fifth in the series, on *Electrons and Photons*, was about to convene, to treat the same subject with greater power and no less controversy. The unexciting title was a compromise. The original held out hope for a reconciliation of classical and quantum concepts, but Ehrenfest and Bohr had made clear to Lorentz, who presided over the Solvay Institute's scientific committee, that no such hope could reasonably be entertained.[2]

The uncompromising conference was significant for the political as well as the intellectual history of science. The two pre-war Solvay meetings had advertised their internationalism. The third, the first after the war, convened parochially in 1921 without any representative of Germany or Austria-Hungary. This ostracism persisted until the conference of 1927. It owed its revocation largely to the astute diplomacy of Lorentz. Attendance at international gatherings was still heavily politicized. Belgian scientists had only just reversed their opposition to inviting Germans (apart from Einstein and other pacificists) to international congresses and German pride and politics had made acceptance of the Solvay invitation problematic.[3]

To help overcome their resistance, Lorentz asked Einstein to join the planning committee and lobbied Planck, who had worried that the committee would choose Germans not for their scientific merit, but for their conduct during the war. Planck instanced his own invitation as proof of prejudice. He had not contributed to the latest developments in physics but had questioned Germany's belligerency and war aims. Sommerfeld had not been invited although he was still a leader in the field presumably because he had favoured the annexation of Belgium. Planck had failed to notice that his attendance at the earlier Como meeting had lent his prestige to Mussolini's government. He came to see that he would be making a political statement whether he accepted Lorentz's invitation or not and decided to help reunite the European physics community.[4] Having thus secured the cooperation of Planck and Einstein, Lorentz faced the problem of limiting the size of the German contingent.

The conference format consisted in the delivery and discussion of pre-circulated reports. The programme drawn up before Einstein joined the committee in April 1926 provided for reports by Compton on light quanta, by de Broglie on matter waves, and by Einstein on new approaches to Planck's radiation formula, the Trojan horse by which the quantum had invaded the world of physics in 1900. Heisenberg was to talk about Göttingen quantum mechanics. Einstein excused himself from preparing a report on the ground that he had nothing new to say and proposed Schrödinger, whose 'truly brilliant extension of de Broglie's ideas' was in press.[5] Lorentz replied that Einstein could report on whatever he pleased and that Heisenberg or Schrödinger might be assigned 'the foundations of the dynamics of quanta'. However, no more than two of the trio Born, Heisenberg, and Pauli could be accommodated.

Whom do you recommend? Heisenberg if only theorists count, and so Heisenberg and Born, 'for it would be difficult to put Pauli before Born', the student before the professor.[6] In the end, all three came. Born and Heisenberg gave a joint report and Schrödinger presented his case, although his name does not appear on the list of participants. Einstein did not take up Lorentz's invitation to talk on any subject of his choice. He again claimed ignorance about the recent 'impetuous' developments of quantum theory although he knew quite enough about them not to like them. 'I cannot accept the statistical way of thinking fundamental to the new theories'.[7]

There was another reason for Einstein's reticence. When Lorentz first asked him for a report, he thought he had something very important to say about the new theories. He had found a way, as he told the Berlin Academy of Sciences on 5 May, to remove the obnoxious probability from quantum mechanics. Perhaps stimulated by Heisenberg's paper on uncertainty, Einstein ascended into the multidimensional space of the ψ waves for a system of many particles, gave it a 'metric' in the style of general relativity, associated the kinetic energy of each particle with a direction in this space, and persuaded himself that he had connected 'completely definite motions to solutions of the Schrödinger equation'.[8]

He sent these good tidings to Ehrenfest and to Born, who passed them on to Heisenberg. In shock, Heisenberg wrote Einstein that 'only after many pangs of conscience did I come to believe in the uncertainty relation, but now I am completely convinced of it'. He earnestly, fearfully, requested a conversation to talk things out. 'Are there experiments that can decide between the two views?'[9] He need not have worried. Einstein had already abandoned his approach. He discovered that it would not allow him to split the many-particle system into two subsystems independent of one another even when separated to great distances. This was to him a fatal flaw. He had laboured to eliminate 'non-local' interactions, which implied unphysical actions at a distance, from gravitation theory and had no wish to reintroduce them into quantum theory. He had stayed with his idea long enough to write a paper about it but had decided against publication before the month was out.[10]

In the end, Einstein did not give a report to the conference. Neither did Bohr. Instead, he substituted a version of his Como talk that replaced his contributions to the discussions, which, for the usual reasons, the rapporteur had trouble following.

At 10 a.m. on Monday, 24 October, the conferees gathered at the Solvay Institute of Physiology in Parc Léopold. They heard first about the latest experimental evidence bearing on the wave-particle conundrum. William Lawrence Bragg reported that he had used his technique of reflecting X-rays from crystals, which had won him a Nobel prize, to show that the distribution of electronic charge in crystal atoms agreed, more or less, with Schrödinger's realistic interpretation of standing ψ waves.

SOLVAY CONFERENCE 1927

A. PICARD   E. HENRIOT P. EHRENFEST   Ed. HERSEN   Th. DE DONDER   E. SCHRÖDINGER E. VERSCHAFFELT W. PAULI   W. HEISENBERG R.H FOWLER L. BRILLOUIN

P. DEBYE   M. KNUDSEN   W.L. BRAGG   H.A. KRAMERS   P.A.M. DIRAC   A.H. COMPTON   L. de BROGLIE   M. BORN   N. BOHR

L. LANGMUIR   M. PLANCK   Mme CURIE   H.A. LORENTZ   A. EINSTEIN   P. LANGEVIN   Ch.E. GUYE   C.T.R. WILSON   O.W. RICHARDSON

Absents : Sir W.H. BRAGG, H. DESLANDRES et E. VAN AUBEL

At 10 a.m. on Monday, 24 October 1927, the conferees gathered at the Solvay Institute of Physiology in Parc Léopold, Brussels.

After Bragg came Compton, with five forceful objections to the wave theory on which Bragg based his analysis. The Compton effect and its discoverer spoke unequivocally for light quanta, or 'photons' as Compton now called them, using a word coined by physical chemist Gilbert Lewis in California for the carriers of the energy previously allotted to electromagnetic waves. 'Do the waves serve as guides to photons or is there another relation between photons and waves? These are new questions, no doubt difficult to solve'.[11]

De Broglie took the stage on the following afternoon to outline an interpretation he had completed six months earlier. He had separated Schrödinger's single equation into two, one essentially Schrödinger's ψ, the other a travelling wave with spikes or singularities in its amplitude. 'Could there exist between the continuous solutions and those with singularities . . . a connection expressible roughly as follows: the continuous solutions would give a statistical picture of the displacement of the singularities corresponding to real solutions and make possible predictions of the "probability of the presence" of a singularity in a given volume of space?' De Broglie had answered, 'yes', the connection being, as usual with him, via phases, which he made equal and raised to 'the principle of the double solution'. For those who might think the formulation unnecessarily abstract, de Broglie had allowed that the singularity of the one solution is guided by the continuous wave of the other. 'One thus conceives of the continuous wave as guiding the motion of the particle. It is a pilot wave'.[12]

At the Solvay meeting, de Broglie explained that although the continuous wave establishes probabilities in Born's manner, it does not imply 'renunciation of determinism of individual phenomena'. We might assimilate Maxwell's electromagnetic waves to de Broglie's electron waves by supposing that neither carries energy and both guide particles.[13] In this scheme, the ψ wave is responsible for all the wave-like effects, such as diffraction and interference, but the particles maintain their integrity as localized, physically real entities.

The recorded discussion of these skirmishes disclosed a strong pre-existing polarization. Heisenberg dismissed Bragg's smeared Schrödinger electrons. Pauli and Born insisted that ψ waves do not exist in ordinary space and cast doubt on de Broglie's electronic trajectories.[14] Wise old Lorentz wondered how X-rays could be strongly absorbed at frequencies that, according to the Bohr theory, were not present in the atom, and how photons could create a diffraction pattern if they entered the experimental apparatus one at a time.[15]

Photons have their uses, Bohr replied, for example in elucidating the Compton effect. But (recurring to a favourite argument) their defining quantity, frequency, has meaning only in the wave theory. 'It seems that there is a logical contradiction there'. And then came a revealing remark, a truly golden text. Bohr admitted that he had tried desperately to do without photons. He had been willing to sacrifice the conservation of energy to the cause. But with the failure (and honourable funeral) of his last best effort, BKS theory, he had run out of alternatives. De Broglie had pointed the way forward. '[He] has removed many of the logical problems of the description [of radiation] in space and time in recognizing similar paradoxes about the nature of material particles'.

Or, perhaps psychologically more pertinent, de Broglie had demonstrated that physicists had been as mistaken about the nature of matter as they had been about the nature of light. A Compton collision, as Bohr now pictured it, involved four wave fields, one incoming and one outgoing for each particle. Schrödinger obligingly drew four packets on the blackboard in different colours of chalk. He might have spared himself the trouble. The interacting packets do not make a visualizable picture, Bohr said, but a 'symbolic analogy'. Progress is painful. We must leave our space-time impulses behind and be prepared for further 'renunciation of intuition'.[16]

Heisenberg and Born now took up the attack against the 'false ideas and ancient prejudices' of the opponents of matrix mechanics. They claimed the Göttingen theory, with its probabilities and uncertainties, as a complete quantum mechanics. '[It is] free from contradictions and predicts unambiguously the results of all possible experiments within its domain'. Yes, it requires some hard mathematics and the admission of a statistical element as an essential ingredient. And why should that be bothersome? 'There is no empirical argument that opposes the assumption of the indetermination [uncertainty] of the microcosm in principle'.[17] Schrödinger's equation? 'A special case', limited in scope, good for calculating, lousy for understanding. Its inventor refuses to acknowledge that indeterminacy results from 'an essential impotence' in our 'ability to comprehend physical phenomena'.[18]

With radical indeterminacy, we avoid ambiguity. With Schrödinger's ψ waves, construed realistically, we embrace futility. With Heisenberg's uncertainty, we see clearly to the foundations of quantum theory. The true meaning of Planck's constant is 'the universal measure of the indeterminism introduced into natural laws by the dualism of waves and particles'. The talisman $\Delta q \Delta p \approx h$ contains it all. 'We hold that quantum mechanics is a complete theory whose fundamental physical and mathematical hypotheses are no longer modifiable'.[19]

Lorentz replied to these open minds that the symbolism of quantum mechanics, which included quantities such as position and momentum named and connected in the same way as their classical counterparts, was a great mystery. And what warrant had the Göttingen group for ignoring good correspondence-principle quantities like the phases of the ψ waves, which defined their instantaneous locations in their characteristic peak-to-trough cycles? Born thought the neglect of phases unimportant. Bohr retorted that, on the contrary, Lorentz's question was of the first importance for 'discussions of the coherence of the methods of quantum theory'.[20]

Schrödinger's turn came on Wednesday afternoon. He struck a blow for realism by suggesting that ψ waves for atoms containing several electrons could be interpreted as realistic standing waves in ordinary space and time. Bohr questioned the project and Heisenberg declared it hopeless. Schrödinger decried the Göttingen alternative. Tricky arithmetic was no answer to problems in physics. 'I cannot yet see as a [legitimate] response to a physical question the assertion that certain quantities are subject to a non-commutative algebra'.[21]

There followed a brief hiatus to allow participants to attend a meeting in Paris to commemorate the centenary of the death of Augustin-Jean Fresnel, the French physicist who had perfected Young's wave theory. De Broglie was to give a lecture at the Société de Physique. The Solvay conference was suspended for a day so that any of its attendees who wanted to celebrate the invention of the theory of light that it rejected could do so. The Brussels conference resumed on the Friday afternoon, when a general discussion took place.[22]

Lorentz's opening declared the articles of the old faith. Clear and distinct ideas, particles that maintain their individuality, effects that follow directly from causes. Perhaps electrons can dissolve into clouds. If so, Lorentz demanded the freedom to ask how it happened. 'If anyone wanted to prevent me from such an investigation by invoking some principle, I would be very uncomfortable'. He did not object to treating ψ probabilistically, but he did object to making such a treatment a requirement. 'Could we not cling to determinism as an item of belief? Must determinism be rejected in principle?' These rhetorical questions called forth a mighty response from Bohr, which, as we know, was omitted from the conference proceedings because the rapporteur could not follow it. Instead, to thicken its linguistic opacity, the editors printed a French translation of the German article that gave the gist of the lecture conceived mainly in Danish and delivered in Bohrish at Como.[23]

Einstein, now roused, excused himself for not having mastered quantum mechanics and tried to embarrass the Göttingen mechanics with a simple thought experiment. 'One can look at the postulate of the domain of validity of the theory from two points of view, which I would like to illustrate with the help of a simple example'.[24] Suppose electrons impinge on a screen (S in Fig. 7) punctured with a very small hole (O). Electrons that pass through are diffracted in the direction of the hemisphere (P), which is lined with photographic film that records the arrival of individual electrons. Assuming that the experiment must involve 'de Broglie–Schrödinger waves' (ψ waves), he offered two possible accounts or 'conceptions' of it.

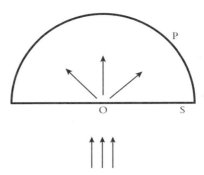

Fig. 7 Electrons impinge on the screen S. Some pass through a small aperture at O and diffract, producing a spherical wave, which spreads out beyond. The diffracted wave then interacts with the hemisphere P which is lined with photographic film.

Conception I—The ψ wave does not correspond to an individual electron, but rather to a 'cloud' or 'ensemble' consisting of many electrons. Quantum theory gives no information about individual processes. According to this conception, quantum theory is purely statistical and the quantity $\psi\psi^{\star}$ or $|\psi|^2$ represents the probability that in unit time *some* electron in the cloud will be detected at a given point on the screen.

Conception II—Quantum theory claims to be a complete theory of individual processes. Each electron is described by a packet of de Broglie–Schrödinger waves that diffracts at the aperture. In this interpretation, $|\psi|^2$ represents the probability that at a given instant a particular electron will be present at a given point on the screen. 'Here the theory concerns an individual process and claims to describe everything governed by some set of rules', for example, the conservation of energy and momentum in individual events, as in Compton collisions.

In conception I, the wavefunction is not itself physically real but instead carries statistical information about the 'cloud', implying, however, that a more exact knowledge of the states of individual electrons allowing exact predictions of their trajectories might be accessible. There appeared to be no way to distinguish experimentally between the two conceptions. The diffraction pattern appears only when many electrons have alighted on the screen and, in both conceptions, the same $|\psi|^2$ determines the probabilities for recording electrons at different places on the photographic film.

Einstein did not regard the choice as a matter of taste. For him, conception II could not stand. It violated relativity theory and common sense.

> I have some objections to make to conception II. The scattered wave directed towards P has no privileged direction. If $|\psi|^2$ were taken merely as the probability of finding a given particle at a certain place and time, it could happen that *the same* elementary process provokes activity *in two or several* places on the screen. But the interpretation that $|\psi|^2$ expresses the probability that *a particular* particle lands at a given point assumes a very peculiar mechanism of action at a distance, which prevents the wave, which is continuously distributed in space, from acting in *two* places on the screen.
>
> In my opinion, this difficulty can be removed only by augmenting the description provided by Schrödinger waves with [some way of] locating the particle during its propagation. I think that M. de Broglie is right to look in this direction. If we restrict ourselves solely to Schrödinger waves, interpretation II of $|\psi|^2$ would seem to me to contradict the postulate of relativity.

If quantum theory applies to an individual electron, what happens when the wave encounters the photographic film? According to Born, $|\psi|^2$ gives the probability that the electron will alight here or perhaps there. But if it is recorded 'here', then we discover simultaneously *and instantaneously* that it is not 'there'. At one stroke the continuously distributed ψ wave 'collapses' to produce one—and only one—mark on the film and is ineffectual everywhere else. That would seem to require an action propagated faster than the speed of light, in contradiction with one of the basic postulates of the special theory of relativity. Or, to anticipate a later change in emphasis, the collapse suggests 'non-local' connections that are hard to describe and bewildering to contemplate.

Apart from recommending de Broglie's line of march Einstein did not say how the theory might be augmented satisfactorily. His failed attempt the preceding May no doubt made him wary. Henceforth he would not attempt a wholesale reinterpretation of the quantum mechanics that Bohr defended but would try to subvert it with counter examples.

Bohr dismissed Einstein's experiment of the diffracted electrons as irrelevant. It made no sense to him.[25] He guessed that Einstein was stuck in a space-time description that demanded what quantum mechanics could not supply. He took ψ too literally, caught up, perhaps, by the snare of language. We must use words taken from older theories, Bohr wrote, we have no alternative. But we must not invest them with the same meaning. The classical concepts we deploy in quantum mechanics are only symbolic!

De Broglie understood the crux of Einstein's argument perfectly well. Either we say that the individual electron always has a definite position and velocity and so a trajectory, or that it is present throughout the diffracted wave train and at the moment of truth 'condenses' to photograph itself. On the first possibility, we can accommodate the ideas of Bohr and Heisenberg by supposing that a law of nature prevents us from determining position and velocity (and hence momentum) simultaneously. Then 'there is no real indeterminacy, but merely uncertainty imposed by the very nature of things'. On the second possibility, we have the problem that 'no mechanism in harmony with our ordinary notion of space-time can . . . explain this instantaneous condensation'. We would have to conclude that a complete interpretation of natural phenomena cannot be accomplished within the framework of space-time. And what then? This much: we must not follow Bohr and Heisenberg. Their approach did not suit the French mind. It was not 'lucid'.[26]

Well, Pauli replied, we certainly cannot follow lucid de Broglie since his 'double solution' does not predict the outcome expected from quantum mechanics for the simple case of an inelastic collision (one which involves an exchange of energy) between a particle and a quantum-sized rigid rotating object. Quantum theory requires that the rotator be in a stationary state both before and after the collision. But de Broglie's theory allows that 'in the individual collision process the behaviour of the particles should be completely determined and may at the same time be described completely by ordinary kinematics in spacetime'.[27] In general that would not leave the rotator in a stationary state. The criticism upset de Broglie so much that he did not hazard another significant contribution to physics for a lustrum. He might not have been so reticent had he known that the Scourge of God ranked his efforts above 'the childish papers of Schrödinger'.[28] In lectures delivered at the University of Hamburg in early 1928, de Broglie signalled his acceptance of complementarity.[29]

The Solvay proceedings contain no direct answer to Einstein's intervention. Instead, Paul Dirac, a Cambridge theorist then successfully developing his own powerful form of quantum mechanics, broke through his habitual taciturnity to observe that a simple argument shows that the deterministic evolution of an isolated system as assumed in classical theory is 'indefensible'. To check the predicted evolution requires a measurement, to measure requires a disturbance, and to disturb a system destroys its isolation. The good news is that the investigator chooses the question to ask. The bad news is that nature chooses the answer. 'The choice once made is irrevocable and will affect the future state of the world'. The experimenter shoulders quite a responsibility! Heisenberg thought that Dirac was too generous in allowing equal authorship of the world to nature. 'The observer himself makes the choice because only at the moment of observation does "choice" become a physical reality'.[30]

Nothing can be more obvious than the participation of human beings in the evolution of the physical world of which they are part. We make harbours, canalize rivers, move mountains, exterminate species, alter climates, sometimes intending the effects we produce, sometimes not. In making measurements, however, we have the clear intent of minimizing our contribution to the development of the universe, to reduce it to zero if possible. Classical physics assumes that the disturbance caused by observing can be made negligible or corrected for. Quantum physics says that nature is not able to collaborate in this way. At least one quantum of action, one h, cannot be assigned confidently to either the micro-entity being measured or the measuring instrument. Whatever the intent and delicacy of the observer, quantum theory prevents him or her from securing the information that would allow exact prediction of the course of events in the microworld.

Lorentz shuddered at his young colleagues' easy admission of this special indeterminacy and wondered at the timidity that conceded without a fight the existence of 'events that we cannot predict although until now we always thought we could predict them'. Why give up so soon?

The question had as much to do with psychology as with physics. As we know, Bohr had his special reasons anchored in his character and Danish philosophy for seeking, accepting, and exploiting apparently irrational elements in physics. His powerful scientific persona and wide-ranging thought gave him the status of a deep-seeing guru to his young German disciples. Other forces acted on them in the same direction when he was out of range. The economic hardships and doubts about traditional values and classical ideals characteristic of the early years of the Weimar Republic were never far from their minds. Their strong and rapid renunciation of the most general assumptions of the wonderfully successful classical physics that Lorentz had helped to build and that fit perfectly with the confidence and security of the world it described, did not come about merely because the concepts of wave and particle did not work in the microworld.

At the very end of the general discussion, after some thrust and parry over the details of photons, Lorentz expressed new reasons for dismay. The ψ wave extends through all space, the atom over a very small portion of it. 'That is very disagreeable'. Wave packets (considered as real physical things) are a charade, they spread, they cannot describe a respectable particle. Lorentz understood Bohr to say that a packet reconstitutes itself after (or during) every measurement. He may have had a statement by Born in mind; in any case, the explanation did not satisfy Lorentz, who wanted particles with everlasting life.[31]

He judged that Solvay V had failed to clarify the murky and sometimes contradictory concepts its participants employed and adjourned the formal proceedings of the conference in an appropriate mood of uncertainty. 'The compromise proposed on several sides of associating waves and point electrons I take simply as a provisional way to resolve the difficulty'.[32] Einstein agreed. 'As for "Quantum Mechanics" [he wrote Sommerfeld a week after the conference], I think that it contains as much truth about ponderable matter as the theory of light without quanta. It might allow a correct theory of statistical laws but not a sufficient concept of individual processes'.[33]

The formal, published, exchanges between Bohr and Einstein at Solvay V hardly constitute a 'debate'. But outside the conference room Einstein had peppered Bohr with thought experiments devised to reveal confusions or contradictions that he deduced from the Göttingen interpretation or from Bohr's complementarity. Ehrenfest witnessed these encounters, which he may have provoked. We know that he enjoyed setting his friends at head-to-head combat, as at the Lorentz jubilee of 1925, when BKS had fallen and 'Einstein triumphed over Bohr'.

This time the triumph was Bohr's. Ehrenfest described the contest to his students: 'At first [Bohr was] not understood at all . . . then step by step defeating everybody. Naturally, once again the awful Bohr incantation terminology. Impossible for anyone else to summarize'. The exchanges between Bohr and Einstein went on like a game of chess. 'Einstein all the time with new examples . . . Bohr from out of philosophical smoke clouds constantly searching for the tools to crush one example after the other. Einstein like a jack-in-the-box: jumping out fresh every morning . . . His attitude to Bohr is exactly like the attitude of . . . [anti–relativists] towards him'. '!!!!!!!BRAVO BOHR!!!!!!'

What impressed Ehrenfest most about Bohr's arguments was his even-handed treatment of light and matter, the great problem of late classical physics. Bohr used the conservation laws (as in the Compton effect) to apply the uncertainty relations to matter particles. '!!!!!!!BRAVO BOHR!!!!!!' Matrix mechanics had arrived in its own way at the same uncertainty relations for particles that classical wave mechanics specified for light. 'Downright undeserved magnificent harmony!!!!' Knowing then that the uncertainty relations preserve the conservation laws, the physicist can consider a collision between an electron and the Moon without worry-ing about destroying time-hallowed principles. Still, in principle, with the Moon as with a molecule, 'conceptual tracking . . . between the moments of observation' is just as false (or true) as 'tracking of a light corpuscle through the wave field between emission and absorption'.

Ehrenfest ended his informal report of the contest of the paladins where Bohr was to end the definitive presentation of his views in *Nature* in 1928. 'Bohr says that we have at our disposal only those words and concepts that yield such a complementary mode of description . . . The famous INTER-NAL CONTRADICTIONS of quantum theory only arise because we operate with a language that is not yet sufficiently revised'.[34]

Ehrenfest was probably right in thinking that Bohr would despair over this account of language. As we know, Bohr refused to alter the mean-ing of ordinary words in applying them to the quantum world. His efforts to devise rules for using them 'unambiguously' where they seemed to contradict would be decisive for defining the separation between the mea-sured object and the measuring instrument. The instrument belongs to the macroworld and the usual concepts in their ordinary meanings apply to it without restriction.

Bohr published two retrospective accounts of his contributions to Solvay V. In the earlier, an important essay about his discussions with Einstein (1949), he gave as the essence of his remarks the necessity of express-ing the arrangement and outcome of experiments in classical terms not only to make clear communication possible, but also to specify where in the imbroglio of instrument and microsystem classical concepts can be ascribed to micro-entities. His criticism of statements that assign classical quantities to micro-entities, such as 'we cannot know both the momentum and position of an atomic object at the same time', is not in conflict with this ascription.[35] As we know, only the aspects of the microworld correlated with the experimental set-up can be, or are to be, described classically; the other aspects are unknowable, not because they pre-existed and are destroyed by the measurement, but because they never existed.

In the later account, a short history of the conferences prepared for Solvay XII (1961), Bohr rephrased the main point of his intervention at Solvay V as 'the renunciation of pictorial deterministic description implied by the new methods'. Uncertainty makes statistics unavoidable. 'From this point of view, the whole purpose of the formalism of quantum theory is to derive expectations for observations obtained under given experimental conditions'. Bohr and his group did their best to convince Einstein of the need for this concession. But Einstein would not accept their demonstration that his efforts to outwit the uncertainty relations were 'futile'.[36]

Measurement and Impossibility

Soon after taking up his professorship in Leipzig, Heisenberg told its philosophers that they would have to give up the idea of causality. You can of course cling to some such concept as 'everything that happens must also happen', he said, repeating a quasi-joke of Bohr's. But if you want to know what everyone who designs a world picture needs to know, namely what physicists are thinking, you must recognize that because of an uncontrollable uncertainty in observations of the microworld, we cannot obtain the information necessary to predict future micro-events precisely.

With this doctrine atomic physics hammered another nail into the coffin of concepts that the philosophers' hero Immanuel Kant had declared necessary for the investigation of the material world. Einstein's relativity theory had proved that we must modify two of the necessary concepts, our ordinary ideas of space and time, to account for a certain domain of phenomena. Quantum theory has enjoined the third: universal strict causality. But philosophers need not bother themselves, young Heisenberg continued, to devise a substitute for causal relations in physics. Bohr has worked it all out from the insight that a causal description is complementary to a space-time description. Both cannot be used precisely and simultaneously, but both remain serviceable if limited by the uncertainty relations that, as everyone present knew, Heisenberg had invented.[1]

Heisenberg brought a more assertive version of this message to physicists and mathematicians assembled in Königsberg (now Kaliningrad) in the autumn of 1930. Causality fails, he said, because of uncontrollable uncertainty. The genius of Königsberg, Kant, had erred in thinking that strict causality is an essential element in human thought. Bohr's complementarity, which reconciles all difficulties in atomic physics, exemplifies how new knowledge destroys false belief. Heisenberg concluded with an easy catechism: complementarity is definitive; the ψ wave contains all the information attainable about the system it describes; atomic physics has no interest in isolated systems, no interest in the interior structure of atoms, and cares only for the process of observation.[2]

In the discussion that followed, the physicist-philosopher Philipp Franck, a good friend of Einstein, objected to reducing physics to an account of observations and asserted that the ψ wave is only a calculational aid, a pure fiction, and no replacement for a robust atom. Heisenberg agreed that ψ is an abstraction but insisted that it contains everything necessary and possible to know about a microsystem. It also contains the secret of uncertainty. Although it develops continuously in time for an isolated system, it suffers a discontinuity—a quantum jump—during an observation. The uncontrollable jump, which gives rise to, or is described by, the uncertainties, must not be placed in the mind of the observer, but in the physical interaction between the measuring instrument and the micro-entity.

A voice that could not be ignored interrupted. It belonged to the mighty mathematician Johann von Neumann, who corrected Heisenberg on a detail of measurement and laid it down that quantum theory had no 'objective validity'. The ψ wave is not observable. It is available to us only when experiment squeezes some information from it or adds some to it.[3] And then it is not the imaginary ψ but the real $|\psi|^2$ that makes the connection between theory and experiment. Von Neumann was one of the young geniuses in Göttingen in 1926, younger even and quicker of mind than Heisenberg, and with a similar penchant for physical and intellectual risk. Like several talented Jewish Hungarians, including his boyhood friend Wigner, von Neumann had been scattered from Budapest to Berlin, in his case via Göttingen, just when Born's school was elaborating matrix mechanics.

Von Neumann had come to Göttingen to work with the prince of its mathematicians, David Hilbert.[4] Together they attended a lecture on matrix mechanics by Heisenberg, which perplexed prince Hilbert because, von Neumann thought, physicists had not succeeded in constructing a theory logical enough for a mathematician to understand. He undertook to purge quantum mechanics of its defects and to explain it in a language comprehensible to Hilbert.

The multiplicity of approaches to the quantum theory aggravated the task. Despite the demonstration by Schrödinger and others that the matrix and wave mechanical formalisms gave equivalent results, they differed greatly in their principles and methods. Matrix mechanics lived in an abstract 'space' spanned by quantities referring to the stationary states and transitions within the microsystem under study. Wave mechanics described ψ waves in a different abstract space, a multi- or infinite dimensional 'configuration space' accommodating all possible physical configurations of the system. Only for the simplest atom, of hydrogen, did this configuration space reduce to the more familiar three dimensions of our classical, Euclidean experience.

Both Dirac and Jordan had treated the matrix and wave theories as special cases of a general 'transformation theory', but in a manner which for von Neumann relied too heavily on some ill-defined mathematical objects.[5] The theories could not easily be brought together: 'That this cannot be achieved without some violence to the formalism and to mathematics is not surprising. The spaces . . . are in reality very different, and every attempt to relate the two must run into great difficulties'.[6]

What interested von Neumann was the relation between mathematical functions in the two spaces and the way the relation figured in the physical interpretation of the theory. He showed that these several functions 'are identical in their intrinsic structure (they realize the same abstract properties in different mathematical forms)—and since they . . . are the real analytical substrata of the matrix and wave theories, this isomorphism [structural identity] means that the two theories must always yield the same numerical results'.[7]

To show the isomorphism, von Neumann cast quantum theory into a space still more abstract, 'Hilbert space', where, with an infinite number of dimensions at his disposal, he axiomatized a general algebra of which the various quantum theories were merely examples. Between 1927 and 1931 he published a series of papers that he reworked into *Mathematische Grundlagen der Quantenmechanik* (1932), a book of befuddling clarity in which (following Born) he claimed to have demonstrated that quantum theory was irredeemably statistical. In so far as quantum theory was true, causality could not operate in the microworld and no set of 'hidden variables'— physical attributes supposedly real but for some reason unobservable—could be introduced into it that could provide the causal space-time description that classical physics demanded.

Von Neumann's 'impossibility proof' was not as dogmatic as those tranquillized by report of it believed. It rests effectively on the premise that, if quantum theory satisfies the axioms on which he founded it, then hidden variables that eluded the axioms cannot be admitted without altering the foundation. Since, he argued, quantum theory had proved to be very competent in its domain, we should not try to alter it. Instead let us act like rational adults and acknowledge that 'in the atom we are at the boundary of the physical world, where each measurement is an interference of the same order of magnitude as the object measured'. The uncontrollable interference cancels causality.

Why should that trouble an unprejudiced mind? After all, there is no evidence for strict causality in macroscopic classical physics whose laws are just averages by which the peculiarities (the 'hidden variables') of a vast number of individual atoms or molecules are washed out. Only at the atomic level can the sort of causality claimed by classical physics be properly tested and in that domain quantum theory shows it to fail. 'Under such circumstances, is it sensible to sacrifice a reasonable physical theory for [causality's] sake?'[8]

As we know, the measurement by which the observer learns the state that experiment and nature have chosen for the quantum system is a messy act. We can judge how Von Neumann tried to clear it up using the problem Einstein had developed at Solvay V, shown in Fig. 7. Before reaching the photographic film, the ψ wave representing the electron develops freely in time in all directions, its motion governed by Schrödinger's wave equation. Von Neumann referred to this continuous, deterministic, and completely reversible evolution as process **2**. Interaction of the ψ wave at the film 'collapses' it into the description of a particle in a definite place. The probability of arriving at that place rises from whatever value of $|\psi|^2$ it had immediately before landing to the singular value 1. Von Neumann referred to this discontinuous, irreversible transformation as process **1**. The processes are distinct and mutually exclusive.[9]

Von Neumann's logic made the collapse of ψ an entirely different process from its undisturbed progression and placed the incomprehensible discontinuity in quantum theory precisely there.[10] And just where is there? Von Neumann identified three participants in every observation: I, the object observed; II, the measuring instrument; and III, the human observer. Quantum theorists can include II with I or II with III, in accordance with what they take the measuring apparatus to be.

Einstein had cast himself as III, as inspector of the combined system of the electron (I) and the photographic film (II). But he might have put himself together with II by including the film, a photon from it, and his own sensory apparatus—'his retina, nerve tracts and brain'. Von Neumann tried to show mathematically that both accounts (I + II/III, and I/II + III) give the same result. In either case we might say that the discontinuity in the measurement process alters the state of the world in so far as perception alters an observer's mind. It would not be safe to say more. 'Indeed, experience only makes statements of this type: an observer has made a certain (subjective) observation; and never any like this: a physical quantity has a certain value'.[11]

Bohr had insisted that the quantum object and classical apparatus be treated as distinct, yet inseparable. But von Neumann saw no reason why process **2** should not be applied to classical situations. The quantum system, governed by the Schrödinger equation, interacts with a measuring instrument and the composite system (I + II) develops smoothly and continuously. The composite system then interacts with the 'observing portion of the world'. Based on conversations he had had with his Hungarian compatriot Leo Szilard, a collaborator of Einstein, von Neumann suggested that component III consists of the observer's 'abstract ego'. Process **1** occurs when the measurement outcome is registered in the observer's *conscious mind*.

The 'unphysical' and acausal conclusion of the act of measurement occurs when the composite system encounters an unphysical object that 'remains outside of the calculation'. Von Neumann did not say just what he had in mind. But his conversations with Szilard concerned entropy reduction in thermodynamic systems through interference by intelligent beings (a variation on Maxwell's Demon).[12] We have here, according to Max Jammer's authoritative account of quantum philosophy, 'the beginning of certain thought-provoking speculations about the effect of a *physical intervention* of mind on matter'.[13] We shall follow some of these speculations later.

Von Neumann's approach appeared to offer sanctuary to those overwhelmed by a theory of the quantum domain riddled with misleading classical references. Borrowing mathematical devices introduced by Dirac, future iterations of von Neumann's formalism—of 'state vectors' in an abstract Hilbert space—would go on to reassure (or confuse) future generations of students. But a microworld of objects whose physical attributes could not be said to *possess* any value left no perceptual anchor or means of visualization for physicists grappling with the theory's mathematical abstractness. A few of those who struggled as students would successfully seek vengeance on the theory for their discomfort. We shall meet these contrarians in subsequent Acts of our drama.

Bohr had very little to say about von Neumann's theory of measurement, other than he felt that it was 'wrongly put'.[14] Nowhere does he discuss von Neumann's 'projection postulate', his process **I** interpreted as the instantaneous 'projection' or 'collapse' of the ψ wave into its final resting state.[15] Bohr had two strong reasons for not regarding measurement as a problem of principle. For one, he did not admit the application of quantum theory to classical instruments. For another, he took ψ as purely symbolic and von Neumann's components I, II, and III as elements in the evolution of our knowledge of the state of the object under study. Since there is no physical projection or collapse, process **I** cannot suffer one. It merely represents the registering of the measurement outcome and the updating of the observer's state of knowledge about it. The observer's 'abstract ego' intervenes only in a mild, passive, and common-sensical way.

Although Heisenberg did not mention von Neumann in any of his early efforts to spread what he understood to be the truths of quantum theory, he certainly knew of von Neumann's work and probably took comfort in it. The declaration by a frighteningly gifted mathematician that measurement presented an impenetrable discontinuity and that no interpretation that invoked hidden variables was available within the imposing edifice of quantum mechanics scared physicists who had hoped that something more satisfying to them than complementarity could be found. And it reassured fellow travellers who inclined towards Bohr and practical physicists who regarded time devoted to 'philosophical' questions as time wasted. These latter virtuosi, who may be exemplified by Slater, were content with an apparently secure dogma to which they could refer their inquisitive students without interrupting more important work.

While von Neumann was building out his logic of measurement, Heisenberg was bringing tidings of what he now called 'the Copenhagen spirit' to the farthest reaches of the globe. For seven months in 1929 he travelled, first to the United States, where he lectured at the University of Chicago and on the East Coast before crossing the continent to Ultima Thule, the California Institute of Technology (Caltech) in Pasadena. On the way he climbed the Rockies and descended into the Grand Canyon. In March and April, he was again in Chicago lecturing and swimming in cold Lake Michigan, then back west via Yellowstone National Park for hiking and a rendezvous with Dirac.

They travelled together from Yellowstone to Yokohama, Dirac as taciturn and Heisenberg as competitive as ever. After landing in Japan, they went for a walk that took them past a pagoda. Heisenberg spontaneously climbed it and, standing on its very apex (width Δq) on one foot in a howling wind, happily maintained an uncertainty Δp too small to knock him over. After some weeks of ground-based lecturing, Heisenberg returned to Europe via China and India and a hike in the Himalayas.[16]

Meanwhile Bohr was distilling the Copenhagen spirit for the more general benefit of humankind. He had always expected that quantum theory would prove important for all forms of thought and he devoted an essay written for the jubilee of Planck's doctorate to float his wider programme. We must be thankful to Planck, he said, for ushering into the world the infant quantum that has matured to support a theory with the 'beauty and self-consistency . . . of classical mechanics'. Abstract and symbolic though the theory may be, it can revolutionize thinking about problems well beyond the physics from which it emerged. The key to the revolution is the recognition that we must observe to learn, that by observing we change the world, and that to describe our observations fully we might have to resort to complementary concepts. 'In the facts which are revealed to us by the quantum theory and lie outside the domain of our ordinary forms of perception we have acquired a means of elucidating general philosophical problems'.[17]

Experience tells us we have free will but reason presupposes that a causal chain prompts our acts. The apparent contradiction dissolves in the attempt to design an experimental demonstration of the necessary causal chain. The investigation must finish in the brain, which, if observed while processing the act of volition under scrutiny, would not be able to finish it. Emotion and volition, thinking about others and preserving individuality, and analysis of a concept and its immediate application, are also complementary pairs. The limitations implied by complementarity arise as usual from interference through observation, in the case of mental phenomena by the complication that the observer is also the observed.[18] These fanciful consequences of quantum theory perplexed its inventor, Planck, to whom they were addressed. He replied that Bohr's essay was 'so deeply thought-out that I shall not now try to comment on its details'. Instead, he sent a recent essay of his own, 'with which, perhaps, you will not entirely agree'.[19]

In August 1929 Bohr told a meeting of Scandinavian scientists that quantum theory had brought 'new artifices' and 'expedient[s]' for attaching meaning to 'words like "the nature of matter" and "the nature of light"', that ordinary space-time concepts did not work in the microworld, and that they should be happy about it. '[It is] an essential advance in our understanding'. Do you have a problem reconciling free will and determinism? Remember that complementarity provides experimenters the leeway to bring out the wave or particle aspect of micro-events. Free will and determinism are also a complementary pair and so are both true, but not, of course, at the same time.

Regarding life and the analysis of physico-chemical systems as another complementary pair afforded another easy solution of an old problem. We can learn much about life from physical intervention if we do not prosecute it so far as to kill the object of inquiry before getting to its living principle. The secret of life will elude us forever. And, also, death. Since consciousness is inseparably connected with life, 'the very problem of the distinction between the living and the dead escapes comprehension in the ordinary sense of the word'. That was news! More was to come, more resignation, more insights. '[T]he new situation in physics has so forcibly reminded us of the old truth that we are both onlookers and actors in the great drama of existence'.[20]

A brief reference to this last expansion of the Copenhagen spirit appears in the printed form of Heisenberg's Chicago lectures, published in 1930 in German and in English. He devoted most of this useful short book to the interpretation of experiments, but also offered as Bohr's opinion that 'indeterminateness' (uncertainty) originated in the arbitrariness of dividing the world into observer and observed, of deciding what the apparatus of observation is to include, and thereby choosing the potential set of attributes of the microworld to actualize. The onlooker is an actor in a quantum drama. You must fit this complicated dichotomy into your conceptual scheme (Heisenberg advised his auditors) before you can judge the consistency of the methods of quantum theory and review the problem of 'separating the subjective and objective aspects of the world'.[21] It was an unfortunate formulation. It invited the conclusion that the Copenhagen spirit had re-invigorated radical subjectivism, whereby every individual creates his or her private world and objects of common experience cannot be specified.

It had not, of course.

Heisenberg's former collaborators in Göttingen also published a textbook on quantum mechanics in 1930. They gave short shrift to Bohr's new emphasis on subject, object, and reality. Their main purpose was to right the imbalance in the few textbooks then available on quantum theory which, they believed, unduly favoured the wave picture. They presented quantum mechanics as matrix mechanics, the good reliable Göttingen way, and offered their book as the promised sequel to Born's *Atommechanik* of 1925. Their derivation of the spectrum of hydrogen using matrices following Pauli's *tour de force* brought the Scourge of God down on them in a published review. After a few unfriendly remarks about their approach, he reached for attributes on which he could be complimentary. And found them: 'The print and paper of the book are superb'.[22]

EPR, Faust, and the Cat

The subject of Solvay VI, which met in October 1930, was magnetism. Bohr and Einstein attended, and Dirac, Heisenberg, Kramers, and Pauli, but not Lorentz, who had died in 1928, or Ehrenfest, whose absence deprived the world of a colourful account of the unofficial proceedings. The official ones contain nothing, in the reports or discussions, that echo the energetic arguments over waves and particles heard at Solvay V. Only when Bohr killed an experiment proposed by Fermi with a blow of uncertainty were objections raised to any manifestation of the Copenhagen spirit, and Bohr carried the argument.[1] As at Solvay V, the action of interest took place informally around mealtimes. But since the organizers had separated the participants, the older in better accommodations than the younger, there were fewer opportunities for group interaction.[2]

At one old-timers' breakfast, Einstein presented Bohr with his latest thought experiment. Bohr's retelling of it in 1949 inspired the title of this book: 'At the next meeting with Einstein at the Solvay Conference in 1930 our discussions took quite a dramatic turn'.[3] The drama was greater than Bohr's version intimated. According to it, Einstein's experiment employed a heavy box hanging from a strong spring and containing a photon and a fixed clock. The clock triggers a shutter that opens just long enough to allow the single photon to escape (Fig. 8).

Let the experiment begin! Weigh the box and its photon before and after the photon escapes and from the mass difference and $E = mc^2$ determine the precise energy of the photon. Now open the box and read the release time t recorded by the very exact clock. Since this measurement is independent of the weighing and can be made as exactly as desired, both E and t can be determined with unlimited precision and the product of their uncertainties ΔE and Δt can be made less than h.[4] No point of principle—no inescapable law of physics—denies us access to the precision required.

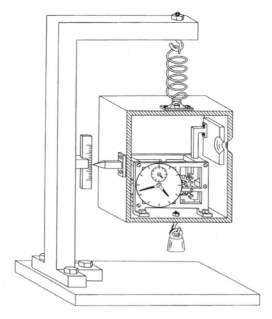

Fig. 8 Bohr's sketch of Einstein's 'photon box' thought experiment.

Well, Bohr, what do you say? In answering, please keep in mind your disciple Heisenberg's dictum that the failure of the uncertainty relations would signify the collapse of quantum mechanics.

Bohr frantically consulted the loyalists at the lesser hotel. Gloom abounded. He wandered between tables at dinner (the participants took this meal in common), mumbling his conviction that Einstein could not be right but offering no proof. Then during a sleepless vigil, it came suddenly to him that since relativity theory (in the form of $E = mc^2$) had produced the embarrassment, it must point to the solution. He worked out that the clock was the culprit.

Bohr accepted the mass difference as a suitable proxy for E, but not the assumption that it could be determined unproblematically by weighing. To bring the box back to its initial position after the loss of its photon, Bohr proposed adding a sequence of smaller and smaller weights until the pointer was returned to within a small distance δq of its original setting. (The δ signifies a classical measurement without quantum monkey-business.) To realign the pointer exactly would require an (unknown) small residual weight δm, representing a (classical) uncertainty. Precise alignment might require a very long time.

Let the time spent fiddling with the box be t as told by a clock fixed to the wall of the laboratory. During the fiddling, the box moves up and down against the acceleration g owing to gravity. This motion, according to Einstein's general theory of relativity that Bohr now ingeniously invoked, introduces time dilation effects. As the clock inside the box moves upwards against gravity, the time it keeps slows down. As it moves downwards, the time it keeps speeds up. The more the box jiggles up and down (the longer the reweighing procedure) the greater the likely disagreement, δt, between the clock in the box and the clock on the laboratory wall. This depends on δq according to $\delta t / t = g \delta q / c^2$, a result that can be derived as a 'weak-field' limit of general relativity.

During the last stage of the weighing, when the box is unbalanced by δm, the patient thought experimenter watches the box move towards rest at δq with a momentum δp, which must be less than $\delta m g t$, the momentum associated with the total fiddling time, or $\delta m g t > \delta p$. Here momentum is of the ordinary variety, mass (δm) times velocity $(= g t)$. If now an impatient theorist exchanges the classical δ quantities for quantum Δ ones, it follows from $\Delta t / t = g \Delta q / c^2$ and $\Delta m g t > \Delta p$ that $\Delta m \Delta t > \Delta q \Delta p / c^2$ (g and t both cancel). With $\Delta q \Delta p \approx h$ and $\Delta m = \Delta E / c^2$, Bohr had the result he wanted, $\Delta E \Delta t > h$. The energy-time uncertainty principle triumphs! BRAVO BOHR!

We have descended into detail about this box to indicate the trickiness of the analysis, which caused much doubt and confusion in its time. Einstein soon redirected the argument, if indeed he had ever urged it in the form Bohr recounted in 1949, and Bohr was never comfortable with his intricate rebuttal.[5] The Δt in Bohr's analysis refers to an uncertainty in the reweighing time, whereas Einstein's Δt refers to the moment of release of the photon. Perhaps most objectionable is the sleight-of-hand that circularly employs position–momentum uncertainty to demonstrate the validity of the energy-time relation. Although most physicists accepted Bohr's answer as brilliant and definitive, he fretted over it for the rest of his life. A rough sketch of the apparatus was on his blackboard the day he died.[6] Although Einstein conceded that Bohr's response was 'free of contradictions', he accused it of 'unreasonableness'.[7]

He might rather have said impertinence, for Bohr's answer did not address Einstein's redirected argument as Ehrenfest explained it to him in 1931.

It aimed at something much deeper than Heisenberg's principle. 'Einstein told me that he had not doubted the uncertainty relation for a long time and that therefore the "movable light-flash box" ... was ABSOLUTELY not designed against [it]'. To work Einstein's new thought experiment, Ehrenfest wrote Bohr, set the internal clock and the laboratory clock to zero and arrange for the shutter to release the photon at 1,000 hours. Weigh the box with the photon inside for 500 hours, then screw it to a rigid frame. At 1,500 hours, when the photon is well launched towards a fixed mirror half a light year away, choose whether you want to establish either the time t at which it left the box or its wavelength λ.

If you opt for time, open the box, which is still rigidly fixed, and compare the readings of the internal and external clocks. This will enable you to compensate for the fluctuations in the internal clock's rate during the 500 hours of weighing and so establish the exact time of release. Alternatively, free the box from the frame and weigh it again for 500 hours. You will now have precise measures of the mass of the box both before and after the departure of the photon from which you can deduce its wavelength from the mass difference, $E = mc^2$, and $\lambda = hc/E$.

The bite of the argument is that the returning photon must adjust its colour, precise or fuzzy, according to the experimenter's choice *while it is in flight*. Quantum mechanics seemed to demand that a micro-entity too distant from the apparatus from which it came to be reached by a signal travelling at light speed must nonetheless instantly respond to the experimenter's manipulations of it. This shocking behaviour, which Einstein thought needed only to be exposed to make all good physicists reject the theory that allowed it, is one of the main intuition-shaking themes of Act IV of our drama.

The argument did not seem sound to Ehrenfest. 'Even if the shutter opens VERY BRIEFLY the double weighing can determine the weight loss very exactly and therefore the energy of the emitted quantum; on the other hand, it is clear that a very short burst of light is very unchromatic'. The burst of light is too short to have a well-defined wavelength. 'I am very ashamed that I am so stupid, but I do not see what I am supposed to think about this "contradiction"'.

But Einstein's new argument did not concern contradictions or detailed measurements or the nature of light quanta, but rather the crazy consequence that the separated photon had to take on one or another property depending on whether the experimenter thought to compare the clocks or re-weigh the box.[8] And the choice could be delayed indefinitely! Admitting uncertainty as a certain consequence of the quantum mechanical formalism, a consequence reaffirmed just before Solvay VI by the demonstration that Heisenberg's inequalities could be obtained directly from the commutation relations,[9] incurred a great cost. The photon and its box remain mysteriously connected, their fates somehow tied together, no matter how far apart they are. Einstein could not make the concession. It would rub out separate, individual objects, essential traits of an acceptable world picture.

The disagreement threw Ehrenfest into one of his manic-depressive states, bleak when considering the situation in physics, excited when anticipating a battle between the titans. He invited them to meet in Leiden in October, to celebrate the award of the Lorentz Medal to Pauli and to talk. 'For ME it is hugely important to find out as exactly as possible how far you are obliged to think the same and where your freedom to go your separate ways begins'. He thought he might help by keeping them from talking past one another, especially by dissipating 'Bohr's terrible clouds of politeness that create so great a barrier to communication . . . Repeat your arguments if you want, Einstein, Bohr of course will be grateful. But if you prefer to do nothing but chat, that obviously is more than enough!!!!' Einstein did neither, since illness kept him in Berlin.[10]

In his account of 1949, Bohr acknowledged Einstein's earlier redirection of the photon-box argument and conceded that it 'might seem to enhance the paradoxes beyond the possibilities of logical solution'. That would have turned up the dramatic tension a few degrees! To relax it, Bohr issued a decree: a mathematically consistent formalism can be convicted of inadequacy only by showing that its consequences disagree with experience or do not exhaust the possibilities of observation. And a verdict: 'Einstein's argumentation could be directed to neither of these ends'.[11]

The Nazis prevented the continuation of the Leiden chats. By the time of Solvay VII (1933), Einstein had fled Germany to the new Princeton Institute for Advanced Study, which would soon also shelter von Neumann. From there, in 1935, Einstein issued a delayed-choice challenge derived from the second version of his photon-box experiment that managed to heighten tension in the Copenhagen camp. This time Einstein was not alone. The challenge came from him and two Russian–Jewish associates of the Princeton Institute, Boris Podolsky and Nathan Rosen. Their entry into the debate over quantum mechanics made an epoch in physics and an episode in the clash of civilizations.

First, the clash of civilizations. Anti-Semitism and the programme of 'Deutsche Physik', which sought to remove 'Jewish' theory from physics, drove many quantum theorists from Europe. Of those in our drama who remained in Germany, Jordan and Heisenberg were the most prominent. They protected themselves from attack for teaching Jewish anathemas in different ways. Jordan joined the Nazi party. Heisenberg 'aligned' in small things, like mentioning favourably the leaders of Deutsche Physik, the Nobel-prize-winning experimentalists Philipp Lenard and Johannes Stark, in his general talks on quantum mechanics. These concessions did not protect him from a serious attack in the weekly newspaper of Hitler's elite guard, the SS.

An article engineered by Stark in its issue of 15 July 1937 accused Heisenberg of being a 'white Jew', a carrier of intellectual bacteria, for teaching the theory of relativity and other triumphs of Jewish science. Heisenberg protested that his teaching had no connection with politics and that he had no sympathy with Jews. He was able to bring the adjudication of the case to the head of the SS, Heinrich Himmler, through their mothers, who happened to know one another. After the additional intervention of a prominent experimental physicist, Himmler called off the attacks and allowed Heisenberg to continue teaching modern physics on the understanding that he would not mention the names or praise the work of the Jewish theorists whose ideas he taught.[12]

Quantum theory was also under attack on the opposite side of the totalitarian seesaw, Soviet Russia, for portraying the microworld as a collection of partially realizable potentialities. Dialectical materialism, the Marxist theory adopted as the official philosophy of Soviet communists, needed a world of rock-solid substance. But quantum theory, especially in the positivist form Marxist theorists associated with Bohr, resembled the frightful gibberish of Mach, which Lenin had taken the trouble to annihilate during the age of the czars.[13] Podolsky, who was born in Russia, and Rosen, a first-generation descendant of Russian emigres, developed strong commitments to the Soviet system.

In 1930 Podolsky returned to Russia with a doctorate in theoretical physics from Caltech to work at the Ukrainian Institute of Physics and

Technology, where he collaborated with important theorists, including the visiting Dirac. He came back to the US in 1933 to take up a fellowship at the Princeton Institute. Einstein rated him highly. Podolsky remained attached to the Soviet cause. He would later be revealed by SVR (the post-Soviet incarnation of the KGB) documents as the spy codenamed QUANTUM, prepared to betray atomic secrets in the hope of obtaining a position in the Soviet Union.[14] His efforts earned him $300, but his lack of access to more important secrets (he was never part of the Manhattan Project) meant that he quickly exhausted his usefulness. Certainly he was capable of indiscreet self-promotion.

Rosen's romance with communism is more easily documented. Six months after the publication of the joint paper (hereafter 'EPR'), he asked Einstein, whose assistant he was, to recommend him for a job in the USSR. Einstein wrote to the Chairman of the Council of People's Commissars, Vyacheslav Molotov, praising Rosen for his talents as a physicist. '[H]e is not only a tireless worker, but also an unconditional adherent to the Soviet political and social system . . . his dearest wish is to collaborate with all his strength in building the new Russia'.[15] Molotov arranged things and Rosen was delighted with his new home. 'It is the only civilized country in existence . . . I have the feeling I am useful and needed . . . I have come to feel that anything one does to help the Soviet Union means furthering the progress of humanity'.

So open and progressive was the country that the leading Soviet philosophy journal had just published one of Einstein's articles against the Copenhagen spirit. It was also a great place to bring up children, 'the only country where there is a future to look forward to'.[16] Soon Rosen had a son. 'I hope [Einstein wrote in congratulation] that he too can help in furthering the great cultural mission that the new Russia has undertaken with such energy'. Rosen did not try the experiment. By October 1938 he was back in the US, having discovered that his research did not prosper in the people's paradise and his conscience would not allow him to remain where he felt useless.[17] We shall meet him again.

Now to the epoch-making EPR paper, a tidal wave in the field of quantum interpretation.[18] It shook the citadel in Copenhagen. Rosenfeld: 'This onslaught came down upon us as a bolt from the blue. Its effect on Bohr was remarkable'.[19] It looked superficially like several earlier sallies by Einstein, including the thought experiment Bohr had defeated at Solvay VI, but it had a clever twist to it that gave him pause.[20] Here is its gist. Two quantum particles, 1 and 2, interact in Copenhagen through a physical mechanism that correlates their properties. They then separate. Particle 1 arrives in Pasadena, where someone, say Caltech's President Robert Millikan, who had a Nobel prize for his dealings with electrons and photons, receives it. Nothing prevents him from measuring *either* its position q_1 *or* momentum p_1 with as exquisite a precision as his apparatus will allow. He hesitates. He knows, because EPR and simple algebra confirm it, that the *difference* in the positions of the two particles, $Q = q_1 - q_2$, and the *sum* of their momenta, $P = p_1 + p_2$, commute: $QP - PQ = 0$. The deep connection between such commutation relations and the uncertainty principle had been established in 1929. That Q and P commute means that they are not constrained by uncertainty. These quantities could be measured precisely and simultaneously when the particles interacted in Copenhagen. So, Millikan's choice about particle 1 would determine what he could know in principle about particle 2.

If he measured q_1 precisely, he could deduce q_2 from Q; if he preferred p_1, he had p_2 from P. Apparently—and this is EPR's argument—without observing particle 2 Millikan could infer more about it than the uncertainty relation, applied to particle 2, appeared to allow. Surely particle 2 must then *possess* precise values of properties such as position and momentum if, at his whim, he could deduce an exact value for the one or the other. Now, quantum mechanics denies that a particle possesses such values. Therefore, the theory cannot be complete.

But inferring is not measuring. EPR tried to overcome this objection with the following criterion:

> If, without in any way disturbing a system, we can predict with certainty (i.e. with a probability equal to unity) the value of a physical quantity, then there exists an element of physical reality corresponding to this physical quantity.

EPR thus declared that a quantity whose value can be predicted with absolute certainty must exist whether measured or not. Like Galileo thinking about falling weights on a moving ship, we can know the answer without looking.

The EPR argument is a subtle elaboration of the basic argument Einstein had put forth eight years earlier at Solvay V. His conception II (ψ is a realistic description of the physical states of individual systems) implies that when Millikan measures q_1, the two-particle ψ wave 'collapses', instantaneously endowing the second particle with the precise position q_2 and eliminating all possibility of investigating p_2. If Millikan instead measures p_1, the ψ wave collapses instantaneously to endow the second particle with the precise momentum, p_2. How does the second particle 'know' Millikan's choice? By signalling it across the vast distance between the particles at a speed faster than light? Or by an action at a distance abhorred by all true physicists?

Another possibility is that the values of q_2 and p_2 were already fixed when the particles interacted in Copenhagen. How might that have happened? Einstein and his collaborators, content to observe that the silence of quantum mechanics on this question proved its incompleteness, volunteered no alternative. One was nigh, however, for anyone not intimidated by von Neumann, in the supposition of hidden variables. EPR would make it respectable to explore this possibility.

It also energized defenders of the Copenhagen way. Pauli alerted Heisenberg to Einstein's new attack, mounted with Podolsky and Rosen, 'in other respects not very good company'. What about their argument? 'I must grant that if an undergraduate had made such objections, I would have regarded him as very intelligent and promising'. However, Pauli went on, the argument rests on an elementary error, a misunderstanding of the way two systems combine in quantum mechanics. Einstein believes that quantum mechanics is correct but incomplete, a juvenile belief, an oxymoron: the new variables supposed necessary for completion would either change the theory and bring it into conflict with proven results, or not change it and so serve no practical purpose.

But (Pauli continued) Einstein is certainly right in suspecting something messy about the notion of measurement, about the composition and separation of the elements of an experiment. This is just the problem of the *Schnitt*, the place where the observer decides to separate, or cut, the object observed from the instrument of observation. You have discussed this problem, Pauli reminded Heisenberg, as has von Neumann, although he with mathematical pettifoggery. So please answer Einstein's trivialities with a definitive account of the operation of the theory and practice of the *Schnitt*.[21]

Heisenberg's draft reply, which he sent to Pauli and Einstein, argues that the only place in the chain of measurement events at which hidden variables can exert their effects is at the cut, the point at which the deterministic evolution of the Schrödinger equation gives way to the deterministic evolution of the classical measuring device. The quantum formalism does not fix the cut, '[which] may indeed be shifted arbitrarily far in the direction of the observer in the region that is otherwise described according to the laws of classical physics'.[22] We can cut between a quantum system A and a classical measuring device B or between $A + B$ and some further system C and expect quantum mechanics to apply to the combined system $A + B$, just as it applies to A. Heisenberg labours to show that the predictions of quantum mechanics are indeed unaffected by the placement of the cut.

Now let a hidden variable determine some outcome with the cut placed *before B*. We must expect the same outcome with the cut anywhere *after B*; but as the system has developed according to the Schrödinger equation and has passed through B (though without a 'measurement'), the hidden variable for the later cut cannot be the same as for the earlier. Since the system must be prepared to give the same pre-determined response no matter what vicissitudes it encounters before measurement by an instrument capriciously chosen, it must be created with a cornucopia of hidden variables to cover every likely eventuality. That is implausible or, perhaps, impossible. That is what Heisenberg concluded: 'the quantum mechanical formalism that is applied to the total system $A + B$ contains no freedom whatsoever, and leaves no place for additional assumptions regarding the effect of A on B'.[23]

Heisenberg's cut was not a physical divide. As he explained 20 years later, the wavefunction is 'completely abstract and incomprehensible, since the various mathematical expressions $\psi(q)$, $\psi(p)$, etc., do not refer to real space or to a real property; it thus, so to speak, contains no physics at all'.[24] It is purely symbolic. We are reminded of Italo Calvino's remark, 'as happens with all true symbols, one can never decide what on earth it symbolizes'.[25] But an abstract symbol cannot annul a physical quantity. Heisenberg's argument appears to us to depend on the judgement that the engineering of a pack of hidden variables to agree with the predictions of quantum mechanics across arbitrary cuts is too far-fetched for serious consideration. If that was his argument, history has proved him wrong. Quantum physicists have given sustained attention to such connections. Much of the drama of our final act arises from taking seriously the possibility that nature permits intricate conspiracies between the measurer and the measured.

Heisenberg withheld his draft from publication because Bohr had already prepared an answer to EPR.[26] Bohr concluded that inferring is not measuring. If, as experimentalists, we insist on making measurements, then by reducing the manipulation of EPR to familiar exercises, such as Young's set-up, Bohr exposed the fatal flaw in EPR's assumption that Q and P could be measured simultaneously using the same apparatus.[27] In order to determine the position of particle 1 in Pasadena the circumstances of their mingling in

Copenhagen could not have been what EPR advertised. For suppose that the particles are electrons and the apparatus a heavy diaphragm with two slits as close together as you please. Then, indeed, $q_1 - q_2$ (the slit separation) and $p_1 + p_2$ (the momentum transfer to the diaphragm as the particles pass through the slits) can be obtained simultaneously. But however exactly Millikan measures the position of particle 1 with respect to his workbench in Pasadena, he will not be able to calculate its distance from the experimental apparatus in Copenhagen. That is because the position of the hanging diaphragm in Bohr's laboratory is undetermined. To secure a reading of position (not of separation!), he would have had to bolt the diaphragm to his laboratory bench; but if he had done so he could not have measured the momentum exchange $p_1 + p_2$. BRAVO BOHR!

In contrast to the Solvay exchanges, EPR's thrust and Bohr's parry took place before the larger community that read the *Physical Review*, the American journal then rapidly becoming the major periodical for international physics. The exchange received some attention, though most physicists who noticed it regarded it as Sabbath sermonizing of little use on workdays. Bohr's church rejoiced, however, at seeing the gospel saved by recourse to the well-known parable of slits in a diaphragm.

But perhaps the most instructive responses came from Carl Wilhelm Oseen, professor of physics in Uppsala, who had followed Bohr's work closely from its beginnings, and Philipp Franck. At last, I understand, Oseen wrote, what you have been saying all along. Before a measurement, an atom's state with respect to the quantity measured is not defined. That was only half of it. Frank understood Bohr to mean that 'physical reality' should not be ascribed to the quantities we associate with quantum entities. Quantum mechanics, as interpreted by complementarity, characterized measuring procedures and results, not the things measured. Bohr acknowledged that that was what he had in mind.[28]

Between the publications of the EPR paper in mid-May 1935 and Bohr's response in mid-October, its foremost critic was Einstein himself. He complained to Schrödinger that 'because of language considerations' Podolsky had written most of the paper after thorough discussion among the authors, but the main point had been 'buried alive, so to speak, by erudition'.[29] No earlier drafts of the paper survive in the Einstein collection and no correspondence suggests that he had seen it before Podolsky submitted it for publication just before leaving Princeton to take up a position, procured with Einstein's help, at the University of Cincinnati.[30] We ascribe Podolsky's incivility to the same impulse to self-promotion that had prompted him to leak the content of the paper to the *New York Times* before its formal publication in *Physical Review*.

The editors of the newspaper were eager to have the scoop. A readership existed responsive to such headlines as 'Einstein Attacks Quantum Theory ... Finds it is Not "Complete" Even Though "Correct"'.[31] Einstein was incensed: he had not been consulted and he deplored the unprofessional practice of announcing new scientific results in newspapers. The *Times* subsequently printed a statement by him to that effect. According to Israeli physicist Asher Peres, Einstein was so upset by Podolsky's indiscretion and impropriety that he never spoke with him again.[32]

The substantive part of Podolsky's sin, the main point that had been buried in the EPR paper, concerned the 'separability' of the two particles. Quantum theory requires that the state arising from their interaction be described by a single, two-particle function, which for simplicity we can denote as ψ_{12}. EPR's analysis assumes that, when well separated, each particle can be described by its own wave function, ψ_1 and ψ_2. '*My* way of thinking is this', he wrote Schrödinger, 'you cannot get the better of the Talmudist without an additional principle: [a] "Separation principle" [Trennungsprinzip]'.[33] This is an assumption later labelled 'Einstein separability'. Particles so separated can be considered as 'locally real'.

The main point was that quantum mechanics appeared to strip interacting particles of their separate identities so that, even after parting company, they remained mysteriously connected, just like the photon and its box. This had nothing to do with ad hoc criteria of physical reality and everything to do with the conflict between separability and completeness. Either the two particles are separable and locally real, and quantum mechanics incomplete, or quantum mechanics is complete and the particles mysteriously bound together, even over vast distances and times. Both cannot be true. It was by putting its disturbing argument in this way that EPR became a watershed.

There remained the problem of the *Schnitt*. Einstein had belittled the bizarre argument that experimenters can place the cut wherever they pleased, not excluding even their own minds. In this extreme, everything downstream, the quantum system, instrument, and the observer's physiognomy, can be described by a single ψ, which only 'collapses' to a definite state when 'measured' by the observer's consciousness. To illustrate the absurdity of the situation, and to persuade Schrödinger of the foolishness of his insistence on a literal (rather than statistical) interpretation of the ψ-function, Einstein proposed an example in which a chemically unstable explosive has an even chance of spontaneously exploding within a year. Such a situation is readily represented in terms of a ψ-function which, after a year, would be described as a blend of not-yet-exploded and already-exploded systems. 'Through no art of interpretation can this ψ function be made an adequate description of a real situation; in truth there is no intermediary between exploded and non-exploded'.[34] Although this was a purely macroscopic example, it prompted Schrödinger to place the macroscopic event in a measurement chain that begins in the microworld. The result was his famous parody of the cat that can dwell simultaneously in states of life and death as long as necromantic physicists want to conjure with it.

Schrödinger exposed the animal's predicament during the storm stirred up by EPR. Let a cruel quantum *Schnittmeister* seal a live cat in a big box together with a closed glass bottle of lethal gas. The box also houses an apparatus consisting of a slowly decaying radioactive substance, a detector of its rays, an amplifying circuit, and a hammer, which smashes the bottle when released by a signal from the circuit. There is a chance of one in two that within an hour the radioactive substance will send a ray to the detector and kill the cat. During that time, the ψ of the experiment includes states in which the cat is dead and other states in which it is alive. To the average mind, it is a question of either-or. To logically consistent quantum minds, of both-and until the opening of the box collapses its ψ and reveals the cat's fate.[35] Both-and is clearly absurd, Einstein acknowledged: '. . . your cat shows that we are in complete agreement'.[36]

Schrödinger embedded this story in a lengthy, three-part article on the state of quantum mechanics inspired by EPR. Regarding its two particles, he made a prescient observation which was to award an enduring label for their mysterious connection: 'If two separated bodies, each by itself known maximally, enter a situation in which they influence each other, and separate again, then there occurs regularly that which I have just called *entanglement* of our knowledge of the two bodies'.[37]

Schrödinger understood Einstein's definition of reality as requiring that every 'real thing' have a *Fähnchen*, a little flag or pin, on the grand map of physics. Do you believe (he asked Pauli) that the case EPR built on this assumption is entirely clear, simple, and obvious? That is what everyone says the first time I describe the experiment to them, 'because they have learned the Copenhagen Credo in one sanctum very well'. Three days later, however, most of these quick responders repudiated their immediate reaction and resolved to think about the question, but none of them had yet come forward to explain to Schrödinger how everything is clear and simple. In his opinion, EPR's point might have been stated better by supposing that a so-called pure state is a collective, the members of which differ from one another in ways unspecifiable by quantum theory. Since the hypothetical variables that specify the differences do not feature in the theory, they are 'hidden' and the theory must be incomplete. And, Schrödinger continued, the theory will remain incomplete if Bohr's clique remains in charge. 'Incomplete because I [if I shared the dogma] will be ordered to regard cases as the same, although I know they are not, because I will be forbidden to inquire further, although I know there is something to look into'.[38]

Pauli replied that Bohr's answer to EPR was entirely satisfactory and that Schrödinger's efforts to do better with his collective were not. Let us understand a pure state as one about which the physicist knows everything necessary to make all possible predictable outcomes of an experiment. (This is how Kramers used the term in his textbook of 1933.) Then the question becomes, what sorts of additional assumptions about the state may be made without changing the statistical predictions? 'The question is very important since in the minds of the conservative older gentlemen (Einstein . . .) there is an apparently ineradicable analogy to the kinetic theory of gases'.

The analogy fails because in that theory the hidden variables—the unobserved positions and velocities of the atoms or molecules of the gas—can be included without changing statistical predictions. Pauli did not favour a crusade against the unbelievers or denial that new variables might be discovered. He insisted, rather, that it is not possible ('as the old gentlemen would like') to declare the statistical predictions of quantum mechanics correct and yet set up a hidden causal mechanism to explain them. Nothing more of substance can be said. 'In my opinion, the only remaining problems in non-relativistic wave mechanics are pedagogical'.[39]

Schrödinger's review of quantum mechanics published in November and December 1935 begins with questioning why the new physics employs the quantities of the old rather than invent its own. Sticking with the inadequate concepts brings in statistics. How are we to construe the position of a particle when we know its momentum exactly? Is it real and unknown or not real at all? The gas theorist would give the first answer, the quantum theorist the second. The meaning of an observation, the most important and interesting point of the theory, is hopelessly confused, as the parody of the cat makes clear.

There is no escaping it, Schrödinger concluded, we must not mince words. Quantum theorists do not understand epistemology. They make no difference between the real state of an object and the totality of what can be known about it. 'In reality there is only detection, observation, measurement'. By definition ψ contains all the information procurable relevant to making predictions about the future state of the system it describes. We are not to sigh for more than ψ can give, which is a catalogue of all possible experimental outcomes with their relative probabilities. And yet, like an incompetent astrologer, it does not predict its own demise, that is, it leaves unanswered the question how a measurement on the system causes the vast evolving catalogue to collapse to a single indelible entry. That is enough to convict quantum mechanics of incompleteness.[40]

One of Schrödinger's students who did understand epistemology, Fritz London, famous for his work on the quantized chemical bond, acknowledged no incompleteness when he summed up the hermeneutics of quantum mechanics for his generation in a widely cited treatise published in 1939.[41] Its centrepiece was the theory of measurement treated in the manner of von Neumann: probabilities enter only when a ψ collapses and measurement is complete only when a human mind takes cognizance of a dial reading. For this mind, system, instrument, and observer constitute the external world. A 'community of scientific consciousness' produces agreement about pointer positions, makes predictions subsequently realized, and saves science from solipsism. The community deploys a minimum of philosophy, '[just] sufficient for its needs . . . It responds to experiment in every detail desirable and says nothing about imaginary questions that experiment cannot answer'.[42] On this understanding many physicists of the 1930s made quantum mechanics objective and freed themselves from further philosophizing.

London's text has another claim on our attention. He wrote it in collaboration with a French colleague, Edmond Bauer, in Paris, where he had

found a temporary position after the Nazis drove him from his professorship at the University of Berlin. Their important treatise was a by-product
of the flight of Jewish intellectuals during the 1930s. London soon left
France for the United States; Bauer eventually climbed the Alps to refuge
in Switzerland.

Meanwhile quantum mechanics had received the endorsement of the Nobel
Institution. Despite uncertainties that would have undermined the chances
of its creators in the days when exact classical experimenters decided the
winners, de Broglie, with 11 nominations, collected his prize in 1929 (the
committee having failed to agree on a candidate for 1928), for ideas that
had precipitated a real discovery, electron waves. In 1931 and 1932 the
committee again could find no one worthy of a prize although in 1932
Heisenberg received 7 nominations and Schrödinger 6. Then jubilation: in
1933 the reserved prize for 1932 went to Heisenberg (10 nominations) and
the current year's prize to Dirac (2 nominations) and Schrödinger (11).[43]

The *Prinzipienfuchser* had conquered the prize machinery. The virtuoso
Born had to wait almost 30 years, until 1954, for his prize for the statistical
interpretation of the wavefunction, and the virtuoso Sommerfeld, although
he received a total of 71 nominations over 20 years (1917 through 1937),
would have waited forever, had he lived that long. Sommerfeld did not
belong to the Bohr or the Einstein group and neither master ever nominated him. Had they done so, he might conveniently have received the
prize for 1934, which was never awarded. Born's omission from the first
tranche of prize winners suggests that the other founders, none of whom
nominated him before 1950, did not think his probabilistic interpretation
sufficiently original or his contribution to matrix mechanics sufficiently
separable from Jordan's or different enough from virtuosity to warrant a
prize. That left Pauli, whose turn came in 1945, for the exclusion principle
he had announced 20 years earlier.

The securing of a Nobel niche for quantum mechanics followed a year of
spectacular discoveries that cleared up several pressing difficulties at a new
frontier of atomic physics, signalled in the reports at Solvay VI. To explain
nuclear charge (Z) and weight (A) using only the bits and pieces they had to
hand, physicists had supposed that an atomic nucleus consists of A positively

The *Prinzipienfuchser* had conquered the prize machinery. L–R: Erwin
Schrödinger, the King of Sweden, and Werner Heisenberg at the Nobel Prize
ceremony in 1933.

charged protons and $N (= A - Z)$ negatively charged electrons. According to this scheme, a helium nucleus ($Z = 2$) consisted of $A = 4$ protons and $N = 4 - 2 = 2$ electrons. The electrons did not fit the part well. It was the fault of the uncertainty principle. Confined to a range of positions Δq equal to the diameter of a nucleus, they would have a Δp implying a non-relativistic velocity greater than that of light. Furthermore, the theory of nuclear electrons also gave nuclei magnetic properties that disagreed with experiment.

In favour of nuclear electrons, physicists pointed to radioactive decays that give off fast-moving electrons (beta particles) apparently originating in the nucleus. But that called attention to the awkward fact that in general the energy of the beta particles was less than the difference in energy between the initial and final states of the nucleus.

As usual, these difficulties discouraged the younger theorists, who wanted to solve problems, whereas Bohr, who liked to poke about for loci for revolution, rejoiced. Perhaps, at last, he had come to the place where he could jettison the conservation of energy and, to accommodate electrons in the nucleus, to sacrifice as much else as possible. But, alas, early in 1932 English physicists discovered the electrically neutral neutron and nuclear physics was saved by replacing the electrons by N neutrons and reducing the number of protons to $Z = A - N$. The helium nucleus now had 2 protons, 2 neutrons, and 0 electrons. Almost simultaneously the experimental detection of positively charged electrons (anti-matter) plugged holes in the quantum relativistic theory that the unerring Dirac had invented a few years earlier. That left beta-decay standing among the fundamental problems identified at Solvay VI.

Pauli had suggested an obvious way out. He supposed that an unknown particle came from the nucleus together with the beta particle to balance up the energy. Why had it not been observed previously? Because, according to its inventor, it has no charge and very little if any mass. Bohr and Ehrenfest rejected this ad hoc solution as a violation of the old injunction against multiplying entities unnecessarily. Their continued resistance to admitting Pauli's 'neutrino' into the nuclear family of particles inspired a hard-hitting, witty parody of Goethe's *Faust* performed in Bohr's institute in the spring of 1932. The plot turns on the seduction of Ehrenfest (Faust) by Pauli (Mephistopheles) via the Devil's favourite creature, the slim gentle Neutrino (Gretchen). Fig. 9 displays a selection of illustrations accompanying the script.

Fig. 9 A small selection of illustrations accompanying the script of Goethe's *Faust*.

Famous physicists are hit off perfectly in caricature if not in verse. Bohr, the Lord, 'Approving views that shatter like a bubble/Sticking your nose into every kind of trouble'. Dirac, preternaturally clever and skinny, 'the revered one-dimensional case'. Some American physicists at a speakeasy (a take-off on the scene in Auerbach's pub in Goethe's play), enthusiastic about Neutrino, 'A shapely and appealing signorina/But tell me, have you been to Pasadena?' Einstein (the Monarch) crosses the stage dragging his pet flea (the general theory of relativity). He is followed by Slater dragging the corpse of a mathematical dragon he has slain, and by Millikan, who tries to bring the theorists to their senses by mentioning Wilson chambers and counting tubes.[44]

The barbs tossed at Ehrenfest-Faust came only too close to the mark. 'All doubts assail me: so does every scruple/ . . . /I grab the eraser, like a frantic pupil'. Pauli-Mephistopheles tries to reassure him. 'Beware alone of Reason and of Science/Men's highest powers, unholy in alliance'. Enter Neutrino-Gretchen (in real life, if the expression be allowed, the girlfriend of one of the physicists) and sings, 'My mass is zero, My charge the same/You are my hero/Neutrino's my name'. Faust falls for her, becomes famous, delights in his fame, and dies. Neutrino is pushed off the stage by her mightier cousin, the newly discovered Neutron, but Pauli-Mephistopheles cleaves to her. No matter that experiment cannot find her, and would not, for over 20 years.

The piece ends with a nod to an ideal world unavailable in any plausible description of the circumstances of 1932. 'Hailed with cordiality/Honoured in song/Eternal Neutrality/Pulls us along!'

Missionaries of the Copenhagen Spirit

In September 1933 a man visited the Institute for Afflicted Children in Amsterdam to see his son, a boy of 15, recently transferred there from a clinic in Germany. An hour later both the patient and the visitor were dead. The man was Paul Ehrenfest. The feelings of inadequacy that had haunted him for years had become unbearable. He had not justified the faith that Lorentz had placed in him, he had not been able even to stay current with the rapid advances in theory and technique driven by younger men. Dirac had tried to console him by praising his role in mediating between Bohr and Einstein. Ehrenfest tearfully acknowledged that the praise meant something to a man who had lost the will to live.[1] It was not consolation enough. In an unsent letter to his students, Ehrenfest declared his intention to end his life 'to leave his place in Leiden free'. He could no longer do research. 'That is the heart of my collapse, of my depression'. 'I am so fearfully wretched', he had written Pauli nine months before his suicide, 'I am completely unable to think about physics'.[2]

The young men whose brilliance had extinguished Ehrenfest's informed one another of his death in short passages embedded in letters about their successful work. 'It makes me personally very sad', Pauli wrote, but then 'Ehrenfest was always very neurotic and suffered from melancholic-depressive attacks'. Pauli knew what he was talking about. His mother had taken her own life a few years earlier. We are all very sad, Heisenberg replied, but it was not unexpected. 'He had written letters of farewell to several friends and spent an hour with his sick son before the end'.[3]

An older man understood the situation better. All worthy academics over 50 experience distress at their inability to master new ideas quickly, or so Einstein, who had crossed the threshold, wrote. The result is 'doubly difficult for a man of fanatical honesty, a mind to which clarity means everything', like Paul Ehrenfest, '[whose] critical sense robbed him of his love for the offspring of his own mind even before they were born'. He succumbed to his sensitivity and want of self-confidence from internal conflict, a rare occurrence, in Einstein's experience, 'possible only in the case of the noblest and morally most exalted personalities'.[4]

Some of the quantum theorists reared in the hothouses of Göttingen and Copenhagen also experienced psychological problems and sought help from the theories of Sigmund Freud and Carl Jung. Jordan reassured himself of the reliability of Freudian psychology by working out parallels between investigation of the id and quantum measurement. Does not the uncontrollable alteration of the quantum system by interaction with the measuring apparatus mimic the spontaneous thrusting of material from the unconscious into the ego? And the restless shifting of the subject–object cut in atomic physics mimics the movable boundary between the inner world of consciousness and the outer world of immediate experience? From these premises and the fact that we can share our outer worlds, Jordan reached the principle that we can also, although with greater difficulty, share our inner worlds.[5] Telepathy would no doubt be very helpful to people with the severe speech impediment from which Jordan suffered. He discussed his insights with Pauli, who urged him to develop them further.[6]

Jordan thereupon discovered an unexpected analogy between a split personality and an electron. Let the personalities be A and B and the answering electronic properties P and Q. The dominance of A (or B) then corresponds to a determined value of P (or Q). The impossibility of simultaneous dominance by both A and B agrees with the complementarity of P and Q. The suggestion that causes the switch from A to B or B to A is analogous to a quantum-physical measurement. The ambiguity of the electron has its counterpart in the ambiguity of personality. Schizophrenia is as natural and inescapable as the 'mutual suppression' of the position and motion of an electron. Telepathy might easily arise among the four personae of two split personalities.[7]

Pauli wound up in Jung's consulting room owing to the conflict between his unrestrained self-indulgence and his unflinching self-criticism, which was inhibiting his scientific creativity. His self-confidence then suffered a more visceral blow. Within a year of their marriage, his wife, a former cabaret dancer, went off with a chemist, and not a very good one. 'If it had been a bullfighter—with someone like that I could not have competed—but such an average chemist!' Pauli took to drink and then to the local shrink.[8] Jung assigned him to an assistant so as not to mar the spontaneity of Pauli's disclosures by observing them. Left on his own, Pauli spontaneously delivered a thousand dreams and visions, and interpreted them, without prompting, in Jungian terms. The analyst at a distance was not surprised. 'Owing to his excellent scientific training and ability [Jung wrote], he did not require any assistance'.[9] Indeed, he might have been in Copenhagen when interpreting his dreams as 'symbolic'.

In Jungian talk, a 'symbol' signifies something necessarily ambiguous, a multivalent 'invisible and untouchable' reality.[10] Although we cannot grasp their full meaning, we can observe the symbols thrown into our consciousness and report our descriptions. As Pauli drew out the evident analogy to quantum physics, the concepts of ordinary, rational mechanics have only a symbolic character when applied to an atom. How the ambiguous thing presents itself in experiment eludes rationality. The modern physicist faces the irrational every working day.[11] Material from the unconscious breaks into the conscious mind as the unknown state of a quantum entity settles into definiteness by measurement.[12] Disturbing mental impulses are as natural as radiating atoms. And so, Pauli recovered, remarried, and returned to physics.

Heisenberg found support in his pals the Pathfinders, in hiking in the woods and communing around the campfire. He retained this form of relief, with its challenges of leadership and hints at homoeroticism, after becoming a professor, and his biographer affirms that his young wife, whom he married when he was 35 and she 22, had to explain to him that she was not a boy scout.[13] He reduced tension further by agreeing with Bohr's complementarity and avoiding the prolonged and agonizing grilling he had experienced over the uncertainty relations.

Heisenberg practised his discipleship so faithfully that he dropped Einstein, de Broglie, and Schrödinger from the history of quantum mechanics. Only those who understood it in Bohr's way qualified as contributors. The others, those too pig-headed to acquiesce in complementarity, puzzled him. 'Perhaps understanding quantum theory is in principle complementary to understanding people who do not wish to understand quantum theory'.[14] Jordan and Pauli also awarded Bohr sole credit for the creative insights that led to the new physics. Pauli let his enthusiasm for rewriting history wax so freely that Einstein and Bohr felt obliged to restrain him.[15]

The true disciples showed their dependency by viewing their master as 'the founder and endower of a religion of complementarity'.[16] Thus Pauli, in explaining his personal philosophy as a mixture of Bohr, Arthur Schopenhauer, and Lao-Tze.[17] Jordan identified rather with Bohr, Mach, and the philosopher David Hume. Klein claimed Bohr as his constant guide in physics and 'human affairs'. Rosenfeld reported a conversation in which Bohr declared his belief that he was called to proclaim the gospel of complementarity, which, 'better than any religion . . . would afford people the guidance they needed'.[18] His mutterings, his parables and riddles, and his struggles to speak no more than the truth, were all marks of his prophetic engagement. 'We all look up to you as the profoundest thinker in science . . . [as] the "Heaven-sent expounder" of the real meaning of these modern advances'.[19]

Bohr easily assumed the prophet's mantle. He had the assurance of the philosophy he had learned from Høffding that all theories end by disclosing their limitations. He had come on the scene when classical physics was encountering the expected boundary alarms. He had devised ways, necessarily irrational, to enter the new land, and he had striven to find a rational generalization that would eliminate the inconsistencies his lonely explorations had forced him to admit. He had endured hardship, illness, ups and downs as he prepared to sacrifice clarity, causation, and conservation on the altar of the new physics he glimpsed but could not grasp. When the eminences of the mother church in Copenhagen, Kramers, Heisenberg, and Pauli, and of the daughter church in Göttingen, Born, Jordan, and von Neumann, found the new physics and proved its coherence and completion, Bohr began to look for places where it, too, would hit its limits.

The next new physics would require more sacrifice, Bohr expected, the surrender of more cherished beliefs. This expectation, and the confusion and frustration of building on shifting foundations that had driven his followers to bouts of depression, did not trouble a mind so ready to renounce concepts most others deemed essential. His followers admired his ability to challenge conventional thinking and to abide by the damage done and freedom acquired, and they understood his peculiar trait of enthusiastic resignation as preparation for recognizing the boundary of the next new physics. As Pauli explained it, Bohr was eager to define the domain of the quantum, to complete and package it, so that he would be related to the quantum theory as Einstein was to relativity.[20] The dean of quantum theorists agreed with Pauli. Planck could see no other reason for Bohr's 'astonishing willingness' to relinquish the search for a deeper and more satisfying atomic physics.[21] Bohr had founded the theory, nourished it, worried that it would not survive, trained the men who would vindicate it and him, and desired to close off the perfected quantum mechanics as soon as he could.

Though much would subsequently be made of the tyranny of the 'Danish priesthood', in truth the missionaries of the Copenhagen spirit enjoyed a success that was limited largely to a few philosophically inclined physicists of continental Europe.[22] Leading British physicists were not persuaded by the new gospel and resisted conversion. Thus, in his widely read textbook, *The principles of quantum mechanics*, first published in 1930, Dirac makes no mention of complementarity. He never liked it, he said, because it gave no new equations. Bohr's longstanding colleague Charles Darwin (the grandson, he liked to say, of the 'real' Darwin) insisted that 'the details of a physicist's philosophy do not matter much'.[23] Frederick Lindemann, professor of physics at the University of Oxford, insisted that the British found no appeal in the renunciation of accepted wisdom. The editors of *Nature* preceded their printing of Bohr's Como lecture with a comment that Pauli paraphrased for Bohr's amusement:[24]

> We English physicists would be terribly happy if the views put forward in the following article should be proved incorrect in the future. But since Mr. Bohr is a nice man, such a pleasure would not be very charitable, and since he is a famous physicist, and more often right than wrong, there is only a small chance for the fulfilment of our hopes.

Those few American physicists who concerned themselves with philosophical questions also discovered they had no need for the Copenhagen spirit. This was no petty provincialism. That the initial development of quantum mechanics had been an entirely European affair might indeed illustrate 'the weakness and provincialism of theoretical physics in the United States at the time',[25] but by the late-1920s American physics was rapidly expanding, transforming, and carving out a substantial role for itself. It had many inherent advantages, not least a much more collaborative culture, within which theorists worked more directly alongside their experimentalist colleagues.

In 1927, Harvard professor of physics Percy Bridgman, who though an experimentalist had worked hard over the previous decade to stimulate the growth of theoretical physics in America, published an influential book, *The logic of modern physics*. Bridgman's philosophy of *operationalism* was unashamedly empiricist: 'The attitude of the physicist must be one of pure empiricism. He recognizes no *a priori* principles which determine or limit the possibilities of new experience. Experience is determined only by experience'.[26] Bridgman mentioned Bohr only in passing, and not in the context of complementarity.

Bridgman's student Edwin Kemble, the first American to write a doctoral dissertation on quantum mechanics, acknowledged this essentially descriptivist approach: 'in the last analysis, the function of theoretical physics is to describe rather than to explain'. '[D]iscarding this question [the origin of wave-particle duality] as ultimately unanswerable, we may address ourselves to the task of describing what we observe in the most compact manner possible'.[27] 'The first answer of the operationalist is that what electrons *are* is not the concern of physics'; 'any attempt to interpret ψ-functions as faithful pictures of external reality rather than tools of calculation leads to hopeless difficulty'.[28]

No mention of Bohr, complementarity, or Copenhagen. The American who had hobnobbed most intimately with the Copenhagen school, Bridgman's student John Slater, disliked its methods as well as its philosophy. He denigrated Bohr's flights of physico-philosophical fancy as 'handwaving', and dedicated himself to intricate calculations about the interiors of atoms and molecules before moving into applied physics.[29]

The psychologically tender J. Robert Oppenheimer spent many hours in Bridgman's Harvard laboratory and his interactions with both Bridgman and Kemble inspired him to become a physicist. He left Harvard in 1927 to hone his style in Europe.[30] His experiences in the Cavendish laboratory in Cambridge convinced him to become a theorist and he went on to Göttingen to complete his doctorate with Born. Compton, meeting him there for the first time, judged him to be a model American theorist, 'one of the very best interpreters of the mathematical theories to those of us who were working more directly with the experiments'.[31] In Leiden (where he picked up the nickname 'Opje', or 'Oppie'), Ehrenfest persuaded him to resist the siren call of Bohrish 'largeness and vagueness' in Copenhagen and instead seek out a 'professional calculating physicist'. Pauli in Zurich would be better placed to 'fix him up'.[32]

Despite his exposure to Pauli, who thought of quantum mechanics as 'The Theory of Complementarity',[33] Oppenheimer would not be drawn into the Copenhagen mysteries. The papers that he wrote while in Europe, some in collaboration with his mentors, were extensions of the mathematical theory to technical problems. On his return to the US in 1929, Oppenheimer took up a joint professorship at the University of California in Berkeley and at Caltech in Pasadena. He proceeded to build one of the strongest and most influential theoretical research groups in America.

Oppenheimer's lectures on quantum mechanics at first baffled his students as much as his manner terrorized them. With a few years' experience he mellowed into an inspirational teacher and worked up a course based largely on Pauli's lengthy summary in the *Handbuch der Physik* of 1933.[34] Course notes, compiled by Oppenheimer's student Bernard Peters in 1939 and issued in mimeographed form, run to over 130 typewritten pages. Peters recorded very little on the correspondence principle, complementarity, and the physical meaning of the wavefunction. Students learned that 'the interpretation of the [light] diffraction effects themselves in terms of photons leads to difficulties whose solution involves the idea of complementarity', but not (from the lectures) how complementarity does the trick.[35] After the war Oppenheimer embraced complementarity with enthusiasm and played a Pickwickian part in reinvigorating the hunt for hidden variables.

While Bohr waited for the new irrationality to show itself, perhaps, he guessed, in an ambiguity hidden in the concept of an elementary particle, he continued to offer complementarity as the cure for age-old problems. He did not have much success with biologists or psychologists. Jordan blunted whatever interest they may have had by converting the uncontrollable play of atoms in the innermost reaches of living beings into a special life force that might give direction to evolution. This was to reintroduce teleology and Lamarckism where science had only recently and profitably rejected them. No matter. 'What wonderful properties the new physics opens for biological research!'[36]

In Jordan's hands, Bohr's hints to biologists became a vigorous vitalism and a physiologically rooted freedom of the will. This too perverted Bohr's teaching, which acknowledged the feeling of freedom along with a belief in determinism as a complementary pair necessary for a complete description of psychological experience.[37] But the reference to freedom or its attribution to uncontrollable atoms did not strike psychologists as useful or desirable.[38]

The German-speaking biologists who made up the audience for these flights of the Copenhagen spirit into their science countered that Jordan's embellishments were not only useless but dangerous. Physicists could amuse themselves philosophizing, but when they were at work they returned to causal descriptions and exact inference, whereas biologists had only a precarious hold on the analytical methods of the physical sciences. It did not help them to reduce life to quantum jumps.

Max Delbrück, a promising physicist of the Copenhagen school before turning biologist, attended a meeting at which Jordan spoke. Everyone had complained about the incursion of quantum theory into biology, Delbrück reported to Bohr, and no one had said a good word about physicists. Nor did the philosophers of the Vienna Circle, the 'logical positivists' who traced their insights to Mach, welcome the imperialism of the Copenhagen spirit. Rather, as several of them warned at a meeting in 1934, Jordan had merely brought good new physics into the service of bad old philosophy. The mischief did not end there. Mysticism and fuzziness and blurred analogies created a mental climate ripe for breeding reactionary ideas.

Some of Bohr's close colleagues joined in disapproval. Dirac and Darwin urged physicists to stick to their trade. Delbrück repudiated Jordan. Planck, Schrödinger, and Einstein, who disliked Copenhagen physics, naturally complained about the effort to impose complementarity on all the sciences. '[That] is not only a mistake [Einstein said], but there is something reprehensible about it'.[39] Bohr went on the defensive. He called a small meeting of physicists to take place in 1936 just before he was to address a conference in Copenhagen sponsored by the Vienna Circle and other positivist leagues. He hoped that these colleagues would stay on to give 'the impression that, as far as physicists are concerned, [complementarity] is not a matter of mysticism but of sober efforts toward understanding the limits of applications of even the most elementary concepts'. His colleagues did not stand in solidarity with him. Why not? Because, Bohr explained to Kramers, whom he had tried in vain to enlist, 'so many of the physicists, who have contributed so essentially to progress in the field, suddenly seemed to be scared of the consequences of their work'.[40]

Perhaps the philosophers would be more amenable. Bohr invited the conferees to his villa in the Carlsberg brewery for a reception the night before they went to work. They were doubtless gratified when he declared that, with Mach, Poincaré, and Einstein, he recognized 'philosophy [as] the logic of science' and the methods of physics as the logic underlying all the sciences. Replying on behalf of the guests, Victor Lenzen, professor of physics turned philosopher at the University of California at Berkeley, reviewed von Neumann on measurement and Bohr on EPR, enrolled

Bohr on the defensive. He called a small meeting of physicists to take place in 1936, just before he was to address a conference in Copenhagen sponsored by the Vienna Circle and other positivist leagues. Above: Heisenberg and Bohr converse over a meal in Copenhagen, 1936. Below: L–R: Bohr, Heisenberg, and Pauli conversing at the Niels Bohr Institute lunchroom, Copenhagen Conference, June 1936.

both in the positivist brotherhood, and insisted on the elimination of all psychological and subjective elements from physical theory. He had come all the way from California to deliver this message.[41]

The purgation continued on the first day of lectures. Bohr devoted his keynote address to repudiating mysticism, spiritualism, and 'anti-rationalistic vitalism'.[42] Frank then stood up to exonerate complementarity from the logical possibility of comforting free-willers and vitalists. A similar note came from the other world, from Moritz Schlick, one of Planck's few doctorands and Mach's successor at the University of Vienna, whose assassination by an unphilosophical student had darkened the opening of the conference.[43] In defending physics from mysticism, positivist philosophers were proving to be more useful allies than physicists possessed by the Copenhagen spirit.

Among the livelier participants in the discussions was Grete Hermann, a socialist organizer by occupation, a Göttingen mathematician by training, and a Kantian philosopher by politics and instinct. On Heisenberg's invitation, she spent time at his institute in Leipzig developing ideas she had set forth in a manuscript she had sent Bohr and Dirac at the end of 1933. 'She is certainly wrong', Heisenberg said on first reading, 'but a fabulously clever woman'. She directed one of her clever arguments to save a shred of Kant's causality. Since after a measurement the history of a particle can be *retrodicted* on classical principles, causality held, she asserted, if the Kantian notion were freed from the requirement that a proper cause must *predict* its effect. Hermann seems to have persuaded Heisenberg that quantum mechanics could be both causal and indeterministic until Bohr straightened him out.[44] No doubt a little causality would have made the more extravagant extrapolations from complementarity additionally implausible.

Pursuing her project, Hermann kicked another strut from under the standard view by convicting her sometime contemporary at Göttingen, von Neumann, of obtaining his celebrated impossibility proof by circular reasoning. Her purpose in re-opening the door to hidden variables was to test whether causality could be so saved. But she decided against them after concluding, in a way difficult to follow, that quantum mechanics had all the machinery it required without them. Consequently, she took her demonstration—the very demonstration that 30 years later would make room for hidden-variable theories—not as a probe for an alternative to the Copenhagen view, but as a position to be defeated to save her neo-Kantian causality. That proved her enduring view, which she urged again at the Copenhagen conference in 1936.[45]

The fabulously clever Hermann had also demonstrated her remarkable prescience three years earlier. During her time in Leipzig she had pondered von Weizsäcker's analysis of Heisenberg's microscope. She seized on the result that following its encounter with the electron, the outgoing ψ-wave takes different forms—spherical or planar—depending on whether the photographic plate is positioned in the image plane or the focal plane (refer back to Fig. 6). Hermann had the idea of omitting the photographic plate entirely, prompting a third description of the ψ-wave as a superposition of the two possibilities: the wavefunction of the electron in a position basis, $\psi(q)$, multiplied by an outgoing spherical wave *plus* the wavefunction of the electron in a momentum basis, $\psi(p)$, multiplied by an outgoing plane wave. '[E]ach of its terms is the product of one wave function describing the electron and one describing the light quantum. Through this linear combination the light quantum and the electron are thus not described each by itself, but only in their relation to each other'. The word used to describe this phenomenon had not yet been coined. But like the photon and its box, the particle pair of EPR, and the radioactive atom and Schrödinger's cat, the light quantum and the electron of Heisenberg's microscope are entangled.[46]

No more than the efforts of the soberer quantum physicists did the circles of the logical positivists contain the contamination of the Copenhagen spirit. Jordan's extravagances were already serving what Frank and several of his colleagues were calling 'a reactionary world view'. Frank had in mind the Nazis' exploitation of vitalism in their race biology. Jordan approved their 'work of renewal in domestic politics' and war as 'the normal way to accomplish something new in history'. As we know, he joined the Nazi party.[47]

Heisenberg also fell into what he later described as a confusion of physics, philosophy, morality, and duty. He had stayed in Germany, he said, and accepted compromises and humiliations, to ensure the survival of physics under a regime that did not prize it. In this he at first had the advice and support of Planck. But the old man would not have condoned Heisenberg's work for the Uranverein (Uranium Club), Nazi Germany's fledgling atom bomb project, or his attempt to justify it with Copenhagen babble. 'Remembering our experience in modern physics, it is easy to see that there must always be a fundamental complementarity between deliberation and decision . . . Even the most important decisions in life must always contain this inevitable element of irrationality'.[48]

In another twist of quantum mechanics in the interest of National Socialism, Bernhard Bavink, a physicist, philosopher, and theologian discovered in uncertainty both a weapon against the red menace and a proof of the superiority of Nazi political philosophy. The weapon destroyed mechanism and materialism, the fabric of Marxism, and with it the 'march of Bolshevistic scepticism'. The coincidence of the rise of communism and the discovery of quantum mechanics could not be fortuitous. 'Is it not imaginable that God . . . may have arranged matters . . . in such a way that . . . the death sentence of materialism should be pronounced before the bar of history, at the very moment when materialism is setting out, from Russia, to conquer the world by force?'[49]

Modern physics gives God room for His mysterious manoeuvres in the freedom that the indeterminism of individual quantum events allows. Consequently, in Bavink's memorable phrase, the real business of the quantum physicist is 'nothing else than counting up the elementary activities of God'. Continuous creation gives Him the opportunity to direct the evolution of world views towards opposition to communism, miscegenation, pollution of the gene pool, and the Jews. Hitler's programme therefore had a double sanction from modern physics: one for proving that no reasonable person could be a materialist, the other for intimating that God approved of National Socialism.[50]

These texts date from 1933. Bavink joined the Nazi party that year, on 1 April, a few days before the new German government issued its law for purging the civil service. He had worked with the party apparatus earlier and had composed a purgation regime that agreed in several points with the new law. Experience of the stupidity and brutality of Nazi officialdom quickly proved his mistake in believing that it had any interest in scientific truth. Although he stayed an anti-Semite and eugenicist, he could not stomach Deutsche Physik and wrote against it as much and as openly as he could in influential articles in the journal he edited.[51] Bavink's doctrine that physics should guide politics worked out no better than Stark's doctrine that politics should guide physics.

Is free will complementary to determinism? Even Anglo-Saxons were tempted to probe for openings for free will and divine intervention in the uncertainties of quantum theory. Compton, who was a religious man, makes a good test case. He expressed himself ambiguously on the matter in commenting on a lecture by Darwin. In it, Darwin had objected 'most strenuous[ly]' against extending physics to philosophy and then contradictorily added that modern physics could not save free will because human actions involve vast numbers of electrons whose average behaviour is perfectly predictable. Compton replied that the doings of a single electron when amplified by the nervous system might produce an entirely unexpected result. '[L]iving organisms are not subject to physical determinism'. But as he wrote these words, some such random quantum event must have happened in his nervous system, because, against expectation, we read, 'this does not mean that the living organism is free to determine its own actions'.[52]

Many voices that did not belong to physicists joined the free-for-all around free will. Some saw that freedom delivered by the uncontrollable behaviour of electrons in the brain was worthless. Others, like William Inge, professor of theology at Cambridge, dean of Saint Paul's in London, and author of too many books, could not decide whether we must concede 'a real wobbling of nature within the limits of "Planck's constant". The utterances of our leading scientists are enough to drive a poor layman to despair'.[53] Many of these despairing laymen got their information from English versions of Planck's popular lectures, which explained that physics was still deterministic because the ψ wave develops in space and time much like any other wave, differing only in conveying information rather than motion. But their main source was the colourful descriptions by a Cambridge professor, practical mystic, and first-class physicist, Sir Arthur Eddington.

Eddington's main contribution to the discussion was the published version of lectures he gave during the first quarter of 1927 at the University of Edinburgh in the famous Gifford series on natural theology. Between presenting the lectures and revising them for a book Eddington became aware of Heisenberg's uncertainty and Bohr's complementarity. Eddington's *The nature of the physical world* (1928) thus has the freshness of a first encounter. For its accessible discussion of physical concepts, striking metaphors, and bold assertions about the nature of science and religion it was widely read, translated, interpreted, and criticized.[54] Writers as far from his Quaker religiosity as Inge and Bavink cited it with approval.[55]

Eddington trespassed outside his subject with the confidence that he brought good tidings. The physicists' world of symbols, shadows, and pointer readings was becoming mystical. Physicists no longer held as a test of the admissibility of a theory that an engineer could make a model of it. '[S]cience thereby withdraws its moral objection to free will' and allows ordinary people to call their souls their own again.[56] The more physicists have set aside the goal of strict causality, the more powerful their theories have become, even though the objects of their measurements, like 'electrons', are in fact unknown in themselves. In quantum physics we know only that 'something unknown is doing we don't know what'. 'We can grasp the tune but not the piper'. The quantum physicist creates an unknowable with every experiment. How then can we get at the world behind the pointer readings? 'It is hard to empty the well of Truth with a leaky bucket'.[57]

Eddington had another bucket to hand in which he mixed pointer-reading physics with the 'spiritual world' of values. This mixing bowl was mind or consciousness, where humans can connect with the world behind the readings. There an unprejudiced mind will find God.[58] The times were inspirational. 'If our explanation should prove well founded that 1927 has seen the overthrow of strict causality by Heisenberg, Bohr, Born, and others, the year will certainly rank as one of the greatest epochs in the development of scientific philosophy'. An epoch in the astronomical sense of a new beginning: 'religion first became possible for a reasonable scientific man about the year 1927'.[59] The concepts of physics had become too important to the human condition to be left to physicists. They cannot decide on the ultimate principles of their science, they wobble, and some, like Bohr's group, beat 'a complete retreat from the major premise of modern thought, the notion of absolute natural law', which Planck and Einstein still defended.[60]

This was too much for Herbert Samuel, liberal politician, atheistic Jewish Zionist, and physicist manqué. Speaking in October 1935 as president of the British Institute of Philosophy and friend of Einstein, he observed the need for a philosophical basis for integrating philosophy and religion with the true gains of modern physics. Unfortunately, the true gains, 'perhaps the most marvellous achievement yet of the mind and hand of man', have been contaminated by the 'singularly unconvincing' extrapolations of Bohr, Heisenberg, and Eddington. To this programme the Bishop of Birmingham replied that although he 'inclined' to reject the principle of indeterminism, he could not plump for its opposite. He could see 'no reason why man should not have some measure of creative freedom', 'some kinship between the mind of man and the mind of God'.[61]

As the 1930s progressed through world-wide depression to world-wide war, most quantum physicists did not stop to draw out the wider lessons of physics for the wider society or rebut the extravagances of the Copenhagen school. Lenzen, who had sacrificed his physics to his philosophy, 'emphatically disclaim[ed] responsibility' for the speculative extrapolations of complementarity and refused to discuss them lest he encourage further misunderstandings.[62] His stand had the benefit of the advice of Aristotle, who, Lenzen observed, had warned against 'argu[ing] about geometry among non-geometers, for the man who argues badly will escape notice'.

Aristotle pointed out further that, just as the professional should not argue his specialty with the uninformed, we must not expect a specialist to know everything. 'We should not, therefore, ask each scientist every question, nor should he answer everything he is asked about anything'.[63] And yet the specialist and the generalist, the technical expert and the politician, the creative physicist and the curious public must and do interact. The outcomes frequently are not encouraging when the ideas in play are as abstruse as the interpretations of quantum theory. Nonetheless, educated laypersons deserve access to these ideas for whatever instruction, inspiration, liberation, and comfort they might afford.

Bohr and Einstein, Heisenberg and Schrödinger, Eddington and Planck believed that enlightening the educated public was as much a duty as straightening out one another.

ACT III

Orthodoxy to Uncertainty

Postwar Hostilities

World War II drove several of the European founders of quantum mechanics to Britain or the US. Bohr spent most of the war in both countries, offering encouragement to physicists struggling to build atomic bombs at Los Alamos and, in full complementary mode, worrying about how to negate their use. His argument that the best way to avoid a postwar arms race with the Soviet Union would be to share nuclear technology did not convince Winston Churchill and US President Franklin D. Roosevelt and others unfamiliar with his unique way of reasoning. To Churchill, it smacked of treason. 'It seems to me Bohr ought to be confined or at any rate made to see that he is very near the edge of mortal crimes'.[1]

Pauli sat out the war at the Institute for Advanced Study in Princeton. Born took up a professorship in Edinburgh. Schrödinger wandered from Oxford (where, he complained, academics did not love women) to Graz (where Nazis behaved even more detestably) to Ireland as the prize catch of the Institute for Advanced Study in Dublin, established in 1940 by Eamon de Valera, the pooh bah of the government of the Republic of Ireland (simultaneously prime minister, foreign minister, and minister of education) and a would-be mathematician.[2] As we know, Heisenberg and Jordan remained in Germany entangled in National Socialism. Rosenfeld went to Utrecht, Kramers stayed in Leiden, and de Broglie continued his physics in occupied France. War work consumed the time of most American, British, and Soviet physicists. The interpretation of quantum mechanics slipped from the agenda.

After demonstrating their competence with such gadgets as radar, the proximity fuse, and the atomic bomb, American physicists were eager to show that they had come of age as theorists too. In 1947 a select group gathered in a domestic version of a Solvay conference at the secluded town of Shelter Island, on the eastern tip of Long Island. Among the participants were men from the age of the founders, Kramers, von Neumann,

Oppenheimer, John van Vleck (Harvard PhD under Kemble in 1922), and Isidor Rabi (Columbia PhD in 1926, postdoc with Sommerfeld). They were joined by several immigrants who had made their reputations in the 1930s and had led groups in the theoretical division at Los Alamos, Hans Bethe (born in Strasbourg, trained by Sommerfeld), Victor Weisskopf (born in Vienna, prominent among Copenhageners), and Edward Teller (born in Budapest, also studied with Sommerfeld). Einstein was invited to attend but declined for health reasons. John Wheeler, Richard Feynman, Julian Schwinger, and former Oppenheimer students Robert Serber, David Bohm, and Willis Lamb represented the younger generation of American theorists. Eleven of these sixteen men were Jewish.

Of their deliberations we need to know little since the conference dealt with pressing technical problems distant from plain 'elementary' quantum mechanics, uncomplicated by relativistic considerations, which was the forum where questions of interpretation were aired. It will be enough to say that the conference organizers consulted Pauli, whose advice to bring in Bohr and the classical theory of the electron with a view to applying the correspondence principle they ignored, and that they took as their chairman Oppenheimer, who ran their conference as if it had been a bomb-making seminar at Los Alamos. They made practical technical advances on a range of puzzles that had accumulated during the war.[3]

Foundational questions returned with renewed energy and little progress in the early postwar years when meaningful anniversaries of the original protagonists came due. Einstein's 70th birthday fell in 1949 and de Broglie's 60th in 1952. Before the volumes published to mark these milestones came out, Pauli launched a pre-emptive strike by inviting several of the contributors to the commemorative volumes to write also for the first issue of *Dialectica*, a Swiss journal founded in 1948.

Among these contributors was Einstein. Pauli presented him to *Dialectica*'s readers with typical generosity. Do not place much value on Einstein's opinions, he said, they rely on an outmoded view of reality, and almost every working physicist follows the views of Bohr and Born. Despite Pauli's efforts to instruct him about the fundamentals of quantum mechanics and the futility of EPR, Einstein had stuck to the antique notion that physics should describe a 'real outer world' existing independently of the observer.[4]

How could such pig-headedness be explained? 'Psychologically speaking [Pauli wrote Bohr, in English], I feel a certain resistance of Einstein to admit an incompleteness in physics . . . inside the whole entity of life. Instead of it he claims an incompleteness of quantum theory inside physics and is blaming the large majority of younger physicists for it'. He has some mental block. 'I believe that [Einstein] is not able any more to follow the arguments of other people'.[5]

Pauli's other contributors stuck to the Copenhagen script. Bohr submitted an account of complementarity that prioritized the 'phenomenon' (the unity of system and apparatus) over the uncertainty relations in explaining the necessity of Copenhagen mechanics. EPR had taught him that phrases like 'disturbance of the object' as used by Heisenberg could give rise to the misleading impression that a micro-object possesses mechanical quantities that are susceptible to disturbance before measurement.

Heisenberg offered a dialectical reading of the history of physics. The subject had advanced by closing one conceptual system and replacing it by another, each perfectly satisfactory within its limited domain: Newtonian mechanics; Maxwell's electrodynamics and special relativity; thermodynamics and statistical mechanics; and non-relativistic quantum mechanics. The closed systems do not contain any completely secure statements about the world of experience. Rather, they reveal the physicist as a wanton idealist, a fertile creator of '*geistiger Strukturen* [mental constructs], whose succession resembles the history of other *geistiger Bereiche* [intellectual fields], like art, for the goal of both is to illuminate the world through *geistige Strukturen*'. De Broglie too joined the chorus, though without the wanton *Geist*. He had long given up on his 'double solution' and was persuaded to join the Copenhagen chorus. With the zeal of the convert, he sang the praises of complementarity and offered a new example of it. Pauli thought little of the example but much of de Broglie's continuing embrace of the Copenhagen spirit.[6]

The volume celebrating Einstein's 70th birthday appeared in *The library of living philosophers* edited by Paul Arthur Schilpp, a philosopher at Northwestern University in Illinois. Each volume of Schilpp's library featured essays teasing out what the life's work of a towering philosopher 'really meant' while the subject was alive enough to answer. Einstein's volume began with 'Autobiographical Notes' ('Here I sit in order to write, at the age of 67, something like my own obituary') in which he returned to the main point of EPR as elaborated in correspondence with Schrödinger, without the unnecessary and philosophically vulnerable criterion of physical reality.[7] 'If one asks: does a ψ-function of the quantum theory represent a real factual situation in the same sense in which this is the case of a material system of points or of an electromagnetic field, one hesitates to reply with a simple "yes" or "no": why?' Physicist A might argue that the ψ-function offers an incomplete description of the real situation of the system since it expresses only what we know based on former measurements, to which physicist B might reply that the individual system has no definite value prior to measurement. The value arises only in co-operation with the unique probability which is given to it by the act of measurement itself: the ψ-function is an exhaustive description of the real situation of the system.[8] How to decide?

In the case of two spatially separated but correlated (or, as we now say, entangled) systems, S_1 and S_2, in which 'the real factual situation of the system S_2 is independent of what is done with the system S_1,' (the systems are 'Einstein separable'), then there is a real and very substantial contradiction. Quantum mechanics predicts that the 'character of [the result for S_2] then depends on what kind of measurement I undertake on S_1'. This is the main point. 'One can escape from this conclusion only by either assuming that the measurement of S_1 (telepathically) changes the real situation of S_2 or by denying independent real situations as such to things which are spatially separated from each other. Both alternatives appear to me entirely unacceptable'.[9]

That was to state the situation with unexceptionable clarity. EPR does not require a definition or preconceived notion of physical reality. Nor, for those few paying attention, did EPR have to do merely with inferring the properties of a distant particle from measurements on its nearer partner. Its argument rested on the practical confrontation of the 'non-locality' implied by entanglement with the local realism demanded by the assumption of Einstein separability. If S_1 and S_2 are assumed to be separable and locally real then they must communicate telepathically via (back to Solvay, 1927!) some 'peculiar mechanism of action at a distance'. Let the experiment be done, measure the properties of *both* entangled particles and see what you get. If S_1 and S_2 cannot be considered as separable and locally real a gaping explanatory hole opens.

Physicists had no way to describe how the initial brief encounter of the particles establishes their perpetual subsequent entanglement. The only reasonable conclusion, Einstein insisted, was EPR's: 'B will have to give up his position that the ψ-function constitutes a complete description of a real factual situation'. Indeed, B should concede the game and look elsewhere.

'I believe ... that this [quantum] theory offers no useful point of departure for future development'.[10] It will take time to find an alternative. Do not rush. Physicists in a hurry risk building on shaky foundations.

Einstein built on sound foundations but of a bygone era. In his interpretative essay in the Schilpp volume, Yale's philosopher-physicist Henry Margenau wrote of Einstein's concept of reality: 'If we apply [quantum mechanics] correctly we must not ask what happens to a system during measurement; but content ourselves with the information given to us in that measurement'. Do not cling to cherished classical principles, do not listen to Einstein's siren song, take the uncomfortable road to the future. 'To travel it, one must leave much of classical physics behind; one must re-define the notion of physical state and accept the more rhapsodic form of reality which it entails'.[11]

Pauli said much the same thing. Summarizing Einstein's contributions to quantum theory, he expressed satisfaction with Bohr's views and 'regret[ted] that Einstein seems to have a different opinion'.[12] There can be no 'regression' to classical ideas, he insisted, rather acceptance that greater deviation from them will be demanded. Born referred to Einstein's contributions to statistical quantum theory and its implications for causality and determinism, both firmly integrated into the received theory and lamented Einstein's opposition to the consensus. It was a tragedy, 'a tragedy—for him, as he gropes his way in loneliness, and for us who miss our leader and standard bearer'. Logic cannot dissolve the disagreement between the lost leader and his former flock. 'It is based on different experience in our work and life'.[13]

Other voices Schilpp selected joined the disapproving chorus. De Broglie reckoned that the younger generation of physicists accepted Bohr's version of quantum mechanics as the only one compatible with the known facts. Frank reminded Einstein of the positivism of his youthful self, of the self that had wrought the special theory of relativity. Bridgman chastised him for abandoning the operationalism of special relativity. The deleterious consequences of forsaking his roots can be seen in general relativity as well as in quantum theory. '[H]e has carried into general relativity theory precisely that uncritical, pre-Einsteinian point of view which he has so convincingly shown us, in his special theory, conceals the possibility of disaster'.[14]

Lenzen identified the force that pushed Einstein away from his colleagues as 'faith'. Einstein adhered to the ancient programme of comprehending the real by pure thought expressed mathematically. This programme cannot be realized simultaneously across all fields of physics. Einstein wanted to

conceive quantum mechanics as some sort of limit to general relativity.
What warrant for such a programme was there? 'The pursuit of truth under
the guidance of mathematical ideals is founded on the faith that a pre-
established harmony between thought and reality will win for the human
mind, after patient effort, an intuition of the depths of Reality'.[15] Einstein
liked Lenzen's account, 'everything said there appears to me convincing and
correct'.[16] But even Einstein's mind could not reach the world harmony he
had supposed. His faith, like many another, turned out to be unjustified.

The masterpiece among the essays presented to Einstein in 1949 was
Bohr's account of their discussions at the Solvay conferences. It took him
two years to write, and then he revised it thoroughly while perfecting
his article for *Dialectica*. During this last stage he visited the Institute for
Advanced Study in Princeton at the invitation of Oppenheimer, its new
director, who helped with the rewriting. So also did a young member
of the Institute, Abraham (Bram) Pais, a Dutch Jewish physicist who had
studied with Kramers and Rosenfeld before the war, survived it hidden
in the apartments of friends, and begun his career after it by serving
as Bohr's assistant.[17] (Later Pais would write influential biographies of
both our protagonists.) Bohr's essays for *Dialectica* and the Schilpp volume did much to demystify the elements of the Copenhagen approach:
unity of the phenomenon; indivisibility of the quantum; ψ as a complete
description; the requirement of classical accounts of experiments and their
results; and the unavoidability of objective probabilities and complementary
descriptions.[18]

To the uncharitable eye, Bohr's essay in the Schilpp volume reads as
an extended summary, not of Einstein's contributions to quantum physics
but of Bohr's achievements as its steward. 'Realizing, however, the many
obstacles for mutual understanding as regards a matter where approach
and background must influence everyone's attitude, I have welcomed this
opportunity of a broader exposition of the development by which, to my
mind, a veritable crisis in physical science has been overcome'. This crisis,
presumably of Einstein's making, was averted by recognition that 'in quantum mechanics, we are not dealing with an arbitrary renunciation of a more
detailed analysis of atomic phenomena, but with a recognition that such an
analysis is *in principle* excluded'.[19]

Einstein's 'Reply to Criticisms', drafted in response to the contributions
of his admirers and detractors, concedes nothing. His critics had lost their
compass, they had abandoned the sound old concept of physics as 'the

complete description of any (individual) real situation (as it supposedly exists irrespective of any act of observation or substantiation)'. Sophisticates who ridiculed his naiveté were still stuck in the Machist mire from which he had escaped as he had mastered his craft:[20]

> Whenever the positivistically inclined modern physicist hears such a formu-
> lation his reaction is a pitying smile. He says to himself: 'there we have the
> naked formulation of a metaphysical prejudice, empty of content, a preju-
> dice, moreover, the conquest of which constitutes the major epistemological
> achievement of physicists within the last quarter-century' . . . What I dislike
> in this kind of argumentation is the basic positivistic attitude, which from
> my point of view is untenable, and which seems to me to come to the same
> thing as [the eighteenth-century philosopher George] Berkeley's principle,
> *esse est percipi* [to be is to be perceived].

There are shades of sin, however, and Einstein placed his admirable friend Bohr's among the venial. 'Of the "orthodox" quantum theoreticians whose position I know, Niels Bohr's seems to me to come nearest to doing jus-tice to the problem'. Bohr had helped to expose the fact that the EPR paradox rests on the incompatibility of two assertions: the quantum theo-retical description in terms of the ψ-function is complete, and the states of spatially separated entangled objects do not depend on one other. Bohr offered no solution or even a signpost towards a reconciliation of these positions.[21]

Oppenheimer did not contribute to the Schilpp volume. Free from edi-torial restraint, he outdid the contributors in Einstein-bashing. Einstein had become 'a landmark, not a beacon', he said, with no 'understand-ing of or interest in modern physics' (page 132). He had established no school of his own. He worked 'all alone [in Oppenheimer's Institute!] with an assistant who was there to correct his calculations . . . He turned his back on experiments, he even tried to rid himself of the facts that he himself had contributed to establish'. No doubt Einstein had made extraordinary contributions to science. 'Of course, I would have liked to be the young Einstein. This goes without saying'.[22] But the old Einstein? No way.

Nor did Heisenberg or Schrödinger contribute to the Schilpp volume. Heisenberg's absence may be ascribed to editorial unease over his collabo-ration with the Nazis during the war; Schrödinger's perhaps to his having overexposed himself as more deviant than Einstein. Whereas the wise sage of the Princeton Institute contented himself with stating his faith that

Oppenheimer claimed Einstein had become 'a landmark, not a beacon', with no
'understanding of or interest in modern physics'. L–R: Einstein and
Oppenheimer writing and conversing.

physicists could do better than orthodox quantum mechanics, the witty
sage of the Dublin Institute indulged himself in ridiculing it.

Schrödinger's opening blast, published in the same year as Pauli's scattershot
in *Dialectica*, appeared in a new journal of the Austrian Academy of Sciences.
Only in Austria, the Viennese Pauli sneered, could Schrödinger have published such ossified stuff; 'it made a strongly neurotic impression on me',
a truly regressive piece, just 'subjective-personal moaning [*Schimpfen*]'.[23]
Pauli's experience with Jung inclined him to think of himself as a profound
psychologist.

    Schrödinger's complaint began with the accusation that physics recently
had repudiated the faith that had underwritten it since antiquity. The old
catechism had contained only two items: natural events can be understood
and there exists a real external world independent of the observer. Now,

he moaned, most physicists accept Mach's positivism, which relinquishes the goal of comprehensible objective explanation. Instead, they aim at ever more precise descriptions that enable reliable inferences about future events. But this is certainly to follow a false doctrine as Mach himself demonstrated by insisting that his principles required the rejection of the concept of atoms.

Most quantum physicists (Schrödinger continued) are so confused that they hold simultaneously to Mach's principle of economical description, which involves some subjective elements, and a microphysics of particles, which assumes greater knowledge about the unseen and unseeable than Mach's principles allow. These confused souls think that they can escape their muddle by the path through which they entered it. 'Many see in Mach's principle a marvellous way out of their dilemma as it releases them from seeking clear ideas about Nature'. The proper way out, of course, is to return to Schrödinger's old ideas about a material wavefunction.[24]

We cannot do science, he declared, if we put ourselves in the world picture. And yet we must acknowledge a connection between mind and matter. How is it effected? The mind makes the world picture but is not in it; like the evil being or non-being in Edgar Allan Poe's *Masque of the red death*, the picture has no mind inside. What then is the connection? Shall we say with Gottfried Wilhelm Leibniz that each of us inhabits our own isolated world, or with the Hindus that all human minds make up a unity? Or with fringe idealists that the world is created when a mind contemplates it? None of this is satisfactory. It is best to conceive of the individual's role in the world picture in analogy to the representations of themselves that painters sometimes slip into their canvases. These intrusions can be ignored without disturbing the overall effect of the painting. Of course, this is but a simile, which expresses the fact that we have not succeeded in creating a comprehensible objective world picture 'except at the cost of removing its viewer and architect from it. The attempt to jam them in results in absurdity'.[25]

Schrödinger sounded the materialistic overtones in this wide-ranging discourse in a popular English journal, *Endeavour*, which caught the attention of some American scientists who saw political potential in Schrödinger's analysis. It might bring a measure of reassurance to government funding agencies to know that the research they paid for related to a real world. A copy of Schrödinger's paper found its way into the Smithsonian Institution's annual report to the US Congress and thence into the Congressional record for 1950. There those interested could learn that they need not accept the views of Bohr's school. The confusion Bohr generated, which dissolved the real world into positivist passivity, all came from the quantum-mechanical concept of a particle. Give waves the role they deserve and all will be better if not well.[26]

In his seminars in Dublin, Schrödinger worked out how quantum processes usually explained by the jumping of particles from one stationary state to another could be interpreted as resonances between waves. The key was to undo the process whereby Bohr had succeeded with the hydrogen spectrum. He had converted an empirical spectroscopic formula expressing discrete frequencies ν_{mn} as the difference between two 'terms' T_m and T_n into an energy equation by multiplying both sides by Planck's constant h. Let us remove the h, go back to the formula, identify the *beats* between the terms T_m and T_n as the observed frequencies, and understand the phenomena that gave rise to the bizarre theory of Bohr, Kramers, and Slater as simple cases of resonance.[27]

For interested parties unlikely to read the Congressional record Schrödinger reworked his annihilation of quantum jumps for the *British Journal for the Philosophy of Science*. To his arguments about resonances and beats he added a lament, in the style of his Austrian article, for the separation of physics and culture in general from their history. He remarked that if twentieth-century physicists understood the work of their predecessors better, they would be less prone to overstate their own achievements and more conscious of their obligation to explain themselves to the wider culture. Let them formulate their science so that its historical roots are clear and their current views comprehensible. What should endure, which will not be orthodox quantum mechanics, will then endure.[28]

This was too much for the old guard. Pauli warned Schrödinger that regression towards classical physics was psychologically unsound. To Bohr he diagnosed Schrödinger's mental state as that of a child crying over a stolen toy. 'Waves are so much more beautiful than particles and these evil statisticians deprive me of my favourite play'.[29] Schrödinger: only 'unadulterated asses' could think so.[30] Born agreed to take on Schrödinger in a debate before the British Society for the Philosophy of Science in London in December 1952. Rosenfeld would come down from Manchester, where he had moved from Utrecht, to second Born. Schrödinger would appear as the lone champion against what he called 'epistemological bungling [*Stümpferei*]'.[31] But he did not appear at all, having fallen ill, and Born had to answer his attack on quantum jumps in print.

Born's response lacked the verve of the stimulus. It pushed the dispute into philosophy which, in the minds of most practising physicists, was synonymous with irrelevance. When it came to computing, Born said, all physicists were on the same page. If we wish to flirt with philosophy, we can assert that Schrödinger's realistic interpretation of the wavefunction cannot be sustained and concede that a quantum-mechanical particle possesses only vaguely the qualities associated with its classical counterpart. Just what 'particle' means in quantum mechanics was a problem in linguistics. Do not worry about it. Bohr had the problem in hand.[32]

Schrödinger continued his campaign in his contribution to de Broglie's anniversary volume. To reach the widest possible readership, he insisted that his contribution appear in English as well as in French. He did not mince his English. '[I]t must have given to de Broglie the same shock and disappointment it gave to me, when we learnt that a sort of transcendental, almost psychical interpretation of the wave phenomenon had been put forward, which was very soon hailed by the majority of leading theorists as the only one reconcilable with experiments, and which now has become the orthodox creed'. They thought they had rid the world of quantum jumps, but instead the orthodoxy grew nuttier and arrived at the notion that the outcome of the jumps produced by measurements depended on some human observer noticing it. To escape from the cage of the solipsist and other disasters, we must admit that the so-called 'particles' created by orthodoxy are not particles at all, not persistent, distinguishable, independent entities, but ghost-like creatures that must be replaced by a consistent scheme of waves.[33]

Impossible, replied Pauli, for all the old reasons, which the master of the waves, de Broglie, had recently reaffirmed. Nature and von Neumann ensured that 'the interpretation of quantum mechanics based on the idea of complementarity is the only one admissible'.[34] Impossible indeed, echoed Rosenfeld; even de Broglie recognized it, and the greatness of his mind may appear more in his surrender of his double solution and his pilot waves to complementarity than in his invention of them.

Rosenfeld's main target was not heretics who rejected the light but a brother-in-arms who had lapsed after kindling it. To speak plainly, Rosenfeld censured Heisenberg's article in *Dialectica* for broadcasting confusion. The article had begun soundly enough, with a dialectical account of the four successive closed physical systems, and yet reached, 'in a grinding dissonance', a subjective, idealistic, solipsistic finale. The *Geist* that overfilled its conclusion had no philosophical significance. Where did the nonsense come from? Rosenfeld diagnosed it not, as Pauli might have done, as a psychological aberration, but as 'the shibboleth of a class [the elite professoriate]. Heisenberg's idealism had to do with sociology, not epistemology. In this domain, the power of his creative thought disguises the poverty of his philosophy'.[35]

As for Heisenberg's historiography, his reduction of the advance of physics to a progression of closed theories, it was sheer 'idealistic humbug'. As a philosophical thinker or interpreter of the history of physics, 'Heisenberg is not worth discussing'. Or so Rosenfeld's friend, the mathematician and positivist philosopher, Martin Strauss, thought. Even Pauli pulled away from the self-proclaimed prophet of the Copenhagen spirit: 'I myself have quantitatively and systematically avoided discussing anything except physics with Heisenberg'.[36]

And thus we arrive at the apparent paradox that most physicists in the first postwar decades accepted the orthodox 'Copenhagen interpretation' even though its main proponents disagreed among themselves about its fundamentals.[37] For example, Bohr located the complementarity he saw between causal and space-time descriptions in the different sorts of experiments needed to bring out the positions and momenta of micro-entities, whereas Heisenberg associated it with the difference between the causal evolution of possibilities expressed in the wavefunction and the definite space-time result produced by a measurement.[38] Nonetheless, adherents of the Copenhagen school appeared to share enough doctrine that Einstein and Schrödinger could characterize its belief and belligerence as a church. No doubt some hesitated over whether Bohr's complementarity or his intellectual son Heisenberg's uncertainty principle or both together constituted the foundation of the faith; but then did the church fathers not waffle over whether the Holy Spirit emanated from the Father, or from the Son, or from the pair combined?

Bohr's 70th birthday, which fell in 1955, gave rise to a festschrift edited by Pauli that provided a perfect place to define core Copenhagen beliefs. The task went to Heisenberg, who was then deep into a shallow assessment of 'physics and philosophy', the subject of his Gifford lectures delivered in 1955/6 at the University of St Andrews. The combined lesson of these lectures and the Bohr festschrift asserts that the 'Copenhagen interpretation' (a term Heisenberg may have first introduced in these presentations) developed in its full form in 1927 from the uncertainty principle, complementarity, and Born's probabilistic account of ψ.

Essential consequences of these core tenets are that nothing definite can be known about the microworld between measurements on it and that what is safely known—the results of measurements—must be uttered in the unambiguous language of classical physics.[39] It is this requirement that saves the Copenhagen interpretation from the charges of positivism, idealism, and subjectivism. Orthodoxy did not demand much more. The nature of the real world, the peculiarities of ψ, and the role of human consciousness in taking cognizance of a measurement were important but undecided questions, adiaphora, as to which the faithful could form their own opinions.[40]

Still peace did not prevail. In a blistering review of the book that resulted from Heisenberg's Gifford lectures, *Physics and philosophy*, Rosenfeld rejected not only the description of the Copenhagen interpretation ('not one of the physicists now working in Copenhagen would subscribe to [it]') but also the deviationist tendency inherent in the term. 'It falsely suggests that there could be other possible interpretations of quantum theory'. For the rest, Heisenberg's lectures were 'hazy [and] superficial' in philosophy, 'worse than valueless' as history, and anathema in their account of the cardinal Copenhagen teaching about the employment of classical concepts in quantum mechanics. Heisenberg declared that the employment was vague, like poetry, whereas Bohr had striven mightily and (as far as Rosenfeld was concerned) successfully to define in complementarity precisely how and where classical concepts can be used in microphysics.

The insertion of fuzzy subjectivity into the measurement process, anathema to Rosenfeld, may have been one of the features of Heisenberg's Copenhagen interpretation that made it 'beautiful' to Pauli.[41] For Pauli, still labouring to bring his physics into parallel with his subconscious, faced a great dilemma. On the one hand, he would argue, the observer cannot be separated from the observed, and to that extent makes his world. On the other hand, he conceded, if the observation were repeated many times the statistical average would not budge, and to that extent an objective world exists.[42]

As the founders traded rhetoric over the subtleties of interpretation, the next generation was being taught quantum mechanics with little attention to them. The success of the Manhattan project and the relentless focus on the applications of physics fit comfortably within an American culture that European observers had diagnosed as less philosophical than that of any other country in the civilized world.[43] The US had continued to foster a deep-seated anti-intellectualism. In his Pulitzer prize-winning *Anti-intellectualism in American life* (1963), historian Richard Hofstadter described the phenomenon as 'a resentment and suspicion of the life of the mind and of those who are considered to represent it; and a disposition constantly to minimize the value of that life'.[44]

As postwar enrolments in physics courses boomed, the student body became more competitive, focused, and conformist. Research students relied on federal money from granting bodies, especially the Atomic Energy Commission and Department of Defense, which had practical missions and mandates. No more than the Commission and the Department did the National Science Foundation encourage investigation of foundational questions. Research advisers sought to steer their students towards projects likely to provide a firmer basis for a career in physics than investigation of its purpose and meaning. In any case, the large classes did not lend themselves to discussions of unresolvable 'philosophical' questions marginal to the profession.[45]

Would-be quantum physicists in the US learned their subject from such textbooks as Leonard Schiff's *Quantum mechanics* (1949). Schiff had served as Oppenheimer's research associate at Berkeley from 1938 to 1940 and, like other graduates of Oppenheimer's course on quantum mechanics who became professors, used it as the basis for his own lectures.[46] His textbook of 1949, which appeared in three editions spanning 20 years, may be regarded as a descendant from the age of the founders with a characteristic exception. In following Oppenheimer, Schiff adopted the logic of its antecedent, Pauli's article in the *Handbuch der Physik*, while omitting Pauli's effusive account of complementarity.

Schiff's presentation of Bohr's ideas occupies two paragraphs from which the puzzled student could learn that complementarity flowed from the uncertainty principle. 'Thus the complementarity principle typifies the fundamental limitations on the classical concept that the behavior of atomic systems can be described independently of the means by which they are observed'.[47] Schiff considered measurement only in the context of the uncertainty principle; he did not mention von Neumann, Einstein's objections, or EPR. There was little here to inform, and nothing to satisfy students who wanted to know the makeup of the quantum world. To the persevering, this dearth could be inspiring. John Bell read Schiff's book during his last undergraduate year at The Queen's University in Belfast. He demanded to know more, was fobbed off, and, as will appear, enjoyed a sweet and devastating revenge.[48]

Though stripped of their philosophical content and largely empty of insight, the references to complementarity gave most students and professors all the information they needed. A few words about Bohr and an insinuation that he had solved all the vexing problems long ago might suffice. David Mermin, a professor at Cornell who has contributed wittily to the discourse on quantum theory's interpretive principles, recalled the situation he faced as a student:[49]

> [I have] vivid memories of the responses my conceptual inquiries elicited from my professors—whom I viewed as agents of Copenhagen—when I was first learning quantum mechanics as a graduate student at Harvard, a mere 30 years after the birth of the subject. 'You'll never get a PhD if you allow yourself to be distracted by such frivolities,' they kept advising me, 'so get back to serious business and produce some results'. 'Shut up', in other words, 'and calculate'. And so I did, and probably turned out much the better for it. At Harvard, they knew how to administer tough love in those olden days.

But, in truth, his professors were 'just indifferent to philosophy. Full stop. Quantum mechanics worked. Why worry about what it meant?' Only in this sense did Mermin regard them as 'agents of Copenhagen'. 'They had no interest in understanding Bohr, and thought that Einstein's distaste for [quantum mechanics] was just silly . . . It was a very unphilosophical time'.[50]

Skirmishes in Princeton

The quantum hermeneutics of Oppenheimer's student David Bohm perfectly exemplified the dialectics of the Marxism he espoused. At first reluctant to accept complementarity, he came to endorse it before devising a synthesis that returned determinism to quantum mechanics. Bohm's adherence to Marxism had developed together with his knowledge of quantum mechanics as he experienced both in Oppenheimer's greenhouse for theorists in the physics department at Berkeley.

Bohm joined it in 1941 after two years of disappointment with the emphasis on book learning and calculation he experienced at Caltech. He had been accustomed to thinking for himself in a vague and lofty way about the nature and the structure of the universe. In his high school in the depressed coal-mining town of Wilkes-Barre, Pennsylvania, and during his undergraduate years at Pennsylvania State University he easily outdistanced his cohort and often his professors in his grasp of the mathematics of physics. He was a typical outsider, comfortable neither at home, where his parents constantly bickered, nor at school, where his physical awkwardness, aloofness, and indigence kept him from bonding with his classmates. His father, an orthodox Hungarian Jew, earned just enough from his second-hand furniture business to enable the family to hover above the poverty line. David escaped into science fiction and mathematics and dreams of a state of society more generous and constructive than his hometown's.[1]

At Berkeley, Bohm could indulge both these dreams with the Jewish contingent gathered around Oppenheimer, who was having his own romance with the left in the person and projects of a communist girlfriend. The contingent included Peters, an anti-Nazi Polish–German émigré who had spent some time in Dachau; Giovanni Rossi Lomanitz, born in Texas and raised in Oklahoma, named after an Italian socialist and, like Peters, an active communist; and Joseph Weinberg, a precocious leftist New Yorker, who became Bohm's closest Berkeley colleague. And we should not omit Melba Phillips, one of Oppenheimer's first Berkeley PhDs and a friend of Bohm, who suffered during the McCarthy era for refusing to answer whether she belonged to the communist party.[2] By 1942 Bohm certainly was a member of a communist cell. Although he soon grew tired of its tedious meetings, he retained his Marxist sympathies.[3]

Bohm's communism kept him from employment at Los Alamos, but it did not prevent Oppenheimer from assigning him a thesis topic on gaseous diffusion whose solution, which Bohm quickly discovered, was instantly classified as a contribution to bomb-making. After finishing this work Bohm joined Weinberg in teaching the absent Oppenheimer's courses and immediately confronted the problem of making quantum mechanics intelligible to others. He and Weinberg had argued vigorously about complementarity, contra and pro, respectively. Neglect of a topic by texts and teachers does not prevent students from airing it, and Oppenheimer may well have taken part. As a 'rather deep and subtle notion which has turned out to be the clue to the unravelling of [quantum physics]', complementarity was a perfect topic for discussion at the sophisticated parties he threw in his Berkeley home.[4]

The responsibility of introducing others to quantum mechanics prompted Bohm to accept complementarity and, like Schiff, to develop Peters' notes into the framework of a textbook that would display the physics behind the mathematics. However, unlike Schiff's *Quantum mechanics*, Bohm's *Quantum theory* delved deeply into foundational issues. Completing his dialectical move, Bohm taught that through Bohr's insights 'it finally became possible to express the results of quantum theory in terms of comparatively qualitative and imaginative concepts'. That is because Bohr's complementarity allowed, indeed required, the use of classical concepts under due restrictions in the description of experiments and their results.[5]

Despite his warming to Copenhagen teachings, Bohm remained Marxist enough to explore the possibility that the radical probabilities of the theory might be removed by some scheme of hidden variables. EPR's analysis fortified the possibility. '[Their assumptions] seem, at first sight, so natural and inevitable that a careful study of the points which the authors raised affords deep and penetrating insight into the difference between classical and quantum concepts of the nature of matter'.[6] Physicists might yet concoct a deterministic scheme that agreed with quantum physics over much of microphysics but not everywhere. If such a scheme existed and experimental technique were properly perfected, an EPR-type experiment might be mounted as a test.[7] EPR's original scheme had involved measurement of the positions and momenta of entangled particles. Bohm proposed a simpler alternative.

In Bohm's version the source is a molecule easily split into two atoms identical apart from their oppositely directed spins.[8] When the atoms separate far enough to have no plausible interaction, a measurement on the spin of one atom allows a correct calculation of that of the equal-and-opposite other. Measuring the spin of an atom is not easy. First a direction must be imposed, typically the axis defined by the poles of an electromagnet through which the atoms must pass. Physicists say that atoms deflected towards its north pole have 'spin-up', those deflected to the south pole 'spin-down'.

Up and down relate to the vertical if the electromagnet has its south-to-north axis aligned floor-to-ceiling (typically defined as the laboratory z-direction). If the axis is horizontal, it lies at 90 degrees to the vertical (the laboratory x-direction). Quantum mechanics does not allow measurement of the spin orientation in both the x- and the z-direction simultaneously. Once the experimenter sets the instrument's axis, the atom has no meaningful spin orientation in any other direction. Whatever the direction chosen, measurement shows the entangled atoms to have opposite spins *in that direction*. An observation of spin-up for atom 1 on the left implies one of spin-down for atom 2 on the right, and *vice versa*.

On EPR's analysis, the delayed choice to measure the spin of atom 1 in the z-direction can have no effect on the spin possessed by atom 2, since this is a quantity fixed in the molecule before disintegration and, by hypothesis, the atoms are 'Einstein separable'. Atom 2 spins oppositely to atom 1 in the z-direction. But what if instead the experimenter chooses the x-direction? How, if the atoms are widely separated, can atom 2 'know' the direction selected and act accordingly? We are firmly back at the EPR experiment.

Bohm took 'potentialities' as the answer. '[A]ll these [directions] exist simultaneously in roughly defined forms', and any of them can be resolved better at the expense of the others by interaction with a proper measuring instrument. But how then does atom 2 deport itself immediately to agree with the measurement on atom 1 when beyond the range of physical interaction, constrained by special relativity to transmission at speeds less than or at most equal to that of light? Bohm explained that the probabilities of the various behaviours of the two atoms, correlated in their original interaction, propagate with them. What could be clearer? 'The present form of quantum theory implies that the world cannot be put into a one-to-one correspondence with any conceivable kind of precisely defined mathematical quantities.' Very probably more general and challenging concepts than those now deployed by quantum mechanics will be required to account for 'the infinitely complex and subtle structure of the world'.[9] These words might have been written by Bohr.

Bohm set the scene for his experimental challenge in a prescient section of his book, 'Proof that Quantum Theory is Inconsistent with Hidden Variables'. It would take 13 years to articulate the challenge precisely, and a further eight years to realize it in the laboratory. But Bohm had few doubts about the likely outcome: '[N]o theory of mechanically determined hidden variables can lead to *all* of the results of the quantum theory'.[10] The context suggests that Bohm had in mind hidden variables that lent locally realistic properties to the entangled particles of EPR, which, in his judgement, would violate the uncertainty principle.

When he published his classic textbook, Bohm was in Princeton. He and Oppenheimer had moved there from Berkeley in the same year, 1947, Bohm to the university as assistant professor, Oppenheimer to the Institute for Advanced Study as its director. Both places sheltered prominent Copenhageners and fellow travellers. The permanent fellow travellers included von Neumann at the Institute and John Wheeler at the university. Both were committed cold warriors unsympathetic to Bohm's political opinions. But Wheeler, who had collaborated with Bohr in the 1930s and picked up some of his mannerisms ('I never stop thinking about physics . . . [and] my relationship to the universe and its laws'), was comfortable overlooking politics to procure an excellent physicist.[11] Especially one with wide views, a respect for Bohr, and a withdrawn personality unlikely to upset a university not yet entirely welcoming to Jews.

Bohr and Pauli came frequently to the Institute. Bohr was there in 1948, polishing his contribution to the Einstein volume and, in 1950, working with Wheeler and correcting proofs of a weighty article on measurement written with Rosenfeld.[12] Pauli visited in the winter of 1949-50. He delivered some notable public lectures, not on physics but on intellectual history, then being sponsored by Oppenheimer.[13] Pauli's contribution was an application of Jung's theories of alchemy to the relations among the astronomer Johannes Kepler, the magus Robert Fludd, and himself. He had the invaluable help of a permanent member of the Institute's School of History, the art historian Erwin Panofsky, in studying baroque occult ideas and Aristotelian philosophy, but his concurrent association of the concept of the field in physics to the unconscious in psychology through a Jungian archetype was his own invention.[14] Although far from being a quantum physicist, T.S. Eliot, a visitor to the Institute in 1948, absorbed enough of the prevailing *Geist* to diagram his emerging play, *The cocktail party*, on a blackboard, and to allow his characters no closer an idea of their future trajectories than a diffracted electron has of its.[15]

The count of Princeton fundamentalists may be rounded off with Wheeler's colleague at the university and von Neumann's old classmate from Hungary, Eugene Wigner, another cold warrior. Wigner's father had converted the family from Judaism to Lutheranism to avoid the stigma and anti-Semitism that derived from the common association of Jews with communists. When a lonely and isolated student of chemical engineering in Berlin in the 1920s, Wigner became fascinated with Freud's interpretation of dreams and the nature of human consciousness, and in the early 1950s he had started on a lengthy quest to 'extend theoretical physics into the realm of consciousness'. His method pressed the logical analysis of measurement to the extreme of requiring a human consciousness to conclude every quantum measurement.[16] Thence arose the paradoxical concept of Wigner's friend, a colleague who observes the outcome of a measurement in Wigner's absence.

The paradox relies on the supposition that the measurement chain extends all the way from the quantum object under study to the human observer doing the studying. Von Neumann had supposed the same and (with tongue in cheek) Schrödinger did so too, as he entangled his cat in the chain. From Wigner's point of view, the microsystem, the measuring instruments, and his friend's mind are entangled in a superposition of the possible measurement outcomes. Wigner interrogates his friend, who has made the observation and seems quite convinced of a particular result. How could that be if, as far as Wigner is concerned, the friend's mind is still entangled with the full range of possibilities?

'[That] appears absurd because it implies that my friend was in a state of suspended animation before he answered my question'. To insist on the absurdity would be ungenerous: 'to deny the existence of the consciousness of a friend to this extent is surely an unnatural attitude, approaching solipsism, and few people, in their hearts, will go along with it'.[17] The measurement chain must therefore terminate—the wavefunction must collapse—on interacting with the *first* conscious mind it encounters.

Physicists cannot rest there, however. Their science is much too difficult for a single human mind, even Wigner's. To continue to progress, physicists must master the psychology of collaboration, the inter-subjective working of human minds.[18]

Against this phalanx stood Einstein, whose acquaintance Bohm sought and whose esteem he won early in his Princeton career. Einstein admired Bohm's textbook as about the best that could be done with the effluvia of the Copenhagen spirit. Though less successful than Schiff's, Bohm's text received many favourable reviews and was adopted at a 'fair number' of universities.[19] At Einstein's request, Bohm visited him sometime in the spring of 1951. Doubts that had begun to creep into Bohm's mind as he had worked on his book now crystallized.

As Einstein explained the basis for his own misgivings, Bohm acknowledged the need to rethink. 'This encounter had a strong effect on the direction of my research', he later wrote, 'because I then became seriously interested in whether a deterministic extension of quantum theory could be found'.[20] 'After I finished [*Quantum theory*], I felt strongly that there was something seriously wrong. Quantum theory had no place in it for an adequate notion of an individual actuality. My discussions with Einstein clarified and reinforced my opinion and encouraged me to look again'.[21] Bohm's transformation from passionate advocate to revolutionary took just four months, accelerated, perhaps, by the political environment.

In May 1949 Bohm's communist associations caught up with him. Some years earlier his Berkeley comrade Weinberg had been recorded betraying atomic secrets in conversations with Steve Nelson, a key figure in the communist party apparatus in the San Francisco Bay Area. The bug planted by the FBI in Nelson's home was illegal and the testimony derived from it inadmissible in court. Hoping to expose Weinberg's betrayal by more legal means, the House of Representatives' Un-American Activities Committee (HUAC) called Bohm to testify. Einstein advised him to refuse, although this might result in imprisonment.[22] But Bohm testified unreservedly except when he pleaded the fifth amendment, a common way to avoid questions that might elicit self-incriminating answers. He had nothing to hide but feared that in an open-ended interrogation he would be obliged to betray his leftist friends. He did the same at a later HUAC hearing in June.

Events over the next 12 months conspired to whip up anti-communist sentiment in the US. Whitaker Chambers, a former Soviet agent and editorial staff member of *Time* magazine, revealed the names of two highly-placed communists in President Harry Truman's administration—Alger Hiss at the State Department and Harry Dexter White at the Treasury. White died of a heart attack in August 1948. Hiss was convicted on two counts of perjury in January 1950 and sentenced to two concurrent five-year prison sentences. As the 'red scare' gathered momentum, the opportunistic

David Bohm reading the latest news in the *Daily Mirror* after refusing to testify whether he was a member of the communist party before the House Un-American Activities Committee, 1949.

Republican Senator Joseph McCarthy launched an anti-communist crusade that became a witch-hunt. Then, on 25 June 1950, the [North] Korean People's Army crossed the 38th Parallel bringing war to the South. Within months, United Nations forces (principally the South Korean Army and US and British troops) faced defeat.

Closer to our subject, Klaus Fuchs, who had worked for the British contingent in the American nuclear project, was unmasked as a Soviet spy in September 1949 and judged guilty in March 1950 of breaching the UK's Official Secrets Act. His crimes, 'thinly differentiated from high treason' (an offence that at the time attracted the death penalty), earned him the maximum sentence of 14 years imprisonment. HUAC would deal severely with physicists. Bohm was arrested on 4 December 1950, charged with contempt of Congress, and bailed for $1,500. He scrambled in vain to secure the support of his Princeton colleagues.

The big man at Princeton was Oppenheimer, rated by some the most brilliant of contemporary American physicists. He had chosen the directorship of the Institute over many other invitations he received after revelation of his leadership of the atom-bomb laboratory at Los Alamos because it enabled him to stay close to Washington, where he continued to serve the government as chief scientific advisor on atomic energy and weaponry.

Though a man of immense mental power known for his ability to synthesize difficult technical material and reach decisions quickly, Oppenheimer suffered as a director from the unexpected infirmity that he often sailed rudderless. He had been kept roughly on track in the mid-1930s by Arthur Ryder, a right-wing professor who taught him Sanskrit at Berkeley; during the war by General Leslie Groves, the commander of the crash project to make atomic bombs; and after the war by Bohr, his guru in physics and politics.[23] But despite these father figures he had occasionally run off the rails, retained communist ties he had formed in Berkeley before the war, misled Groves's security officers, and betrayed his leftist students.

The President of Princeton University, Harold W. Dodds, ordered Bohm not to set foot on campus. Dodds regarded him as a communist and communists as slaves who had abandoned 'their rights as persons, made in the image of God'. Since Princeton hired only such persons, in June 1951 it fired its most promising young physicist. Wheeler had done nothing to help; he was off in Los Alamos developing the hydrogen bomb and, in any case, 'found it hard to accept Bohm's decision to shield those who adhered to communist ideology at a time when the Soviet Union was suppressing its own people and threatening world peace'. And, possibly, Wheeler had decided that the sacrifice would not be too great for Princeton to bear. 'As I became better acquainted with [Bohm's] work, I came to realize that it was based more on intuition than the rock-solid foundation I had at first imagined'.[24]

Only Einstein tried to help. He proposed to take Bohm on as his assistant, 'a clear mind of . . . a rare independence in his scientific judgments'. But Einstein's administrative superior, Oppenheimer, vetoed the appointment.[25] Employment with Einstein could only exacerbate the situation, for Einstein's support of intellectuals persecuted by Congressional committees had earned him the sustained attention of the FBI and inspired rabid anti-communists to seek his deportation from McCarthyite America.[26]

As his life turned upside-down, Bohm struggled to concentrate on his physics. Confronted with the task of correcting the page proofs of his new book, he found it 'hard to concern myself with getting all these [mathematical] formulas correct'.[27] In fact, Bohm had done no wrong and after further questioning and a short spell in prison he was acquitted in May 1951.[28] Unable to find a position at a US institution, he left for the University of São Paulo in Brazil in October 1951. Einstein had written him a letter of recommendation ('a very gifted and original theoretical physicist . . . [and] a lovable personality') as had Oppenheimer, that 'linear combination of Jesus Christ and Judas', pleased thus to remove in one stroke a burden on his conscience and a possible embarrassment to his career.[29]

Amid this mental turmoil, lonely and depressed, the character that supported counter-cultural political identification did not shy from opposing scientific authority, and a month after his firing from Princeton Bohm sent to the *Physical Review* the challenge to the Copenhagen spirit that he had foreshadowed in his textbook.[30] The immediate inspiration for this new dialectical turn may well have been some new attacks against Bohr by Soviet physicist-philosophers refreshing the hostilities that Lenin had begun with Mach and positivism in 1908. There had been attacks in the 1930s, but they had not been sustained. In the immediate postwar period, however, Soviet physicist-philosophers noticed that complementarity as then taught did not admit a well-defined microworld before a physicist measured its properties.[31]

Andrei Alexandrovich Zhdanov, a party philosopher, propagandist-in-chief, and possibly Stalin's successor-in-waiting, initiated the skirmish in 1947 by condemning the dissolution of matter into waves, ascription of free will to electrons, and other absurdities he attributed to the Copenhagen school. Effective condemnation lasted in the Soviet Union until 1960, but traces lingered on there and in the West, where as late as 1967 the inverse effect, an accusation of communism against a physicist who rejected the Copenhagen interpretation, might be levelled.[32] Between the renewal of hostilities in 1947 and their decline towards 1960, Soviet physicists scrambled to defensible positions, either condemning quantum mechanics altogether if they could do without it or materializing it in various ways. One of these ways was to admit hidden variables in the manner Bohm would adopt.[33]

Bohm made clear the connection he saw between his politics and his physics in a letter to his one-time girlfriend Hanna Loewy. He had discovered a connection between plasma theory and quantum theory that suggested to him that matter comes in an infinite set of qualitative, interconnected levels, the 'particles' at any level being determined in part by the collective in which they participate. The goal of physics is to move through the levels to gain an 'absolute understanding of the laws and properties of matter'. These laws are causal; understanding them is essential; absolute understanding is the only way to save society. 'Humanity as a whole . . . has the possibility of expanding its knowledge without limit; and thus, by understanding the causal laws, to go beyond all conceivable limitations . . . [I]f we fail to understand causality and necessity, both external nature + our own natures will enslave us'.[34]

Bohm spared Hanna the mathematical manoeuvre he used to set up his hidden variable theory. In fact, it is not at all difficult. Bohm assumed that the wavefunction can be decomposed into two functions, R and S. R is an amplitude, S a quantity taken from a classical model in which the paths of particles are defined as perpendiculars to abstract waves, in analogy to the relation of light rays to Young's waves. Bohm assimilated the abstract S waves of the classical analogy to Schrödinger's waves and the associated particles to a micro-entity with definite position q and momentum p.

When written in terms of R and S, the Schrödinger equation could be split into two parts, one involving only real terms, the other only those multiplied by i—even quantum theory does not allow equating a real number with an imaginary one. The simpler of the two equations, which involved only $|\psi|^2 = R^2$, had the form of a classical hydrodynamical condition of conservation. This conservation was encouraging, since, regarded as a probability, $|\psi|^2$ should equal unity when summed over all possible states at any time. The other equation could be understood as the description of a particle moving under a classical force or potential augmented by a 'quantum potential' proportional to h^2 and thus negligible in physical systems where Planck's constant h was too small to bother about. This quantum potential has the further agreeable property that it depends only on $|\psi|^2$. The amplitude of the solution to Schrödinger's equation exercises a force on the particle and, at the same time, gives the probability that it will be found at a given place and time.[35]

This conclusion might seem entirely at odds with Bohm's objective. If his scheme defined a particle's trajectory, it should have no need for probabilities. He had another master to serve, however. His version should not give results different from established experimental outcomes that agreed with quantum mechanics. He had to incorporate Born's probability interpretation and Heisenberg's uncertainty relations. To do that he conceded the practical (not radical) impossibility of securing initial data about the particle, its starting position and velocity, to as great a precision as desired owing to the disturbance of the slight particle by the weighty instrument of measurement.

Here recourse to probability is born of ignorance of the initial conditions, much as statistics arises in the kinetic theory of gases. The probability Bohm invoked was the likelihood that if chosen at random from a very large collection or ensemble of electrons similarly situated, the one under investigation has given values of position and momentum. But he had to pay a price for aligning his interpretation with standard quantum mechanics. Since $|\psi|^2$ determined both the quantum potential that helped to fix the trajectory of the particle and the probability of having its trajectory thus fixed, the particle's state 'look[ed] like a mysterious dependence of the individual on the statistical ensemble of which it is a member'.[36] The price was surrender to non-locality. 'Thus, the "quantum-mechanical" forces may be said to transmit uncontrollable disturbances instantaneously from one particle to another through the medium of the ψ-field'.[37] Einstein did not like this solution. No wonder: the price paid was admission of action at a distance.

The trick of disassembling the Schrödinger equation into a hydrodynamic and an energy part was known in the era of the founders. Its first discoverer, Erwin Madelung, Born's successor as professor of theoretical physics at the University of Frankfurt, also deduced the odd dependence of the individual on the ensemble and, regarding the connection as implausible, dumped the entire approach. And we know that de Broglie discovered the trick and used its consequences in his 'theory of the double solution'. According to a historian of this prehistory, theories of Madelung's and de Broglie's type 'foundered on the shoals of nonlocality'.[38] As we know, Einstein gave up his attempt of May 1927 because it produced non-local effects. Replacement of action at a distance by local fields had secured many a triumph in classical physics; readmitting non-local effects was reckoned weak-minded and retrogressive.

Bohm did not tremble before non-locality or probability or the dual role of the wavefunction in his formulation. And a non-local hidden variable theory did not contradict the prescient conclusion he had drawn in his textbook. No doubt ψ's double dealing, as a field of force and as a measure of probability, was a disquieting consequence of the theory. Perhaps a different quantity might be taken to describe the statistical ensemble. But that would not change anything, for, according to Bohm, thermodynamics would eventually drive any arbitrary form of probability the theorist might impose as an initial condition to $|\psi|^2$. In sum, Bohm gave the particle a trajectory in space and time and a probability of performing it. The epistemology of this probability differed from that of the Copenhagen account, but the mathematical equivalence of the equations guaranteed that Bohm's results would coincide perfectly with those of the conventional quantum formalism.[39]

Bohm could hardly credit his good fortune. 'I can't believe that I should have been the one to see this'. He was not, as he soon learned, but he was the one who put the Madelung and de Broglie approach back on the agenda of physics, though far from its top. Why he? His answer: 'Most people . . . suppose that they do not have the necessary passion and courage to act in a truly creative way'. Bohm attacked all the ideas that interested him, whether in physics, philosophy, politics, or art, with the 'vibrant tension and passionate energy' they deserved.[40]

Juvenile Deviationism

Bohm had sent a draft of his challenge to the Copenhagen school to Pauli, whose earlier praise of his textbook suggested fairmindedness. Pauli reacted with orthodox indignation. This was 'not even new nonsense'.[1] Bohm writes me as the pastor of a sect (*Sektenpfaff*), he scoffed, 'to convert me to a warmed-up theory of the pilot wave'—a theory Pauli believed he had stamped out in Brussels a quarter of a century earlier and from which he had been trying to save Bohm.[2] With his usual passionate energy Bohm replied to Pauli's criticism that he had answered all the old objections to the pilot theory. He had found the diamond de Broglie had been obliged to discard, cleaned it up, and now claimed it. Your counter-argument at Solvay, Bohm wrote, rested on 'excessively abstract assumptions' about the ψ-function of a free electron. Replace it with a suitable wave packet and the argument fails. No-one, he concluded, had been able to find an inconsistency in his treatment.

Von Neumann's famous 'impossibility proof' fared no better than Pauli's destruction of de Broglie. Bohm thought that he had located von Neumann's error in the omission of the possibility of hidden variables in the instruments, which, when pitted against the hidden variables of the system under study, can provide the chaotic situation that collapses the wavefunction. Bohm's argument turned out to be wrong, but it served briefly to deflect a serious objection to his model. Von Neumann was good enough to interrupt his cold-war commissions to look over Bohm's draft and pronounce it logically consistent.[3] By December 1951 Pauli too had conceded that Bohm's approach could not be rejected by logic alone. Pauli invited Bohm to let him know when any discrepancies between the predictions of hidden variable theory and standard quantum mechanics cropped up. Until then he wanted no more of Bohm's correspondence. He was busy working out a parallel between the Neoplatonic doctrine of the successive emanations of world souls and radioactive decay.[4]

In December 1951 Bohm criticized Pauli for claiming to prove, in his old *Handbuch* article, that the equations for quantum mechanics must have a certain linear form. Relax that, Bohm asserted, and thermodynamics justifies my expression for the probability, $|\psi|^2$, which, of course, is the same as yours. Since you cannot fault my approach logically, 'it seems to me that your desire to hold on to the usual interpretation can have only one justification; namely, the positivist principle of not postulating constituents that do not correspond to things that can be observed'.[5] And your insistence that a good physical theory must include the human mind! We know nothing about the functioning of the mind. It would be far better to extend quantum theory my way, which permits us to conceive of many improvements, rather than the Copenhagen way, '[which] does not permit us to conceive of anything at all'.[6]

This zeal provoked Pauli to recover a nice image from 1927. We used to joke, he wrote Swiss physicist Markus Fierz, that de Broglie's double solution amounted to a nursemaid (the wave) perambulating a baby (the particle). 'A fool like Bohm cannot be helped'. He was dangerous, however, as he might lead others into heresy, not in the US, apart from a few lefties, but among the many communist physicists in France stirred up by the Soviet condemnations of complementarity.[7]

The prince of French physicists, Louis de Broglie, who stood far from the left, was surprised one spring morning in 1951 to find in his mailbox a revision of his old theory of the double solution by an American physicist. He showed it to his communist assistant, Jean-Pierre Vigier, who perceived in it half of a grand synthesis of quantum mechanics and general relativity. Although de Broglie was not yet entirely convinced that Bohm had shielded the synthesis Vigier perceived from Pauli's old arguments and von Neumann's powerful proof, he expressed sufficient interest in Bohm's deviance to agitate Bohr's Praetorian guard.[8] Pauli sounded the alarm in January 1952 as Bohm's challenge appeared in print: 'Catholics and communists in France are uniting against complementarity!' A letter from de Broglie in February, stating that Bohm's work had returned him, with Vigier's help, to the double solution, confirmed the threat. De Broglie had recovered his youthful ardour, his 'dazzling poetic vision', that is, he had relapsed into his 'shadow physics beer-idea wish dreams' under Bohm's 'devilish influence'.[9]

Pauli went to Paris in March for scheduled lectures and took the opportunity to throw every argument he could think of against Bohm. He repeated this critical collection in his unfriendly contribution to de Broglie's festschrift. Bohm's theory is useless, Pauli thundered, it introduces asymmetries between q and p, messes up the attributions of statistical relations established by Bose and Einstein and by Fermi and Dirac, and requires a period greater than the age of the universe to produce thermodynamic equilibrium. Pauli's diatribe, which in Born's opinion 'slay[ed] Bohm not only philosophically but physically as well', did not change many minds in Paris.[10] Pauli complained that his audience could not take the trouble to understand complementarity although, he assured Bohr, they regard you most respectfully, though as a great master hanging in a museum, 'the Rembrandt of contemporary physics'. The later Rembrandt, that is, who painted as you speak, in chiaroscuro. What could be expected of communists raised on the outmoded clear-and-distinct push-pull physics of Descartes?[11]

Rosenfeld entered the lists energized by Pauli's worry that Cartesians, Catholics, and Communists would join forces in misleading French youth and confounding physics. All the young people in Paris oppose Bohr 'under the banner of Marxism'.[12] They thus sinned against Marx as well as against Bohr. Since for Rosenfeld, who claimed to belong to 'the almost extinct species of *genuine* Marxists', complementarity was a necessary consequence of experience, it could not clash with dialectical materialism. Had not Friedrich Engels, co-author with Marx of *The communist manifesto*, made agreement with experience the fountainhead of materialism? The modern form of dialectics *was nothing but* complementarity.

Before directing his attention to the sins of the French, Rosenfeld had perfected his polemics against the Soviet pseudo-Marxists in a fight in Bohr's headquarters. Inspired by the Russians and supported by the Danish Marxist press, a physicist nephew of Bohr had opposed the Copenhagen spirit right in Copenhagen. Why could these errant Marxists not see that scientific advance (read complementarity) always took place dialectically (read classical physics against the old quantum theory)? Rosenfeld devoted his article in de Broglie's festschrift to setting forth this truth and to warning that the physicist who does not grasp it 'abandons the rational attitude of the man of science for . . . that of the metaphysician'.[13]

Pauli left Paris with a glimmer of hope because, as he reported to Rosenfeld, de Broglie had distanced himself from Vigier's throwbacks to Descartes. Rosenfeld was not comforted. He urged his brother in physics and politics,

Frédéric Joliot-Curie, to pit his Nobel prize against de Broglie's in the fight for the souls of immature French physicists. 'It is my duty to inform you . . . that they are quite sick . . . They are under the spell of a scholasticism . . . as opposed to the genuine spirit [of Marxism] as is the blackest Catholicism . . . it is up to you, my dear Joliot, to take the initiative'.[14] Joliot had other things to do.

Having done what little they could for the French, Pauli and Rosenfeld proceeded to instruct one another. While exchanging views about the history and concept of materialism, the Scourge of God and the Square Root of Bohr × Trotsky (Pauli's nickname for the old Marxist Rosenfeld) brought to light new complementarities between idealism and materialism, individual and collective, unity and plurality, and the atomic world of quantum mechanics and the unconscious of psychology. To go further, Pauli had felt the need for a 'unified psycho-physical language'. He was already searching for this philosophers' argot.[15]

Meanwhile Oppenheimer denounced Bohm's scheme of hidden variables as 'juvenile deviationism' and advocated a conspiracy of silence: 'if we cannot disprove Bohm, then we must agree to ignore him'.[16] Others despised the scheme as Marxist-inspired. Nor, as we know, did Einstein like it. Despite the hopes he had based on Bohm's 'clear mind' and 'rare independence', he judged Bohm's route back to determinism to be 'too cheap'.[17] 'As for Pais and the rest of the "Princetitute" [Bohm wrote] what these little farts think is of no consequence to me'.[18] They belonged to the old country, the land of the pork barrel, of 'cynicism and cowardice'.[19] And the Copenhageners? 'Pauli has opposed every new idea that was ever suggested'. Rosenfeld and Bohr offered 'almost childish' objections; and Heisenberg, the opportunist positivist, well, 'may his name be cursed'.[20]

Bohm did not find Brazil a paradise. He knew almost no-one. He had to learn the language quickly from scratch to be able to lecture. The academic standards of the institute, both moral and scientific, fell short of Princeton's. And the US Embassy confiscated his passport. He corrected for these weaknesses by becoming a Brazilian citizen, improving discipline at the institute, studying Portuguese, and collecting an entourage interested in advancing or suppressing his ideas. It was tough going. To Bohm's surprise and disappointment, not even fellow Marxists from Oppenheimer's old Berkeley group or the physicist-philosophers of the Soviet Union supported him.[21]

Rosenfeld came to São Paolo to do battle, armed with the information that Bohr thought Bohm's approach 'very foolish', but conceded at last that

Bohm's views, however undesirable, were logically tenable. At a meeting of foreign physicists in Brazil in 1952, several Americans criticized Bohm for pushing a barren theory. Only Richard Feynman, drawn to Brazil less by physics than by bongo drums, could vouchsafe a good word. Stronger support was coming, however, in the form of Vigier, who arrived in Brazil in 1954 to help develop Bohm's ideas. He was just what Bohm needed: a man of action, a hero of the French resistance to the Nazi occupation, a champion of the Vietnamese communists and a friend of Ho Chi Minh, but withal a resourceful theorist who had managed, despite his communism, to gain the full confidence of de Broglie.[22]

Bohm and Vigier secured Bohm's theory against Pauli's repeated objection that $|\psi|^2$ could not be taken as equivalent to Born's probability and simultaneously serve as the physical field guiding electrons. The boys in Brazil argued further that the ψ field could be modelled as a fluid and an atomic particle as an inhomogeneity in the fluid. Suppose that the fluid is subject to many small fluctuations on the scale of the particle. Owing to these vicissitudes, any arbitrary statistical distribution of inhomogeneities will tend towards $|\psi|^2$. To achieve this demonstration, Bohm and Vigier had to invoke possible explanations and reasonable postulates too often, perhaps, to convey full conviction to their readers.[23] The general idea, however, that the random fluctuations of a wave might guide the behaviour of a particle had a high pedigree. Newton's fertile brain had contemplated an ethereal medium whose vibrations put particles of light into 'fits' of transmission or reflection at the surfaces of transparent substances.

Bohm justified his sorts of fits in a small book for a wide audience, *Causality and chance in modern physics*, published in 1957. There he sketched a possible connection between an electron and the ψ wave via a 'quantum force' that nudges the electron towards regions where $|\psi|^2$ is large while random fluctuations in ψ, arising perhaps from a subquantum level in the manner that Brownian motion does from impacts at the molecular level, ensure that it also spends a little time in regions where $|\psi|^2$ is small.[24]

Bohm was easily a match for Bohr in vagueness. Although he blasted the Copenhagen school for its arrogant insistence that 'the current quantum theory must apply, uncontradicted and without approximation, in every domain that will ever be investigated',[25] he could oppose only vagueness to its dogma and 'future music' (as Einstein would say) to its results. When will the music be audible? Bohm typically answered that his inability to deliver his opus was not a fatal objection. Proponents of atomism had held out for 2,500 years before the vagueness of Democritus gave way to Rutherford's nuclear model. Let us wait and see.

Bohm did not want to do the waiting in Brazil, for he had no place there to fix the lever with which he would move the world. He aimed for a position in Israel at the Technion in Haifa and again enlisted Einstein's support. Einstein had recently recommended his old collaborator Rosen to the Technion as able and energetic, 'very effective and beloved as a teacher', 'successfully active in the service and the cause of Israel'. And as a scientist? 'He has also produced independent scientific work although nothing of outstanding value'. To Rosen he now wrote of Bohm. 'He is, in my opinion, very gifted and a solid scientific thinker. He has a systematic mind and is at the same time open minded with respect to every theoretical possibility. His style is lucid and inspiring and shows always originality and independence'.[26] And stubbornness. Einstein had been trying to bend Bohm to his own way of thinking. Keep up the cause, he wrote, but remember there are no macro and micro laws, '[only laws] of general rigorous validity, laws that are logically simple'. This was Einstein's last advice to the man some took to be his intellectual son.[27]

We suppose that Bohm and Rosen had time for chats about the state of the communist world. Both had become disenchanted with it. Bohm was studying Hegel with other people in Israel, reforming his ideas about Marx and Engels, and recognizing the evils of the Stalinist version of the communist state. With the help of Nikita Khrushchev's condemnation of Stalin and the Russian invasion of Hungary in 1956, Bohm and Rosen understood that the road to a wholesome socialist state did not run through the Soviet Union.[28] The realization threw Bohm into one of his frequent depressions. In 1956 he married an Israeli, Sarah Woolfson, who raised his spirits and eased his work for the rest of his life. Most of this they spent in England, to which they emigrated in 1957, first to the University of Bristol and then to London's Birkbeck College. On the way to Bristol, Bohm stopped in Copenhagen, where he found Bohr affable but unreachable, and in Paris, where he worked with his more obliging colleagues, de Broglie and Vigier. The visits to Paris became annual until, in the late 1960s, Bohm himself lost interest in hidden variables.[29]

About the time that Bohm's revolutionary papers appeared, Bohr's collaborator in the creation of quantum theory via the correspondence principle and Heisenberg's collaborator in the decisive step towards matrix mechanics, Hendrik Kramers, died. A man of great humanistic as well as scientific culture, who could soothe Bohr by playing Beethoven when calculations went badly, Kramers was a cosmopolitan at home in several languages and many countries, and a fine representative of the highest attainments and aspirations of classical Europe.[30]

He regarded the quantum theory to whose development he had devoted his career as one of these attainments. It ingeniously incorporated into a complete and coherent description of the microdomain the very classical concepts that quantum mechanics was devised to overcome. From this point of view, complementarity brilliantly formulated the adjustments necessary to speak about nature in ordinary human language. This declaration of the Copenhagen spirit comes from Kramers' textbook on quantum mechanics, first published in Germany in 1933 and given a long life in the US through the American Alien Property Custodian, who seized the copyright in 1944 and reprinted the text. There anyone still in doubt could learn that hope for a classical description of micro-events in space and time is as forlorn as 'the opinion that electromagnetic phenomena can be explained by a mechanical model'.[31]

Kramers was the first of the old guard to fall. Einstein followed, in 1955, engaged almost to the last in arguing the incompleteness of quantum mechanics with his old friend Born. The occasion for this last passage at arms was a festschrift to honour Born on his retirement from his professorship at the University of Edinburgh in 1953. Einstein contributed a 'little nursery song about physics' to the tune of the correspondence principle. Let us do the quantum mechanics for a tiny macroscopic ball bouncing back and forth between two perfectly reflecting walls. A solution for ψ describes a particle with equal probability of being found anywhere on its itinerary, which can hardly describe the classical case of a ball that is in a definite place moving in a definite direction. So ψ cannot describe an individual case but only the behaviour of an ensemble of similar cases, in which the ball is as likely to be here as there. Nor can de Broglie's pilot wave, 'developed with great astuteness by Bohm', apply to the individual; for Bohm's expression for the velocity of a micro-particle gives, for the ball, the absurd result of zero.[32]

A very clear argument, Born replied, but plainly wrong. 'Forgive my cheek, you have chosen an incorrect solution'. The quantum-mechanical solution requires a wave packet, which, for a particle with large mass, behaves like a classical ball. Even if you are right about ψ, Einstein rejoined, the packet would spread and the ball be lost unless you want to invoke Oppenheimer's 'cheap way of calming down one's scientific conscience' and assert that the time required for substantial spreading was close to the age of the universe.[33] Born replied by solving the quantum mechanics of a minute billiard ball in all detail and rigour, and also in great discomfort, in an 'ice-cold hotel (as they all are in [England])', and by reaffirming the Copenhagen line. 'Bohr's expressions are frequently nebulous and obscure. I am simpler and possibly clearer'. Not for Einstein. 'Dear Born, your concept is completely untenable . . ., incompatible with the principles of quantum theory . . . irreconcilable with the superposition principle'.[34]

Einstein continued his corrections: 'every system is at any time (quasi-) sharp in relation to its macro-coordinates'; ψ cannot deliver this information; therefore, quantum mechanics is incomplete. Born: 'Your starting point is untenable'. Listen to people who know! Even Schrödinger agrees with me; ask around Princeton, ask von Neumann![35] Instead Einstein asked Pauli, once again visiting Princeton. Your argument is correct, Pauli wrote Born, but it does not meet Einstein's, which does not call for a deterministic account of a micro-object but a 'realistic one'. He just wants a theory in which a body has a place before it is found there. Once again, according to Pauli, Einstein erred by ignoring the problem of initial conditions and the measurements necessary to fix them.[36] Born acknowledged his misreading, considered that Pauli might have been a greater genius than Einstein, and rewrote his analysis of the oscillating ball for Bohr's festschrift in 1955.

Exactly a year after Einstein attacked the application of ψ to individual systems in the Edinburgh festschrift, Born received the Nobel prize 'especially for his statistical interpretation of the wavefunction'. He had long fretted that he had been passed over in the Nobel awards of the early 1930s and wondered why now, 20 years later, he was honoured. Probably, he wrote Einstein, I was chosen as a physics offset to the chemistry prize-winner, Linus Pauling, since both of us have opposed the use of nuclear weapons.[37] Curiously, Wigner had a similar suspicion about his Nobel prize, awarded in 1963. 'I thought that the prize might have come partly as a reward for my political acts'.[38] But whereas Born associated his prize with his efforts towards peace, Wigner associated his with his belligerent anti-communism. It is hard to plumb judgements from Stockholm.

Passing the Torch

In 1956, a precocious young American, Hugh Everett III, drafted a thesis on the foundations of quantum mechanics that offered a far-fetched but logically permissible solution to the problem of measurement. At the age of twelve, Everett had written to Einstein about the force that held the universe together. Einstein judged him to be 'a very stubborn boy who has forced his way victoriously through strange difficulties created by himself'.[1] At the Catholic University of America in Washington, Everett attended to the less strange difficulties of chemical engineering before migrating to Princeton for a year of mathematics (including military game theory), followed by graduate work in physics.

Everett's attention turned to foundational questions in quantum mechanics sometime in 1954 under the indirect influence of Bohr, then visiting Princeton with his assistant Aage Petersen. One evening Everett, Petersen, and Everett's roommate, Charles Misner, aided by a 'slosh or two of sherry', ascended to philosophy. As Everett recalled the event to Misner, 'You and Aage were starting to say some ridiculous things about the implications of quantum mechanics and I was having a little fun joshing you and telling you some of the outrageous implications of what you said'. Physics amounted to more than calculation at Princeton.[2]

In 1955 Everett took the foundational question of measurement for his thesis. He was probably inspired, and certainly guided, by the English translation of von Neumann's formulation of quantum mechanics published that year by Princeton University Press. By the winter, he had completed a draft dissertation with the exotic title, 'The Theory of the Universal Wave-function'. His advisor did not think the subject would yield the speedy degree Everett sought and encouraged him to seek another: 'You can get out quicker with Wheeler'.[3] He did so.

Everett spotted his opportunity in von Neumann's dual scheme, whereby ψ evolves continuously and dependably in time according to the

Everett's attention turned to foundational questions in quantum mechanics
sometime in 1954 under the indirect influence of Bohr, then visiting Princeton
with his assistant Aage Petersen. L–R: Charles Misner, Hale F. Trotter, Bohr,
Hugh Everett III, and David K. Harrison converse probably in Princeton chapel.

Schrödinger equation (von Neumann's process **2**) if the quantum entity it
describes does not interact with anything, but is projected or collapses suddenly and inexplicably on encountering a measuring device (process **1**).
Instead, Everett proposed to assume the universal validity of process **2**:[4]

> The general validity of pure wave mechanics, *without any statistical assertions*, is
> assumed for *all* physical systems, including observers and measuring apparata.
> [The erroneous Latin plural (*apparatus* is wanted) reveals that Everett was
> reading Wigner, who makes the same mistake.] Observation processes are
> to be described completely by the [wave] function of the composite system
> which includes the observer and his object-system, and which at all times
> obeys the wave equation (Process 2).

Everett would allow ψ to go on placidly, never collapsing, to realize all
its potentialities, not just the outcome recorded by the observer. The other
possible outcomes define other worlds, accessible mentally, perhaps, but not

physically. Everett did not propose that the observer enters into a super-position of worlds, experiencing all outcomes simultaneously, but rather 'splits' between different states, experiencing different outcomes in each. He did not make clear the nature or cause of the 'split', which he interpreted literally as a physical phenomenon acting on (or promoted by) a real wave-function. 'We are then led to the novel situation in which the formal theory is objectively continuous and causal, while subjectively discontinuous and probabilistic'.[5] Although Wheeler's physics was full of fancy, Everett's many worlds went beyond the unusual. 'Split? Better words needed', he scribbled on Everett's presentation of it, and recommended careful rephrasing to avoid 'mystical misinterpretations by too many unskilled readers'.[6]

Wheeler nevertheless thought that Everett's fundamental idea did not conflict with ordinary quantum mechanics and might, if couched in less extravagant language, appeal to Bohr as the capstone to complementarity. But the language had to change. Under Wheeler's guidance and during long nights in Wheeler's office, Everett eliminated 100 pages from his 137-page draft dissertation and titled the result, 'On the foundations of quantum mechanics'. It was published in January 1957 in *Reviews of Modern Physics* as '"Relative state" formulation of quantum mechanics'. Wheeler published a companion paper intended to make Everett's ideas more digestible. By then, Everett was no longer in Princeton. He had joined the Pentagon's Weapons Systems Evaluation Group, a civilian organization that could appoint scientists on terms more generous than those typical of government employment.[7]

Wheeler sent Everett's thesis to Bohr and visited Copenhagen to drum up support for it. Despite his efforts to make Everett's language more acceptable, he was obliged to retreat under a barrage from Petersen and Rosenfeld ('hopelessly wrong ideas'). Rosenfeld accused Everett of a double misunderstanding, since von Neumann's theory of measurement was in itself 'wrongly put', 'a distorted and largely irrelevant rendering of Bohr's argument'. It was simply wrong to seek to fix the problems von Neumann had created by replacing it with a 'theory' based on 'quite untenable assumptions'.[8]

Wheeler arranged to send Everett to Copenhagen to fight his own battle in March 1959. Petersen defended. But all was in vain. Bohr, ever mindful of the use and misuse of language in physical descriptions, had no wish even to discuss 'any new (strange) upstart theory'.[9] Everett's approach challenged too many Copenhagen teachings, such as complementarity and the interpretation of quantum probability. Rosenfeld as usual was most outspoken: 'He [Everett] was indescribably stupid and could not understand the simplest things in quantum mechanics'.[10] Although Everett spent six weeks in Copenhagen, he did not see that much of Bohr and his (or Wheeler's) project of a Princeton plank in the structure of complementarity was 'doomed from the beginning'.[11] Everett took the rebuff hard and rejected Wheeler's later attempts to lure him back to academia. He preferred calculating nuclear bomb kills for the Pentagon.[12]

The staunch old defenders of the Copenhagen spirit sought to stifle the rebellions of Bohm and of Everett in their infancy with the help of the indifference of a physics community more interested in 'getting to the numbers' than inspecting the foundations underpinning their quantum calculations. Bohm and von Neumann saw in this hegemony some similarity to the hierarchical arrangements of the Roman Catholic Church.[13] The analogy has this much to recommend it: the pope and cardinals of the Copenhagen church could not reproduce themselves and were not immortal. This they proved during the decade after the grim reaper harvested Kramers. Von Neumann died in 1957, Pauli in 1959, and Bohr in 1962. They took with them their opponents Einstein (1955) and Schrödinger (1961).

Schrödinger had remained vigorous to the end. 'Complementarity', he declared, 'is a thoughtless slogan'. And its inventor? 'If I were not thoroughly convinced that the man is honest and really believes in the relevance of his—I do not say theory but—sounding word, I should call it intellectually wicked'. He supported this charge with a quote from *Faust*: 'For just where concepts and logic are at the end of their tether/You are sure to hit on a *word* to help you in your troubles'. Bohr and his opponent agreed in this respect, however, that the reality did not distinguish between subject and object.[14]

The obligatory obituaries (apart from Schrödinger's) provided final opportunities to recommend the Copenhagen interpretation. Thus, Bohr praised Kramers for breakthroughs using the correspondence principle; Pauli praised Einstein for advancing correct quantum physics and grieved over his inflexible opposition to the belief of most theoretical physicists, 'the "Copenhagen interpretation" ... founded by Bohr'; and Bohr praised Pauli for his masterful handling of the correspondence principle and the foundations of quantum mechanics.[15]

The founders reinforced their story by brief formulaic renditions of quantum mechanics for general audiences. Here Heisenberg once again outdid his colleagues. During the 1950s he wrote and rewrote literally scores of pieces reaffirming the Copenhagen interpretation. We already know his contribution to the Bohr festschrift of 1955 and his Gifford lectures of 1955/6, *Physics and philosophy,* which attacked Bohm and other opponents of the Copenhagen interpretation individually before disposing of all competitors, past and future, as seekers of the 'impossible'.[16] Heisenberg's outpouring reached flood level in 1958/9 with articles for the centennial of Planck's birthday and notices of Pauli's life and death.[17]

Bohr made some notable contributions to this literary glut. In 1958 he published a collection of articles demonstrating the application of complementarity to 'many fields of human knowledge' outside physics. He regarded the collection as a sequel to a book he had published in 1934 and reissued in 1961 with a strong reassertion of the Copenhagen interpretation. Under the titles *Atomic physics and human knowledge* and *Atomic theory and the description of nature,* respectively, these slim books may be regarded as Bohr's fullest statement of his belief in the need to renounce causal space-time descriptions for the microworld and the great value of the freedom thus achieved. For him, accepting the principle of complementarity in microphysics was not, as he had written when introducing it, a concession or renunciation, but the recognition of a route to a general liberation of human thought.[18]

To Pauli this liberation confirmed the possibility of a fruitful reconciliation of the spiritual and the physical, East and West, the unconscious and the conscious, Jung and Bohr.[19] Quantum physics demonstrated the fuzziness of natural processes and exposed the 'regressive hopes' of people who would bind nature to human notions of rationality. Give them up! Accept the licence that quantum physics affords; acknowledge that different 'thought forms' apply to different domains and that the domain claimed by quantum mechanics has no place for the concept of strict causality.[20] From this point of view Pauli hunted for earlier thought forms resembling complementarity and, as we know, seized on alchemy as understood by Jung: a successful union of the spiritual with the physical, of the quest for self-purification with the pursuit of the philosophers' stone. His historical investigations led him to the discovery that the complementary union represented by alchemy perished during the early seventeenth century in the debate between Kepler and Fludd.[21]

Jordan went even further than Pauli by seeing in the Copenhagen interpretation a turning point in world history. In lectures delivered over the radio in the early 1950s, he urged the replacement of the history of kings and queens, battles and treaties, politics and diplomacy with something more progressive and decisive, the true history of civilization: the history of science and technology. This he supplied, from the Greeks onward ('truly modern history can be written only by historians who are at the same time natural scientists'), to arrive at our epoch of 'momentous insights' won by quantum physicists against 'gigantic difficulties'.[22] We have learned that mechanism cannot stand, that the atom is 'inalienably free and unpredictable', and that 'new perspectives [can open up] for our reflection on the problems of nature and man, world and God'. Freedom from materialism and strict causality have renewed religion, rehabilitated creation from nothing, condemned irreligion as a disease, and suggested that life originated via the 'free and individual decisions of micro entities'.[23] There was something at stake in the survival of the Copenhagen interpretation!

In these ways, the old guard magnified their achievements as they began to depart from the contested world of the quantum. They were rightly proud of their creation. Their grandiose claims seemed confirmed by the magnitude of their struggle and the suddenness of its resolution, so many hopes thwarted, calculations aborted, models discarded, brain power expended. Then in the space of only two years, between Heisenberg's ecstatic vision on Helgoland and Bohr's indecipherable whispers at Como, everything came together. The physicists around the Copenhagen–Göttingen axis had solved the problem of the quantum domain in principle and in perpetuity, finished its foundations, and on the way annexed chemistry. Not even the great brain of Einstein could convict it of inconsistency or incompleteness. Anyone who knew the story of the struggle in its details could not help but entertain the Copenhagen interpretation of its outcome. It remained to enlist historians in the telling.

In 1958 Bohr gave a lecture in memory of Rutherford in which he discussed his quantization of the nuclear atom more vividly than in the many short, formalized accounts he had given of his invention. In writing up the lecture for publication with Rosenfeld's help, he expanded it with descriptions of his false starts and glitches. His old friend Darwin, to whom he sent a draft, objected that Bohr had cluttered the narrative, 'with all of the difficulties being brought out all the time', rather than give a clear and logical overview of his discoveries about the hydrogen atom. Bohr replied that after much deliberation he could not follow Darwin's suggestion without doing violence to history: 'I have striven to use the opportunity to revive the development in a factual and detached manner'. It was a struggle to reconstruct the struggle. 'It has been quite a difficult task, and I have often been scared by the length of the manuscript, resulting from the many different points that presented themselves to give the whole story a reasonable balance'.[24]

Bohr's last prepared historical work was a lecture on the history of the Solvay conferences on physics delivered in October 1961, on the fiftieth anniversary of Solvay I. He said, rightly, that the reports of the conferences are a most valuable source for students of the history of science. More reliable, indeed, than the account he gave then of the origin of his atomic model. Now he followed Darwin's advice, though in a manner that allowed for the complexities in the Rutherford lecture. He had read the reports of the first two Solvay meetings, and perhaps of others too, from which he had learned for the first time that the experts on radiation and the quantum who gathered in Brussels in 1911 did not once mention the nuclear atom. Nor did it receive much attention at the second meeting, in October 1913, which was devoted particularly to atomic structure; Bohr's first paper on the subject, which had appeared in June that year, did not enter the discussions at all. Like other people, physicists sometimes lose the compass they need to find their way. 'It is illuminating for the understanding of the general attitude of the physicists at that time that the uniqueness of the fundament for such exploration given by Rutherford's discovery of the atomic nucleus was not yet generally appreciated'.[25]

Heisenberg was among the celebrants of the Solvay semi-centennial in 1961. Bohr tried to enrol him in the new set of problems. 'In recent years I have become increasingly engaged in historical questions', 'dwelling more and more on thoughts of the great adventure which we all experienced'. Won't you help? An American team sponsored by the 'Academy in Washington' (the US National Science Foundation) proposed to interview Bohr at length and many inquirers had asked for archival material. The more he tried to peel back the layers, the more he experienced a problem as challenging as measurement in quantum theory. No longer was it a question of a single observer and the observed: 'the difficulty [lies in] giving an accurate account of developments in which many different people have taken part'.[26]

Between 1961 and 1964, a wide-ranging project, 'Sources for History of Quantum Physics', directed by Thomas Kuhn, interviewed nearly one hundred people involved in the development of quantum physics. In the summer of 1962, the project staff, which included Paul Forman and, as assistant director, John Heilbron, left Berkeley for a year's stay in Copenhagen. Bohr provided office space. The arrangement had come about through Wheeler, who spearheaded the project through the US funding agencies and chaired the board that oversaw it. Wheeler gave the project a sense of urgency. Death was at work among the principal actors, depriving posterity of the true history of adventures and struggles 'so much to the glory of the human spirit'. '[T]he immortality of [the] heroes is at stake'![27] The need for urgency was quickly confirmed. Bohr died a few weeks after interviewing began, just after reaffirming that complementarity was 'an objective description' of the atomic world, 'the only possible objective description'.[28]

Heilbron remembers the excitement of the project, the lure of Copenhagen, the loyalty and erudition of Rosenfeld, the quiet powerful personality of Bohr, the complacent manufacture of memories by Heisenberg, the confident reserve of Jordan, and the impersonation of Oppenheimer by Oppenheimer. The rehabilitated leader of Los Alamos had come to Copenhagen to gather material for an obituary of Bohr. During several lunches he took with the project staff, he never put aside the character of an Eastern sage. He did not disdain to parade his knowledge of the Hindu scriptures before indifferent junior members of the academic proletariat.

The project remained in Copenhagen after Bohr's death and redoubled its efforts to collect correspondence and manuscripts. Bohr's extensive archive of professional correspondence provided material that helped in preparing oral histories. Abundant archival documentation of the great adventure Bohr had led was assembled, some of which, and all his publications, appear in his *Collected works* (1986–2007). Physicists and historians then made available the writings, published and manuscript, of many of Bohr's close associates. We now have editions of the papers of Kramers (1956), Ehrenfest (1959), Pauli (1964), and Heisenberg (1984+); the invaluable correspondence of Pauli (1979–1993); and the papers, correspondence, and manuscripts of Einstein (1987+). Like quantum physicists of the

Copenhagen persuasion, but in their own slow way, historians have helped to shape the story much as Bohr would have wanted it told.

As the founders fell silent, two gatherings on either side of 1960 resuscitated face-to-face discussion of foundational questions and reopened old hostilities. The earlier gathering took place at the University of Bristol in 1957, the year Bohm beached up there. The latter took place at Xavier College in Cincinnati in 1962, a year after Podolsky became its professor of physics. The British conference had a single funder, the Colston Research Society of Bristol, named for its benefactor, the seventeenth-century merchant Edward Colston whose wealth had been founded partly on the slave trade and whose statue was toppled in the Black Lives Matter protests in June 2020. The American conference had three, one private and two governmental agencies—NASA and the Office of Naval Research—vigilant to keep the foundations of science out of Soviet hands.

The Bristol event was a mixture of physicists and philosophers, that at Cincinnati a pure ensemble of physicists. At the former, the ferocious Rosenfeld continued his defence of the Copenhagen view. At the latter, and not very enthusiastically, the more gentlemanly Wigner ('so generous that it is most embarrassing to work with him'), defended.[29] The other main speakers in Bristol were Bohm, Fierz, and Vigier among physicists, the polymath Michael Polanyi, and among philosophers Karl Popper (in absentia) and Popper's tough-talking, no-nonsense disciple, Paul Feyerabend. In Cincinnati, Bohm's student and collaborator Yakir Aharonov, then at Yeshiva University, Oppenheimer's former student Wendell Furry, Cambridge's Dirac, and the surviving two-thirds of EPR were conspicuous.[30]

Bohm began the business in Bristol with a resumé of the model that he had worked up with Vigier, in which an electron swims in a sea of chaotically moving sub-quantum particles. Naturally, Rosenfeld rejected such attempts: 'To object to a lesson of experience by appealing to metaphysical preconceptions is unscientific'. In full fighting mode, Rosenfeld argued that although experiments had favoured Copenhagen 'with merciless definiteness', its critics still had not felt the death thrust. They are alive because they do not fight genteelly. Is it not a hit below the belt to accuse the Copenhageners of positivism? To be sure, many misstatements and exaggerations by Heisenberg and Jordan, and a few by Bohr, might give colour to the charge. But anyone acquainted with the history of positivism from Newton to Mach will know that the charge is baseless. That certain relations between classical concepts cease to be meaningful in quantum theory states a fact, not an emotionally ambiguous 'renouncement' or 'resignation'. 'Some critics seem to take the invitation to "renouncement" as an attempt on their personal freedom: the right to indulge in metaphysical dreams is not disputed; only, this activity is not science'.[31]

Surely the Copenhagen teaching about the use of ordinary language and classical concepts in describing experiments and their results is enough to refute the charge of positivism? So said Rosenfeld. Feyerabend challenged the teaching. Rosenfeld: 'Isn't that quite an obvious thing?' Fierz gave a counterexample from relativity and joined Vigier in objecting to Rosenfeld's rough-and-tumble dialectical dogmatism. They effectively sidelined him, which we shall do as well, for a time.[32]

Feyerabend had decided against Copenhagen with the help of Popper, under whose direction he worked at the London School of Economics in 1952/3. By then Popper was an experienced campaigner in the muddy field of quantum mechanics. Aware already in the early 1930s that the subject was too difficult for physicists, he explained that Heisenberg's uncertainty relations indicated not a radical probability in nature but a statistical spread or scatter in measurements on an ensemble of particles. Nature did not prevent an adroit experimenter from simultaneously obtaining q's and p's to an accuracy better than Heisenberg allowed. Popper described such an experiment. It bought him the honour of a correction from Einstein and an invitation from Bohr to breathe the clarifying air of his Institute after the positivist conference in 1936 at which Hermann, another of Popper's critics, would distinguish herself. Popper's chagrin over his error and retreat under 'the tremendous impact of Bohr's personality' removed him from active quantum combat until the mid-1950s, when he devised a novel statistics that made all the puzzles of complementarity vanish. This 'propensity' statistics was the burden of the paper he submitted to the Bristol proceedings.[33]

Feyerabend had come to Popper after a complicated education that included a visit to Copenhagen in 1952, just when Bohr discovered Bohm. The leading spirit of Copenhagen had not been worried, just puzzled, by the challenge from America. Popper made it clear that philosophers should not be puzzled, but seriously worried, by Bohr; like all positivists he was wrong and his complementarity, with its subjectivist tendencies, was positively dangerous. Feyerabend embedded himself in Popper's system by mastering Bohm and probing for weaknesses in von Neumann's proof. His lecture at Bristol consisted of a lengthy cancellation of the problems of measurement and a few words in support of Bohm against von Neumann.[34]

The Bristol participants sought relief from such hard teachings by playing with Schrödinger's cat. According to the standard view, Bohm said, we not only cannot say whether the cat is dead or alive before we look, we cannot even say that it is dead or alive. Evidently the ψ account is incomplete. Some hidden parameters must kill or spare the cat before our observation. Vigier: this is correct, something happens at the macro level we cannot describe. Georg Süssmann, from the Institute for Theoretical Physics in Munich, recalled that ψ does not describe the state of the cat but our knowledge of its state. On this understanding, there is no paradox. In general, we know nothing about contingent events without direct sensory experience or reliable report. But perhaps the cat knows something that ψ does not? 'Of course [Bohm continued], the cat could in principle make another wavefunction, which describes itself as either dead or alive'. A feline physicist? '[W]e are getting into very difficult paradoxical problems'.[35]

Like the old astrology, the young quantum mechanics needed a death parameter, that is, a way to kill off all possibilities but the one realized in measurement. But it had none; therefore, it was incomplete. So what? Three participants sided with Süssmann, denying the existence of a paradox, whereas Süssmann, though sticking with ψ as a mere descriptor, agreed that it does not convey the exact nature of the state before measurement. A fresh PhD from Queen's University in Belfast, Uuno Öpik, son of the famous Estonian astronomer Ernst Öpik, summed up the inconclusive discussion. '[T]he wave function . . . gives us all the information which we can obtain. The cat knows more, sure enough'.[36]

The Xavier College symposium could plumb greater depth because (if for no other reason) all its participants were physicists. They began with a bowl of hard nuts. Is there action at a distance in quantum mechanics? How does the wavefunction collapse? Is it purely subjective? Why do particles carry electrical charges of 0, or precisely balanced positive and negative units only, and what secrets lie in the mystic combination $2\pi e^2/hc = 1/137$, thought to be a pure number in all systems of units irrespective of the values of the atomic constants that define it (but now less romantically evaluated to be $1/137.03599913$). The transcript's answer to these questions admits us into

EPR figured prominently in the discussions at Xavier College. L–R: Aharonov, Furry, Rosen, Podolsky, and Wigner.

the workshop of the quantum mechanic, noisy with disagreements, shouts, and table banging, but also full of good humour, chuckles, and laughter.

Furry and Wigner, who described themselves as orthodox mechanics, conceded that a new catechism might be necessary. The most heterodox, Aharonov, allowed that Bohm's world was still under construction. Dirac urged leaving what he called 'class I puzzles', questions like the status of causality, the subtleties of measurement, and the enigma of 137, on whose solution the progress of physics did not depend, to posterity and a better theory. That is what most physicists do. 'Many people live long and fruitful lives without ever worrying about [them]'.[37] The symposiasts did not agree with Dirac, at first.

EPR figured prominently in the discussions. Furry started them with an imperfect analogy. Let there be two boxes, A and B, four numbered envelopes, and two sets of bi-coloured cards, one black and red, the other blue and yellow. An experimenter in Harvard tears a black-red card in half, puts one part in envelope 1 and the other in envelope 3, and repeats the exercise for the blue and yellow card, putting the corresponding pieces in envelopes 2 and 4. Then he places envelopes 1 and 2 in box A and ships it to Pasadena. The apparatus has one dangerous peculiarity. When an envelope in either box is opened, its counterpart in the other spontaneously explodes.

The Harvard physicist (still presumed male) opens box B and envelope 3. He finds a half card, either red or black, and from it can predict reliably what his counterpart in Pasadena would find in envelope 1, either black or red. He can say nothing about the blue–yellow distribution because his decision to open envelope 3 has destroyed all knowledge, for him, of the contents of envelope 2.

Now let the red-black represent momenta and the blue-yellow positions of the interacting EPR particles. The experimenter determines which properties of the situation he wants to realize and thereby excludes the possibility of determining other properties, just as in Bohr's explanation of the experiment. The preparation of the boxes and envelopes (the initial entanglement of the particles) produces the outcomes. There is no question of action at a distance. Although the analogy fails by assigning fixed states to the cards, which would not be the case for quantum-mechanical properties between measurements, it is good enough to pinpoint the problem or discomfort caused by EPR. 'Now this is hard to say', Furry said, 'there is a statistical relation, between these two particles, *no matter how far apart they get*'.[38]

A voice cried out from the audience of physicists. What is the best answer to EPR? Podolsky ducked the question. Rosen conceded that the argument could not be settled by 'any operational procedure of measurement' and so belonged to philosophy or even metaphysics. Aharonov gave Bohr's answer, the only consistent way, he declared, to translate the mathematics into words. Furry: But you are a student of Bohm! Why not use his ideas? Aharonov: 'Bohm . . . has not solved the paradox yet'; his model still obliges acceptance of a disagreeable action at a distance. Rosen, showing flexibility: I have not changed my mind for 25 years; I always picture electrons with definite positions and momenta and never discuss these things in elementary classes in quantum mechanics. Podolsky: if ψ is realistic, EPR implies an interaction faster than light. Aharonov: we are asking too much of the experimental apparatus. We ask it to collapse the wavefunction in a way we do not understand and now we ask it to operate faster than the speed of light. 'It is not a consistent or satisfactory picture to see all these things happening without reason'.[39]

The audience then learned from Furry that very few of the things that theorists treat as measurable are so. In practice only energy, momentum, both linear and angular, and position. These were too many for Wigner. How do you measure position? Furry: By Heisenberg's gamma-ray microscope. Wigner (correctly): 'You don't measure position with that'. Does measurement occur when the photon is emitted, scattered, or received? Aharonov: 'Better give it up and measure energy instead'. Wigner: 'Good idea, most, maybe all, measurements are made on stationary systems'. Aharonov: 'Why do we fuss so about measurement? It is a special case of a microsystem meeting a macrosystem, which happens all the time whether we look or not'. Wigner: No. '[T]he collapse of the wave packet takes place (excuse me for [provoking] laughter) only through the act of cognition'. Rosen countered that a machine record would be tantamount to an act of cognition. Wigner had some trouble responding. 'The piece of paper on which the machine was supposed to write it down would be in a linear combination of two states . . . eh, uh, eh . . . it is very difficult to say things . . . well, eh, uh, uh'.[40]

Furry observed that in the usual story, the interaction between micro and macro is so complex that it annihilates the overlapping bits that each contribute to their joint ψ and throws the microsystem into a defined state. Looking at a printout cannot do that. Except, Wigner replied, the machine cannot produce a definite answer; for as Heisenberg has written, 'the conception of objective reality evaporat[es] into the mathematics'. Furry: that's just one of his philosophical crotchets. A real measurement is so complicated that it reduces the wavefunction of the coupled micro and macro system to a state indistinguishable from a mixture, that is, without detectable quantum-mechanical interference effects. Rosen then challenged Wigner's assertion that a machine cannot print an unequivocal answer to the question whether an electron spins one way or another. Would not the brain observing directly also be in two states? 'Where is the decision finally made?' The audience chuckled. Wigner: 'A very pertinent and very disagreeable question'. The audience laughed. Furry went to the blackboard to explain himself to Wigner's staccato commentary: 'No, no', 'No, no, no', 'No!' Aharonov broke in: '[T]he point is that the theory is not very satisfactory'.[41]

True, but Furry desired to drive another nail into its coffin. He observed that cosmic rays are continually ionizing the air around us. What does it matter to the trails they leave whether we set up an apparatus to make them visible? 'But according to the point of view that puts all the emphasis on cognition, they aren't even in the cloud chamber unless you take a picture!' He began to shout. 'And they are not even in the cloud chamber or on the picture unless you look at it!' A preposterous idea. And not the only one that made Furry furious. He enriched his diatribe against Wigner with a swipe at another oddity from Princeton whose relevance Rosen had suggested in the first session of the symposium. The oddity was Everett's relative state formulation of quantum mechanics.

Everett had flown to Cincinnati from Washington to participate in one of the 'limited-attendance sessions' at Xavier.[42] Podolsky threw down what he supposed was a challenge: 'It looks like we would have a non-denumerable [uncountable] infinity of worlds'. Everett: 'Yes'. Furry: 'I can think of various alternative Furrys doing different things, but I cannot think of a non-denumerable number of alternative Furrys'. Podolsky laughs. Rosen is reminded of a short story by O. Henry, 'Roads of destiny', in which a would-be poet comes to a fork in the road. He has three ways to continue his journey, each of which involves him in a different disastrous adventure. He always dies the same death, from a bullet from the same gun.[43] Interpretations differ, they are stories, but the ending, the computed probabilities for the world we live in, is the same.

The symposiasts at Xavier were not allowed much rest. On Wednesday evening, they listened to Dirac discuss ways he had devised to stamp out the plague, then infesting the theory of photons and electrons (quantum electrodynamics). One possibility was to endow the electron with some size and shape. As he worked out the consequences of his fat electron on the blackboard, Wigner jumped on him. Their exchanges demonstrated physics in the making. You are wrong, said Dirac, 'that's not essential'; 'No, no', Wigner replied, 'that [concession] would be fatal'.[44] It is nonsense, said the one; it is not nonsense, said the other. They must often have reasoned thus together, as they were brothers-in-law.

The final talk at the symposium, a critical summing up, fell to Abner Shimony of MIT. Shimony had graduated in philosophy and mathematics at Yale, earned a master's degree in philosophy (on linguistic symbolism) at the University of Chicago under Rudolph Carnap, one of the founders of the Vienna Circle, and a doctorate in philosophy at Yale before completing a second PhD in physics at Princeton under Wigner. Shimony devoted most of his talk to psychological problems. Was it possible for the human mind to collapse a wave packet? Answer: no known account of the mind allowed it. And if possible, it would make physicists solipsists. Dirac: 'What is a solipsist?' Podolsky: 'If I were a solipsist, I would think you only a product of my imagination', and therefore, Aharonov added helpfully, 'you wouldn't mind destroying him, because it's only an effect of the imagination'.[45]

Shimony then took on Wigner's friend. Why, Wigner, do you have to look at the entire system, micro-object, macro-instrument, and your friend to collapse the system's ψ-function when by hypothesis your friend has already looked and pronounced? And what if you had many friends, say the

500 co-authors of the latest experimental paper in particle physics? Must an individual suspend judgement no matter what the other 499 report? And what guarantees that they will all agree? Either we must concede that the macro interaction resulting in a permanent mark collapses ψ or we must believe in a degree of pre-established harmony at which even Leibniz would have baulked. As for Everett's way out of the impasse, Shimony deemed it too spendthrift of worlds for serious consideration. Are we therefore wasting our time, as Dirac had said and Furry recalled, in pondering problems that may well be too hard for us? No, replied Podolsky, in words suited to close the symposium, 'There is no way of telling what path we have to take in order to get to the kind of theory we want to have'.[46]

The torch had been passed to a new generation of physicists and philosophers. A few used it to rekindle the old debates. New voices joined a familiar chorus. By the early 1960s, general interest in foundational questions had warmed from the ice-cold depths of postwar irrelevance to a state of tepidness. Let us hear again from Princeton, from Freeman Dyson, an English physicist installed by Oppenheimer at the Institute for Advanced Study. In Dyson's experience, a student took six months to learn quantum mechanics and spent the next six agonizing over it. 'He works very hard and gets discouraged because he does not seem able to think clearly ... it is strenuous and unpleasant'. And when enlightenment comes, he realizes that he has been wasting his time. 'I understand now [he says], that there isn't anything to be understood'.[47]

ACT IV

Productive Inequalities

The Theorem of John S. Bell

There was little talk of Bohm's hidden variables at Xavier College. Rosen brought up the quantum potential, which he had hit on independently and published obscurely in 1945, but he was so badly pummelled by Wigner over it that he ran to the more defensible position that Einstein had adopted at Solvay V: ψ deals with ensembles, not individual particles.[1] Elsewhere Bohm's ideas had enjoyed some attention since the Colston symposium. The mostly positive reviews of its proceedings and Bohm's *Causality and chance*, the rhetoric of Rosenfeld and Wigner, and work by Bohm himself kept his views conspicuous within the small circle of interested parties.[2]

Besides *Causality and chance*, Bohm's principal contribution in this period was a close analysis of EPR in collaboration with Aharonov, carried out when both were at the Technion in Haifa. This was published in *Physical Review* in 1957. They had discovered an old experiment capable of testing a simple suggestion made by Einstein and others to resolve the paradox. Does quantum mechanics apply when the two EPR particles are far enough apart that they cannot act directly on one another or mediately through the apparatus?[3]

The old experiment (it dated from 1950) did not exploit the then impractical play with oppositely spinning electrons in atoms that Bohm had discussed in his textbook and elaborated further with Aharonov. Instead, the experiment of 1950, conducted by Chien-Shiung ('Gigi') Wu, 'one of the most talented and most beautiful experimental physicists',[4] and her collaborator Irving Shaknov employed electron-positron pairs or, rather, the two photons produced when a negative electron (e^-) collides with a positron (e^+). The high-energy photons created in the annihilation of the e^-e^+ pair have spins (or polarizations) that are necessarily correlated. Wu and Shaknov tested these correlations only. They had not thought to look for connections of the EPR-type.

Bohm and Aharonov managed to squeeze the answer they sought from Wu and Shaknov's results by comparing the reported correlations to calculations of the behaviour of the coupled photons if quantum mechanics did not account for the situation at large separations. The answer seemed decisive. Standard quantum mechanics did not break down anywhere and the puzzling correlations at a distance in the EPR scenarios had to be accepted as a fact of life. As Bohm and Aharonov put it, 'the aspects of the quantum theory discussed by Einstein, Rosen, and Podolsky represent real properties of matter'.[5] The correlated photons were not Einstein separable or locally real. Since Bohm's hidden variable theory based on the quantum potential was non-local, it survived this analysis of the experiments of Wu and Shaknov.

Soon Bohm's circle widened to include another powerful analyst, who enlisted after attending a seminar Bohm gave at University College, London, as preparation for his move to England from the Technion. The recruit, John Bell, then a fresh PhD from the University of Birmingham, was startled by Bohm's counterexample to von Neumann's proof of the impossibility of hidden variables. Subsequent reading of Bohm's papers proved a 'revelation'. 'I saw the impossible done'.[6]

It would be another decade before the revelation worked a revolution in the foundations of quantum mechanics. This was not because Bell lacked confidence. When convinced of his position he could be passionate, 'quite heated', defending it against slower colleagues or battling obtuse nature; 'you can almost see the steam rising from him as he grappled with some problem'. The risks he took were not only intellectual. He climbed mountains in the dark and motorcycled over potholes with a recklessness that might have ended his days long before he undertook to unriddle the paradox that EPR had created. The undertaking was illuminated by the embers of a youthful flirtation with logical positivism, broken off when Bell glimpsed its terrifying logical end in solipsism. His view that women belong to a different species from men, which he defended with his usual ardour, further illustrates his willingness to run great risks.[7]

Like Bohr, whom he sometimes commended and would at other times lambast, Bell practised physics with strict morality. Both insisted on getting to the bottom of things, whether the things existed or not. We are told that Bell could speak with the ferocity of an Old Testament prophet against sinners who served their science superficially. This intolerance spiked early when, as an undergraduate, he had questioned his lecturer,

Robert Harbinson Sloan, about the uncertainty principle. Sloan was more comfortable with spectroscopic stamp collecting ('All the atoms in the periodic table') than with the tortuous twists and subtle arguments of the rhetorical crutches fashioned by the founders. Alas, he could not answer Bell's questions with the clarity Bell demanded. 'I was getting very heated and accusing him, more or less, of dishonesty', Bell later recalled. 'I can imagine Bell at nineteen', wrote one of his biographers, Jeremy Bernstein, 'his hair probably redder than it is now, his Irish temper flaring, because Dr Sloan could not explain the uncertainty principle clearly enough'.[8]

Again, like Bohr, and Bohm too, Bell worried about the meaning of words and proposed to proscribe ones that in days of yore Francis Bacon had called idols of the marketplace, coinages rendered useless or misleading by promiscuous exchange. Bell's list included items that, if struck from the accepted lexicon, would have muted theorists, philosophers, and historians alike: system, apparatus, macro-, micro-, measurement. 'System' and 'apparatus' imply a distinction that Bell regarded as untenable; 'microscopic' and 'macroscopic' have no definable border; and 'measurement', the worst of them all, should be banned altogether. Replace 'measurement' with 'experiment', recognize it as a tool, and cleave to the understanding that the fundamental aim of science is understanding. 'To restrict quantum mechanics to be exclusively about piddling laboratory operations is to betray the great enterprise'. The great betrayal, Bell told his colleagues, was to accept the low standard of 'satisfactory for all practical purposes'. For this sinful practice he introduced the dismissive enduring neologism, 'FAPP'. Quantum mechanics might be good enough FAPP, but it was 'rotten at the core'.[9] Or, as Dirac put the point with less affect, 'People are . . . too complacent in accepting a theory which contains basic imperfections'.[10]

Bell was another intellectual heir of Einstein. He would have liked to invent a realistic world that met Einstein's expectations and to do so as the fulfilment of a duty to the science he professed. He rendered judgement about ideas not in the usual professional terms—good, poor, interesting, very interesting—but in moral categories like sound and rotten.[11] His combination of probity, brilliance, and audacity enabled him to discover in 1964 his profound theorem about the EPR experiment. Its formulation did not require anything not known to thoughtful physicists during the 30 years since EPR or, indeed, the half century since the age of the founders. It awaited someone with special insight and commitment.

Bell before his blackboard. The equations and drawing on the blackboard refer to his theorem and its implementation through experiments devised in the 1970s by Alain Aspect.

In identifying 'Bell's moral character [as] primarily responsible for his discovery', Shimony implied that most physicists lacked it. A moral character, Shimony continued, that Bell expressed in many ways, in his vegetarianism, speech, demeanour, and political views. 'He was passionate in his pursuit of clarity, understanding, and social justice'. The French theorist and philosopher Bernard d'Espagnat, who had spent time in Copenhagen and other international centres, put the point in another way equally unflattering to the profession. Most physicists, he said, are too caught up in their calculations to think. '[T]hey tend to consider genuine thinking as quite an obsolete activity'. Bell loved to indulge in this discredited activity. It gave him an allergy to positivism and to attempts to restrict science to mere description of experience. 'To John Bell this idea was absolutely repellent'.[12]

The real world began for Bell in July 1928 in Belfast, in a Northern Ireland then only recently separated from the Republic by border posts. His father eked out a living trading horses and his mother worked as a dressmaker. Bell's family, like Bohm's, was pinched but not impoverished. The boys grew up in working-class neighbourhoods, awakened their minds by wide reading in the local public library, and made it through higher education with the aid of scholarships. Both broke with traditional religion in their early teens, as had Bohr and Einstein. Bell suffered from migraine as Bohm did from depression. Although they enjoyed taking radios apart, they were inveterate intellectuals. Bell carried bookish learning so far as to teach himself dancing, swimming, and skiing from reading alone.[13] To these accomplishments he added the FAPP version of quantum mechanics as taught at Queen's University in Belfast.

On graduating in 1949 he decided to enter public service and soon found himself in a second-class job in the theory division at Britain's atomic energy research establishment at Harwell, working under the direction of Klaus Fuchs, about to be exposed as a first-class spy. Bell chose remunerative work rather than pursue a doctorate, as his teachers had suggested, because he wanted to be able to contribute to his parents' income.[14] Bell retained his civil-service appointment until 1960, working primarily on the design of particle accelerators, when he and his wife Mary, also a vegetarian accelerator physicist, obtained positions at the European Centre for Nuclear Research (CERN) in Geneva.

During his time in public service, Bell had obtained a doctorate from the University of Birmingham under the direction of Rudolf Peierls, a theorist trained in the age of the founders (he had studied under Sommerfeld and Heisenberg), co-architect with Otto Frisch of the memorandum that had launched Britain's atom bomb project before it was folded into the American enterprise.[15] Peierls came to admire Bell although he did not approve of sniffing around foundations. Some bits from their later correspondence give a good measure of the divide between the last of the old believers and the new followers of Einstein. 'I am not aware [Peierls wrote] of any other description [than Bohr's] of what the wave function is for'. Bell: what about de Broglie, Bohm, Everett, the Soviet physicist-philosophers? Peierls: only the Copenhagen interpretation makes sense, it does all that is needed, '[it] gives us a set of correlations between successive measurements'. The wavefunction does not describe the microsystem but our knowledge of it. Bell: Einstein maintained that we should not settle for less than an account of the real world. And Einstein was right.[16]

When still at Harwell and buffeted by the contrary winds of Peierls and Bohm, Bell collaborated on a project that made plain to him the grip of the Copenhagen orthodoxy on the presentation of quantum theory in student textbooks. The project was checking technical aspects of the translation of the major Russian textbook on the subject. One of its authors, Lev Davidovich Landau, had worked at Bohr's institute and, after some disagreements with the master, had accepted the Copenhagen version of events. He and his collaborator, Evgeny Mikhailovich Lifschitz, stuck so closely to Bohr's views that Bell imagined that their volume came close to what Bohr himself would have written had he attempted, against his nature, to write a textbook.

In fact, 'L&L', as Bell abbreviated them, made few explicit references to complementarity, but propounded its chief assertions. Classical concepts are necessary to the formulation of quantum mechanics; the measuring instrument must be considered a classical object; quantum mechanics is peculiar among fundamental theories in requiring another theory, classical mechanics, for its formulation and limitation; the classical performance of the instrument during the registration of a measurement takes place independently of a human observer. L&L repeated this last proposition several times. Bell noticed the repetition and joked about their 'inhumanity' in leaving out the observer, which L&L doubtless did to escape the sort of attacks Marxists directed at Bohr for his 'subjectivity'. Bell graded L&L as adequate FAPP, 'when used with good taste and discretion'.[17]

Since Bell's job at CERN concerned accelerator design, he had to pursue his interest in quantum foundations as a hobby. He enjoyed a little more freedom in 1964, when he took a year's sabbatical, most of which he spent in California at the Stanford Linear Accelerator Center (SLAC), an hour's drive from Berkeley. Since he was supported there by the US Atomic Energy Commission, he could not focus all his energy on foundational questions. He continued to work on particle physics and quantum field theory.[18] But he made a breakthrough: he had put his finger on the flaw in von Neumann's impossibility proof—the same flaw that, unknown to him, Grete Hermann had identified in 1933. The great mathematician had subjected states defined by hidden variables to a peculiar condition obeyed by quantum-mechanical states. But why should the entities of a theory, which at some level might conflict with quantum mechanics, be expected to obey all its rules? The flaw did not lie where Bohm had put it, in the neglect of hidden variables in the apparatus, but in gratuitously imposing 'a quite peculiar property of quantum mechanical states' onto states governed by hidden variables.[19]

While at SLAC, Bell submitted the subversive paper in which he rang down the curtain on von Neumann's proof and several updated forms of it to the *Reviews of Modern Physics*.[20] An unfortunate compounding of errors ensued. The editor of the journal requested revisions; the journal's office misfiled Bell's revised manuscript; Bell returned to CERN and the SLAC mail room (which denied knowledge of his existence) returned further correspondence from the journal editor. The comedy came to light when Bell made his own enquiries of the editor, and the paper came out in 1966, two years after its submission.[21]

Removing the obstacle of von Neumann's proof was a prelude to re-opening the door to hidden variables and legitimizing questions about the consequences of their introduction. 'It would therefore be interesting, perhaps, to pursue some further "impossibility proofs" replacing the arbitrary axioms . . . by some condition of locality, or of separability of distant systems'.[22] Bell mulled over these consequences during his sabbatical year and had them persistently in mind during the weeks before he interrupted his stay at Stanford for short visits to Brandeis University in Waltham, Massachusetts, and the University of Wisconsin at Madison.[23]

Bell adopted the version of EPR based on the correlation of spins that Bohm had first discussed in his textbook and had further developed with Aharonov in 1957. Two atoms *A* and *B* emerge in a state that preserves their opposite spins (the manner of their preparation is unimportant). *A* moves to the left and *B* to the right. Each atom passes through the poles of a 'Stern–Gerlach magnet'. This device, first used in a mind-twisting experiment performed by Otto Stern and Walther Gerlach in 1922, demonstrated the 'space quantization' of angular momentum. To the surprise of many physicists at the time, including the *Prinzipienfuchser* Bohr, Einstein, and Ehrenfest, the asymmetric field created by the Stern–Gerlach magnet split a beam of silver atoms into two and only two components, one corresponding to angular momentum along, the other to angular momentum opposed to the direction of the magnetic field. Electron orbits offered no useful purchase for understanding the behaviour of the silver atoms. The splitting is a quantum effect owing to the spin of the atom's exposed outermost electron.[24]

Let us assume that in a Bell–Bohm experiment the magnets line up in the laboratory z-direction and that atom A, after suffering its Stern–Gerlach ordeal at magnet 1, has its angular momentum or spin aligned along the z axis ('spin-up'). Then its correlated companion B, after passing its magnet 2, must be in a 'spin-down' state. There are two possibilities: A up or down, B down or up, which we label $+-$ and $-+$. Fig. 10(a) shows the situation with A $(+)$ and B $(-)$. In this situation, quantum mechanics assigns a probability of $1/2$ to each of the two possibilities, although it forbids us from knowing in advance which result we will get for any particular pair.

The business becomes much more fruitful when we ask, with Bell, what happens when the axes of the magnets make an angle θ relative to one another. The spinning atoms are still obliged to produce results dictated by the orientation of the poles of the magnets. But, as the magnets become misaligned, the probabilities relating to these results change. Let atom A be measured with spin-up $(+)$. It no longer automatically follows that B will appear with spin-down $(-)$; for if θ were 180°, the north pole of magnet 2 would lie in the direction of the south pole of magnet 1 (Fig. 10(b)) and B might therefore also register as spin-up $(+)$. Misaligning the magnet axes opens the possibility for joint results $++$ and $--$. Since the probabilities for all the outcomes must sum to 1, those for $+-$ and $-+$ must be less than the value each had $(\frac{1}{2})$ at full alignment, when $\theta = 0°$.

None of this is contentious. We can enliven the discussion by supposing that the spins of the atoms are fixed at parting by an unspecified hidden variable and that the travelling atoms are Einstein separable and locally real. Measurements made on A then cannot influence the outcomes of any measurements made on B, and *vice versa*. Bell's derivation of what logically follows went through many refinements, later simplified by himself and others.[25] Since the business involves only logic, we can follow his reasoning by what we may call a 'parabell', an analogy that does not invoke quantum mechanics.

We have a collection of all the brass (b) and steel (s) screws in the universe. That does not exhaust their interest; they also possess the additional non-overlapping character pairs, right-handed/left-handed screw thread (r/l) and new/used (n/u). It is then a fact of this universe that the number of new

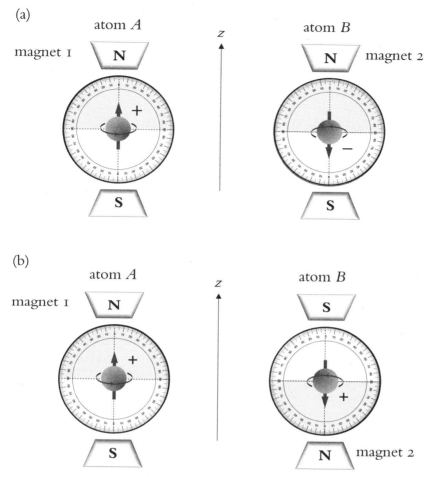

Fig. 10 In Bell's version of Bohm's thought experiment, the correlated spins of atoms A and B are detected as spin-up (+) and spin-down (−). If the magnets are aligned as shown in (a), the only possible outcomes are +− (as shown here) and −+, each with a probability of $1/2$. If we invert magnet 2, as in (b), the same orientations of the spins would instead register either as ++ (as shown) or −−, again each with a probability of $1/2$.

brass screws $N(b, n)$ is less than or equal to the number of right-handed brass screws $N(b, r)$ plus the number of new left-handed screws, $N(l, n)$. Or $N(b, n) \leq N(b, r) + N(l, n)$.

Obvious, no? Just add in the missing elements. The number of new brass screws $N(b, n)$ must include both right-handed and left-handed varieties, $N(b, n) = N(b, r, n) + N(b, l, n)$. Likewise, the number of right-handed brass screws will include both new and used screws: $N(b,r) = N(b,r,n) + N(b,r,u)$. And, finally, the number of new left-handed screws will include both brass and steel varieties: $N(l,n) = N(b,l,n) + N(s,l,n)$. The sum of $N(b,r)$ and $N(l,n)$ therefore consists of four terms, two of which add up to give $N(b,n)$, plus two 'extra' terms $N(b,r,u)$ and $N(s,l,n)$. If our universe has conspired to exclude used right-handed brass screws and new left-handed steel screws, $N(b,r,u)$ and $N(s,l,n)$ will both be zero and $N(b,n)$ exactly equal to $N(b,r)$ plus $N(l,n)$. But if our universe has any used right-handed brass screws or new left-handed steel screws, $N(b,n)$ must be less than the sum of $N(b,r)$ and $N(l,n)$. In our ignorance we hedge our bets and settle for the result $N(b,n) \leq N(b,r) + N(l,n)$.[26]

This relation is what we want but we have not justified it for the case of spinning atoms, for our inequality refers to three properties of a single item, whereas the atoms come in coordinated spinning *pairs*. A little consideration, however, shows that the same inequality holds. We first fix on the specific correlation we are going to study with our Stern–Gerlach magnets, $++$, $+-$, $-+$, or $--$. It will serve our purposes best (as it did Bell's) to select $++$.[27] Now, instead of counting objects possessing three different properties we count the number of pairs of atoms that record $++$ results for three different sets of orientations of our magnets.

Spinning atoms are not screws, but we can assume that, like screws, they possess properties (call these 'factory settings') that are fixed before their encounters with the magnets. Let a local hidden variable fix the spins, necessarily in opposite directions in space, the moment they form and maintain these orientations as the atoms move apart towards their respective magnets. And let the atoms be Einstein separable: the passage of A through magnet 1 does not in any way affect what happens to atom B (no action at a distance). The spins can enter their respective magnetic fields at any angle. In our model, spins pointing into the 'northern hemisphere' of the field are recorded as up $(+)$, those pointing into the 'southern hemisphere' as down $(-)$.

Fig. 11 illustrates the situation. We begin with magnet 1 at an angle $\alpha(=0°)$ to the laboratory z-direction and magnet 2 at an angle $\beta \ (= 45°)$. Define the difference, $\beta - \alpha$ as $\theta_1 = 45°$. Measurements made on many pairs, N, in this configuration will give a $++$ correlation in a portion $N_{++}(\alpha, \beta)$ of the pairs equal to N times the probability that both the spins lie in the northern hemispheres of their magnets (the shaded areas

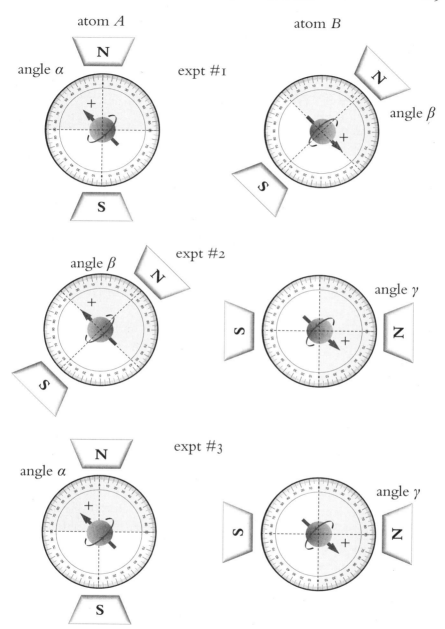

Fig. 11 Bell deduced that experiments performed on the correlated spins of two atoms would provide a direct test of the quantum-mechanical predictions against all alternatives based on local hidden variables.

in Fig. 11). Call this probability $P_{++}(\alpha, \beta)$ and let $N_{++}(\alpha, \beta) = N \times P_{++}(\alpha, \beta)$. On the assumption that the spins of successive pairs distribute randomly but uniformly over the whole sphere, we can see that this probability depends on the *overlap* of the northern hemispheres imagined superimposed one on top of the other. The smaller the overlap, the higher the probability $P_{++}(\alpha, \beta)$.

For experiment 1, our assumptions make $P_{++}(\alpha, \beta)$ equal to θ_1 divided by 360°, or 0.125. In experiment 2, the angles are $\beta = 45°$, $\gamma = 90°$, and $\theta_2 = 45°$; $P_{++}(\beta, \gamma) = 0.125$. In experiment 3, the angles α and γ retain their values, $\theta_3 = 90°$, and $P_{++}(\alpha, \gamma) = 0.250$. We can now reach for our inequality, $N_{++}(\alpha, \gamma) \leq N_{++}(\alpha, \beta) + N_{++}(\beta, \gamma)$. But in all three experiments we measure (in thought!) the same number N of pairs, which we can cancel from our inequality leaving only the probabilities. We therefore expect that the inequality $P_{++}(\alpha, \gamma) \leq P_{++}(\alpha, \beta) + P_{++}(\beta, \gamma)$ will hold. And it does, since 0.250 is less than or equal to $0.125 + 0.125 = 0.250$. We can expect these results if the spins retain their factory settings throughout the experiments.

Nothing in this analysis depends on quantum mechanics. Indeed, a proper quantum-mechanical treatment gives a probability for a ++ result as half the square of the sine of half the difference angle. For experiment 1, $P_{++}(\alpha, \beta) = \frac{1}{2}\sin^2(\frac{1}{2}\theta_1)$, or 0.073 and, since $\theta_2 = \theta_1$, the same probability holds for experiment 2. For experiment 3 the probability is 0.250, as before. But now we sense an arithmetical disaster. According to quantum mechanics our inequality requires that $0.250 \leq 0.146$, which, we can agree with Bell, 'is not true'.[28] Quantum mechanics violates what we can now call *Bell's inequality*. It does not permit the possibility of factory settings.

This wide-ranging conclusion does not depend on the details of the hidden variable theory provided it allows the particles A and B to be separable and locally real. Since the hidden variables make the theory local, Bell could package his results in a neat epigram: 'If the [hidden variable] extension is local it will not agree with quantum mechanics, and if it agrees with quantum mechanics it will not be local'.[29] This is *Bell's theorem*. The extent of the correlation between two entangled particles can sometimes be greater and sometimes less than *any* locally real hidden variable theory will allow. Bell's analysis did not require anything not known to the founders and, as we know, 20 years earlier Hermann had told the Copenhagen school how to elude von Neumann's proscription of hidden variables. Pondering, not calculating, created Bell's inequalities, proof of the assertion that Bohm had made in his textbook of 1951: quantum theory is inconsistent with (local) hidden variables. Bell: 'Probably I got that equation into my head and out on to paper within about one weekend'.[30]

Nature is uncannily resourceful. Might it not have invented a place for hidden variables even if a Bell–Bohm experiment found unequivocally for quantum mechanics? Bell's subtlety spied the possibility of a conspiracy between the two sets of magnets and the apparatus used to detect the atoms. Perhaps naive experimenters had allowed the instruments 'to reach some mutual rapport by exchange of signals with velocity less than or equal to that of light'. Bell proposed a way to thwart their machinations: 'experiments of the type proposed by Bohm and Aharonov, in which the settings are changed during the flight of the particles, are crucial'.[31] Since the particles cover metres in nanoseconds, realization of Bell's crucial experiment would not be easy.

The simpler Bell–Bohm experiment, though far from easy, was practicable in the late 1960s and enticing to the bold or reckless. Their efforts made an epoch. The old *Prinzipienfuscher* had been obliged to bolster their arguments with thought experiments. Thinking real ones beyond reach, they could not move the discussion outside their heads. Bohm, the Marxist materialist, had pulled at the thread of EPR's thought experiment and cast it into a simpler form almost accessible to experiment. Re-analysis of the Wu–Shaknov experiments had suggested previously unlooked-for correlations between distant entangled particles. Bell, the moralist crusader and Einstein's quantum heir, now offered a direct test.

Many physicists later misunderstood Bell's work as a direct experimental challenge to complementarity and Copenhagen. But as Bohm and Aharonov had pointed out in 1957, Bohr's point of view could not be challenged in this way: 'no paradox can arise in the hypothetical experiment of [EPR] . . . the question of how the correlations come about simply [have] no meaning'. The challenge to Copenhagen concerned the grand old question about the closure of a domain of science. 'We can show that there is no inconsistency in the quantum–mechanical conclusion that such correlations exist, but there is, in this [Bohr's] point of view, no way even to raise the question of what is their origin'.[32] The domain is closed against the inquiry. But is it not commonsensical to entertain the question whether Bell's inequality passes the test of experiment?

Interested experimenters had first to discover the question and Bell had not made it easy for them. A few physicists at Brandeis and at Madison heard him talk about his inequality and he tried to reach a wider audience through the *Physical Review*; but the journal demanded payment that ordinarily came from the author's research grant or institution. With his fine sense of propriety, Bell did not want to ask his hosts at SLAC for the money and, unwilling to pay the charges himself, submitted his paper instead to a new journal, *Physics, Physique, Fizika*. It did not work like other journals: its founders intended it to be an outlet for unspecialized articles and it paid authors for their contributions.[33] No doubt Bell's paper fit the purpose of the journal, but the journal was not fit for purpose. It expired after four years.[34]

When he returned from his sabbatical, Bell sought to help people over the hurdle of his ideas through familiar talks at CERN. The nature of his homilies may be inferred from his later invocation of an odd habit of a

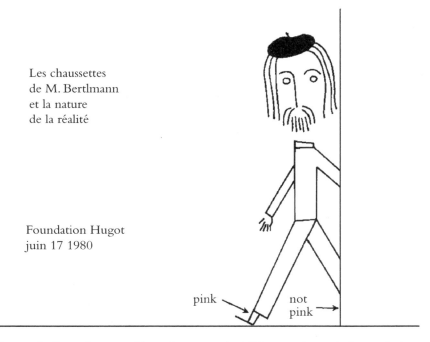

Fig. 12 Bell sought accessible analogues to the EPR experiment in 'everyday' examples of spatially separated objects with correlated properties. He found one such example in the dress-sense of his CERN colleague Reinhold Bertlmann.

younger colleague, Reinhold Bertlmann, who worked at CERN in the late 1970s. Bertlmann had taken to wearing unmatched socks as a symbolic contribution to the student protests of 1968. It was a perfect analogy. Although the colour he would wear on either foot could not be foreseen, his practice permitted a reliable prediction. If he wore a pink sock on one foot, a knowledgeable observer could predict with certainty that his other sock would not be pink (Fig. 12). 'There is no accounting for tastes', Bell wrote, 'but apart from that there is no mystery here. And is not this EPR business just the same?' Bertlmann's socks pre-exist, their colours pre-determined at the factory, before observation. No one needs to look at them to 'collapse' the colour of each sock into existence. Bell worked up his own parabell, involving washing Bertlmann's socks at different temperatures, to provide another derivation of his inequality.[35]

Lest it be mistakenly supposed that Bertlmann was the first to lend his socks to science, we declare that the honour belongs to Robert Symmer, who anticipated him by two centuries. Symmer wore a pair of silk stockings, one black, the other white, on each foot during the winter. When he removed and separated them, they crackled; when he released them, they floated off to dance against the wall of his room. He deduced that stockings of different colour when rubbed together took up contrary electricities and replaced negative charge, conceived by Benjamin Franklin as merely the absence of positive charge, with a substance as real as its counterpart.[36] By giving rise to the concept of the two electrical 'fluids', Symmer's socks laid the foundation of the physics of imponderables, the Grand Unified Theory of 1800.

Bell Tests and Protests

Experiments involving appropriately disintegrating molecules, rotatable Stern–Gerlach magnets, and perfect detectors were easier to conceive than execute. Bell had devised his inequality in an abstract, ideal world, where instruments perform perfectly and cost nothing. In the ordinary world physicists had too little interest in foundations and too few dedicated resources to try to overcome the material impediments to tests of Bell's inequality. And those who might have been interested in Bell's ground-breaking paper had first to disinter it from its burial in an obscure journal.

Fortunately, Bell took the trouble to alert Shimony, to whom he sent a copy of the manuscript he had written at Brandeis. Shimony shared it with his advisor. Wigner liked it and devised a simpler derivation of the inequality. Although Wigner clung to his subjective interpretation, he took a supportive interest in the work of Bell's champion d'Espagnat and the speculations of Wheeler, who had not yet abandoned Everett.[1]

Shimony had been teaching a course in quantum foundations at MIT and was familiar with its susceptibility to crank theories. 'Here's another kooky paper that's come out of the blue'. Bell's manuscript was off-putting, badly typed, duplicated in smeared blue ink, and not free from arithmetical errors. Overcoming these obstacles and reading more closely, Shimony realized that the kooky paper was not only correct, it was also brilliant. Bell had proved that 'nonlocality was not just a feature of one particular model that Bohm was clever enough to think of. It was inevitable in order to recover the quantum mechanical predictions'.[2]

The 1950 Wu–Shaknov experiment had involved the measurement of the polarizations of photons arising from electron-positron annihilation rather than the spins of correlated atoms. In principle, this made their experiments more accessible (no unwieldly Stern–Gerlach magnets), but because the energy of the annihilation photons is too high for analysers suited to the determination of their polarizations, they had had to resort to less sensitive Compton scattering. Furthermore, as Shimony recognized, their experiment could not check Bell's inequality. Wu and Shaknov had placed the Compton detectors either parallel or perpendicular to one another, precisely the angles ($\theta = 0°$ and $90°$) at which local hidden variables and quantum mechanics predict the same results. (In the thought experiments pictured in Figs. 10 and 11, results for hidden variable theories and quantum mechanics agree at $0°$ and $180°$; in the real experiments of Wu and Shaknov, at $0°$ and $90°$; the discrepancy arises from the difference between the intrinsic spin properties of our atoms and their photons.) No one had done experiments at the 'magic' angles that would most readily expose a violation of Bell's inequality.

Shimony approached Aharonov, who was also by now familiar with Bell's paper, seeking support for experiments to test it. Aharonov declined. He had already addressed the question with Bohm in 1957 and had other things to do. Failing in his quest to interest anyone in his proposal to perform a variant of the Wu–Shaknov experiment and other possible Bell tests, Shimony shelved the project.[3] But he did not abandon it. In the summer of 1968, he moved from MIT to a position in physics and philosophy at Boston University, which gave him a place and a mandate to pursue projects like Bell's. Soon he would have some collaborators.

John Clauser discovered Bell's 1964 paper in the library of NASA's Goddard Institute for Space Studies (GISS) in New York City. He had graduated from Caltech in 1964 and completed a master's degree in physics at Columbia University before starting to work on a PhD at GISS on the recently discovered cosmic microwave background radiation. Columbia required at least a B grade in Advanced Quantum Mechanics to continue in the doctoral programme. Clauser could do no better than a C before earning a pass on his third try. 'I really kind of need a model', he later explained, 'or some way of visualising something in physics'. The quantum formalism as it was taught could not give him what he needed, and his brain 'kind of refused to do it'.[4]

He developed deep misgivings about the formalism he had such difficulty mastering and pitched on Bohr and the Copenhagen interpretation as the source of his problem. He gave up on his teachers, retreated from his fellow students, sat in a corner, and tried to work it out by himself. He turned to the dissidents, consulting de Broglie's *Non-linear wave mechanics*, first published in French in 1956 and in English in 1960. He found EPR, Bohm's papers, and Bohm and Aharonov. Then, in 1967, he found Bell. His thesis advisor, Patrick Thaddeus, advised him that he was wasting his time with this off-topic obsession and risked ruining his career. 'He really got quite angry with me'. It did not matter to Clauser: 'once I got involved in the quantum mechanics stuff, everything else paled in comparison'.[5]

The time and place were propitious for adventure: the revolutionary year 1968 and Columbia University. The student revolt demonstrated the power of wayward thinking, though Clauser was inclined to defend the US government on its involvement in Vietnam and mostly kept his head down. 'I went back and did physics, and/or went sailing'.[6] And he approached Madame Wu. Her group at Columbia was still engaged in correlation experiments that Clauser realized could be adapted to test Bell's inequalities. Clauser: Have you thought to measure coincidences at intermediate scattering angles that are neither parallel nor perpendicular? Wu: No. She suggested that he talk with Len Kasday, who was working on an improvement of the Wu–Shaknov experiment in place of Shaknov, who had been killed in action in Korea in 1951. Kasday had not seen Bell's paper, but once informed agreed to make measurements at intermediate angles.[7]

In January 1969 Clauser gave a seminar on Bell's theorem to a research group at MIT he hoped to interest in Bell tests. In the audience was Carl Kocher, a postdoctoral associate newly arrived from Berkeley. He had earned his PhD under Eugene Commins, a much sought-after mentor in the physics department who advised his students to work only on fundamental questions.[8] Clauser now learned about the Kocher–Commins experiment based on the 'cascade' emission of two photons from an excited state of the calcium atom. The configuration of atomic states involved in the cascade correlates the polarizations of the emitted photons. And, importantly, unlike the high-energy photons studied in the Wu–Shaknov experiments, the cascade produces pairs of low-energy photons in the visible region, one blue and one green, analysable using conventional polarizing filters and detectable by conventional photomultipliers.

Clauser dashed back to New York to consult the paper that Kocher and Commins had published in 1967. There was no doubt. They had built a system perfectly suited to testing Bell's inequality. Although they had acknowledged the relevance of their system to EPR, they had made no reference to Bell and, like Wu and Shaknov, had measured the polarization correlations only for the uninformative angles $\theta = 0°$ and $90°$.

Clauser wrote to de Broglie, Bohm, and Bell. Did they know of any experiments to test Bell's inequality? Did they agree that a repeat of the Kocher–Commins experiment with improved polarization analysers and intermediate angles would be convincing? How did they view the importance of such a test? He received unanimous replies: 'no' to the first question, 'yes' to the second. Bell thought that the previous successes of quantum mechanics suggested little doubt about the outcome. Still, the game was worth the candle, if anyone would pay for it, and Bell nourished hope in the 'slim chance of an unexpected result, which would shake the world'.[9]

The remote eventuality of shaking the world was more than enough to motivate Clauser. 'To me, the possibility of experimentally discovering a flaw in quantum mechanics was mind-boggling'. Revenge would be sweet. He believed that the test would find for local hidden variables. 'I was certainly a maverick, no question'.[10] Desperately wanting to be the first experimentalist maverick to check Bell, he staked his ground in an abstract submitted to the spring meeting of the American Physical Society, to be held in Washington D.C. in April 1969. Its exaggeration revealed its author's passion. 'Such an experiment must rule out all local-hidden-variable theories governing the polarization of photons or disprove the Copenhagen interpretation and predictions of quantum theory'.[11]

Shimony learned about the Kocher–Commins paper from an experimentalist colleague at Harvard: 'This is what we need. Don't look anymore'. He showed the paper to his student, Michael Horne, who had come to Boston after graduating from the University of Mississippi. Shimony soon discovered that Frank Pipkin at Harvard possessed precisely the apparatus required to repeat the Kocher–Commins experiment with photons from a cascade in mercury atoms instead of calcium. Pipkin agreed that the experiment should be tried and his student Richard Holt made ready to do it.

Before they could start, they ran across Clauser's abstract. 'We felt pretty sick', Shimony recalled. Wigner advised him to contact Clauser and suggest a collaboration. Clauser held back until Shimony explained that they already had access to the necessary apparatus and a student lined up to perform the experiments. Clauser, Horne, and Shimony formed a fruitful collaboration and a lasting friendship.

Their first task was to generalize Bell's original inequality and fit it for testing in an imperfect world. They faced the problem of gathering enough photons. The cascade sprayed from the excited atoms in all laboratory directions. Optical filters used to isolate them to the left and right would not transmit all the photons incident on them. The polarization analysers would sometimes mis-identify polarizations by amounts that might depend on their orientations. Counting of coincident detections of the photons left and right would be complicated by stray coincidences. By August 1969, the group had replaced Bell's inequality with something more suitable for non-ideal situations: the 'CHSH inequality', so named for its four authors, Clauser, Horne, Shimony, and Holt.[12]

Clauser was not invited to join Holt at Harvard.[13] After completing his PhD with Thaddeus, he secured a postdoctoral position at Berkeley with Thaddeus' former thesis advisor, Charles Townes, the inventor, in 1951, of the maser, a microwave forerunner of the laser. Clauser was hired to work on a project in radio astronomy, which did not interest him much. He remained intent on shaking the world, and Townes generously allowed him time to work on his obsession. He also helped to cajole Commins into providing funds to resurrect the Kocher–Commins apparatus and a PhD student, Stuart Freedman, to work on the project.

By 1970, the international community of foundationalists had grown large enough to inspire the establishment of a journal for their subject. Its founders were two German-born physicists with rich and varied careers, Henry Margenau (Yale) and Wolfgang Yourgrau (University of Denver). Both ended with joint appointments, Margenau in philosophy and physics, Yourgrau in philosophy and history. Their editorial preface to the first volume of *Foundations of Physics* (March 1970) sounds an alarm. There are 'infelicities' in our scientific understanding, they warn, which we must not ignore; 'the risk of overlooking [them] is great in our day when the need and actualities of public support place an excess of emphasis upon the pragmatic aspects of science'.[14] We should clarify our ideas before we go rushing ahead to implement them at public expense.

Following this philosophy, the journal began by rushing backwards, with articles by de Broglie (the double solution and the pilot wave) and Wigner (physics and psychology). The first issue also carried an article on measurement by Hans Dieter Zeh, from the University of Heidelberg, with his own spin on Everett's interpretation.[15] Margenau and Yourgrau correctly anticipated the burgeoning interest in foundational questions.

Their journal continues to flourish and to list, among pressing unsolved problems, 'the nature of measurement in quantum mechanics'.[16]

By 1970 the foundationalists had sufficient traction and institutional resources to hold a jamboree ('the Woodstock of quantum dissidents') on the shores of Lake Como.[17] The Italian Physical Society sponsored the gathering at its Villa in Varenna, some 40 kilometres north of where Bohr had uttered the first edition of complementarity over 40 years earlier. The society appointed d'Espagnat to run the school and choose its lecturers.[18] Everyone who had contributed to the subject since Bohm was either present or represented at Varenna that summer. It was hardly a gathering of fringe unbelievers.

Wigner opened the meeting with the announcement that its subject straddled the boundary between physics and philosophy. In this borderland, five notions about the nature of measurement—the core concept of quantum mechanics—struggled against Bohr's formulation, which, in Wigner's opinion, was 'entirely satisfactory'. If it referred to only one particle! But the world is more than an electron and measurement more subtle than reading dials. Neither positivists nor realists have the answer, Wigner insisted, and the only approach still standing was that of his friend von Neumann: the observer effects the collapse of the wavefunction. Unfortunately, this insight did not amount to an explanation. We know too little about the mental acts in question; 'we explain a mystery by a riddle'. We must unriddle psychology to do physics. 'The self-contained nature of quantum mechanics is an untenable illusion'.[19]

The prominent role of Wigner among the lecturers accounts for the absence of Rosenfeld, the natural choice to represent the interests of Copenhagen. The merest whiff of Wignerism was to him a sickening odour to be cured (he advised d'Espagnat) only by breathing 'the pure air of Copenhagen'.[20] Instead of Rosenfeld, Jørgen Kalckar, much less acerbic and outspoken, represented Bohr's Institute.

Bell described his test of salvation by local hidden variables, which he still hoped would find for them but rather expected to rule them all out except for the non-local variety. And experiment did just that, according to Kasday's report of the work that he, John Ullman, and Wu, now explicitly engaged in foundation studies, had almost completed.[21] Their paper would not be published until 1975, but Kasday was already confident that his results would provide 'no evidence for a breakdown in quantum predictions'. In addition to his collaborators, Kasday acknowledged five other physicists,

American and European, for their direct help. By 1970 there was much more heterodoxy in the world than the picture of a dominant Copenhagen church acknowledged.[22]

One of Kasday's five additional advisors was Josef Maria Jauch, professor of physics at the University of Geneva, who had earned his PhD in the US and spent the war years teaching at Princeton. Jauch offered an axiomatic version of Bohr's 'profound', 'straightforward and immediately obvious common-sense notions'. No challenger had been able to predict any observed result contrary to them. What choice have we (Jauch asked) but to deal with quantum mechanics as its founders left it and concede that the reasonable desire of most physicists for a realistic account of the world has been dealt a fatal blow by the recent work of Wu's group? Another problem was still alive, however, a very deep problem, measurement, for which the Schrödinger equation does not permit even the possibility. We can only say with Bohr that the apparatus manages the trick of collapse and against Wigner that consciousness has nothing to do with it.[23]

Continuing along this line, Shimony supplied details of the Bell tests under way at Harvard and Berkeley. It was still too early to say how these would turn out: 'we should know within a few months whether we come to bury local hidden variables or to praise them'. Shimony thought it useful to give his colleagues a little lecture against taking the epistemological situation of their subject too seriously. The empirical situation might be unique, but the problems it presented went back to Plato and Aristotle. According to Shimony, Wigner's position resembled Mach's and might be compatible with Leibniz's. Bohr's was more like Kant's though less solid in argument; a consistent but miserly phenomenologist who gives no other compensation for renouncing insight into the order of being than escape from the problem of measurement.[24] This left the realist position, which several familiar voices supported.

De Broglie gave in absentia another evocation of the past in a clear resurrection of the theory Pauli had sought to destroy in 1927. He recalled that he had proposed two solutions to the Schrödinger equation, ψ and χ (Greek chi), which differed only by a multiplicative constant. The familiar ψ carries probabilities but has no physical effect; χ is a guide wave for a particle whose velocity it determines. The particle has an internal clock that keeps it in phase with χ and within a region of exceedingly small dimensions. This picture, updated by Bohm and Vigier's ideas about fluctuations of the particle around its trajectory, illustrated the present state of wave mechanics.

When perfected, it would reveal 'the true nature of the coexistence of waves and particles'.[25]

That was an old-fashioned way to put the realist agenda. For those who wanted something bolder Bryce DeWitt, one of Wheeler's protégés and collaborators, championed Everett's alternative. DeWitt was a convert: he had sent Everett a lengthy critique of his relative states in 1957 and been won over by the answer. At the time, DeWitt needed a way out of the fix in which his struggle to combine quantum mechanics and general relativity had landed him. The struggle had eventuated in the Wheeler–DeWitt equation, which implied a wavefunction for the whole universe; a grand ψ that did not lend itself to measurement, since nothing can exist outside the 'system' (the entire universe!) to collapse it. Everett's interpretation offered a way out and DeWitt sought to raise its profile. Although 'bizarre', DeWitt conceded, in the crazy world of quantum mechanics it was straightforward. For it did without the metaphysics, external observers, human consciousness, collapsing wavefunctions, contorted epistemology, and micro-macro puzzles invented by the *ancien regime*.[26]

DeWitt was not entirely fair to his predecessors. 'According to the Copenhagen interpretation of quantum mechanics [he wrote in *Physics Today*], whenever a [wavefunction] attains a form like that [a superposition of wavefunctions pertaining to measurement] it immediately collapses'. This was nonsense. The Copenhagen interpretation has nothing to say about the collapse of the wavefunction or von Neumann's theory of measurement. Against his version of the Copenhagen interpretation DeWitt pitted 'one that pictures the universe as continuously splitting into a multiplicity of mutually unobservable but equally real worlds, in each one of which a measurement does give a definite result'.[27] Thus he repackaged Everett's relative state formulation as the *many worlds interpretation* and raised it from one of the most obscure to one of the most controversial interpretations of quantum mechanics.

The advantages of Everett's approach appealed to Zeh, another Varenna lecturer, whose variation on the theme was beginning its journey to the *many-minds interpretation*. But despite the enthusiastic acceptance of Everett's theories by a few physicists and an imaginative public, the many worlds interpretation did not become competitive. Wheeler later allowed that Everett's interpretation 'carried too much metaphysical baggage along with it' and made 'science into a kind of mysticism'.[28] As he distanced himself from it, however, he acknowledged Everett's contribution as one of the most original and important in decades.[29] Although today it retains the support of a few well-known vocal theorists, it remains for most one of those speculations that, if too frequently entertained, 'could irreparably damage one's intellectual health'.[30]

Bohm, the modern instigator of the new ways, took 57 pages in the Varenna proceedings to argue that the situation called for an entirely new 'order' in physics. The classical order had been nibbled away by relativity and quantum physics to the point that we know our ideas are inadequate but not how to escape from our old ways of 'thinking, using language, and observing'. We remain at the stage of Galileo awaiting our Newton. We cling to old ideas (here Bohm referred to Kuhn's *Structure of scientific revolutions*) long after we have discovered their shortcomings and glimpsed their replacements.

Take the infamous problem of measurement. We separate system and apparatus knowing full well that they are tied together, for everything is connected; the whole is without parts, 'a total order is contained in some *implicate sense*, in each region of space and time'. In what sense? Under everything physics investigates lies a 'holomovement', an unbroken and undivided reality, also, unhelpfully, 'undefinable and unmeasurable'. Still, we can know that 'electron' is a 'name by which we call attention to certain aspects of the holomovement'. The new order will be founded on wholeness, 'a thoroughgoing wholeness, in which there is no division between research and any other aspect of life'. And on 'communition', that is, careful talking and sympathetic listening, through which system and apparatus negotiate a measurement and an inspired colloquist 'relevates certain aspects of the holomovement'.[31] Although a champion jabberwocky talker, Bohm did not persuade many of his colleagues to attempt to relevate the holomovement.

In addition to the familiar spokespersons on foundational questions, the Como participants heard also from their hosts, led by Franco Selleri of the University of Bari. Whereas his colleague, Giovanni Maria Prosperi, took a position between Bohr and von Neumann so complicated mathematically that perhaps only Wigner understood it, Selleri was a clear and uncompromising idealist. If he were not, he said, he would be a split personality, 'forced . . . to believe in reality as a man and to decide that reality does not exist as a researcher'. A possibility of escape existed in the formulation of the measurement process. People assumed that the process coupled a definite state of the system to one of the apparatuses with a certain probability.

This was the assumption then being tested by Clauser et al., which Selleri expected to fail.

He was not alone in pointing to the formulation of the interaction between system and apparatus as a major problem; d'Espagnat took a similar tack, although without anticipating experimental results. Selleri was unique, however, in attaching the question of realism in physics to the demands of everyday life. 'In this time when the social responsibility of the scientist is so strong, where the destruction or survival of the world depends on him, it is important to develop a science not in basic contradiction with the social reality'.[32]

Selleri put his finger on the driving force of the Varenna conference. The pursuit of experimental Bell tests, the likelihood that non-locality would win out, and the inadequacies of the Copenhagen orthodoxy were not the preoccupations of the student body. Selleri had been turned off particle physics by its faddism and barrenness (as he had experienced it) and on to social problems by the student uprisings of 1968, the discovery that many fundamental problems in physics were still open ('it was really fantastic'), and the attractions of dialectical materialism.[33] Many of the students, graduates, postdocs, and lecturers gathered at Varenna had strong leftist commitments and believed with Selleri that physicists had lost their way and their virtue by cravenly following bosses corrupted by the capitalist system and fraternization with the military.[34]

The choice of Wigner to present the main issues and interpretations confirmed the students' Marxist-informed disenchantment. Although he stated the various interpretations fairly, his intemperate advocacy of nuclear armaments exemplified the military-scientific complex that the students identified with the troubles of their profession. Wigner's decision to throw a party on the fourth of July to celebrate American Independence Day, which fell during the school's brief term, exacerbated his sin, for the students charged the United States, still entangled in Vietnam, with leading the subversion of science in the interest of the warfare state.[35]

During the evenings dissidents assembled to write a manifesto, 'Notes on the connections between science and society'. It followed the Marxist line on capitalism and the rise of science, the tyranny of the ruling classes, and so on, and applied it to physics. The manifesto taught that modern society encourages extreme competition, faddism, specialization, and a hierarchical structure of professors and their would-be successors. It rewards research that leads to new technologies and empty reputations and, by wastefully multiplying useless research, imitates the non-productive investments that sustain capitalism. The manifesto ended with utopian demands: a turn to humanistic values; a realistic, communicable physics; and an end to US dominance of European science.[36]

Selleri and other leaders of the Italian Physical Society decided to strengthen their case by scheduling a school on the history of quantum physics and its relation to the wider society. He had recognized that when it came to assessing fundamentals his colleagues suffered from something worse than a tendency towards positivism, idealism, and indifference. They had no time for the history and philosophy of science. About the struggles of the founders of quantum mechanics, most of them knew nothing; a grievous lack, Selleri thought, for a reason that we might have set down ourselves. 'The most serious damage caused by rejecting the history of physics is ignorance of the dramatic elements of the struggle that began with and continued after the birth of quantum mechanics, and of the fundamental problems over which the struggle took place'.[37]

The school inspired by these reflections opened in the Varenna villa in August 1972. Famous physicists, leftist students, and obscure historians participated. The historians' lectures traced the development of the old quantum theory, the early years of quantum mechanics, and the involvement of physicists with the military. As in 1970, the students organized evening sessions to rail at the establishment, boycotted the showing of a film about Fermi and the atomic bomb, and issued a short manifesto. John Heilbron was one of the historians. He remembers the heady mix of physics and politics, generational conflict, national differences, and friendships formed at the interfaces.

The student manifesto blamed the use of laser-guided bombs in Vietnam on 'a systematic application of *scientific* discoveries for military purposes'. Five Nobel-prize winners had collaborated with the Pentagon's Institute for Defense Analysis' Jason Division: Luis Alvarez, Murray Gell-Mann, Donald Glaser, Townes, and Wigner. 'The operational use of scientific knowledge in the Indochina war is of particular concern to us as participants in the 1972 Varenna Summer School in the History of Physics . . . [I]t is no longer possible to separate our attitude on these issues from our professional activities. This is why we express . . . our condemnation of those colleagues who have willingly involved themselves in the waging of this war'.[38]

In April 1972 the *Physical Review* published the results of Freedman and Clauser's Bell test. As the first definitive experiment of its kind, it deserves detailed description. The experimenters used ultraviolet light from a deuterium arc lamp to irradiate a beam of calcium atoms, thus exciting the cascade of entangled photons. This was not an efficient process; only 7% of the atoms thus excited decayed via the state producing the desired cascade. To detect the photons' polarizations, Freedman and Clauser used a pile of ten thin glass sheets tilted at an angle to maximize transmission and rotatable through increments of 22½° (Fig. 13). The setup limited their measurement to photons of one type of polarization, vertical or horizontal.

Their measure of merit was the coincidence rate (counts per second) for two-photon detection with the polarization analysers set to different difference angles, $R(\theta)$, roughly normalized by division by R_o, the coincidence rate with both polarization analysers removed. Fig. 14 presents their results, which took 200 hours to accumulate. The error bars on each point indicate a single standard deviation, conservatively estimated. The curve drawn through the points is the quantum-mechanical prediction, suitably modified to account for instrumental inefficiencies. For example, at $\theta = 0°$, perfect correlation (no inefficiencies) corresponds to $R(\theta)/R_o = 0.5$, and

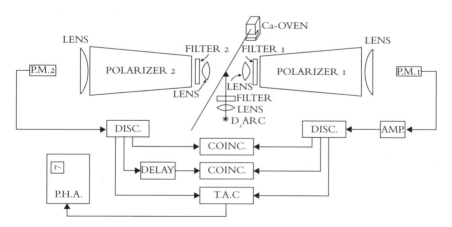

Fig. 13 Schematic diagram of the apparatus used by Freedman and Clauser to perform the first experimental tests of Bell's inequality. Photons from the calcium atom cascade pass through differently coloured filters and 'pile-of-plates' polarization analysers before being detected by photomultipliers (PM1 and PM2). Signals from the photomultipliers pass through discriminators (DISC) to coincidence counters (COINC), a time-to-amplitude converter (TAC), and a pulse–height analyser (PHA).

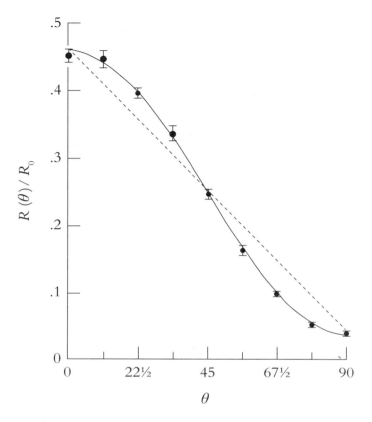

Fig. 14 Freedman and Clauser measured the variation of the normalized coincidence rate $R(\theta)/R_o$ with the difference angle, θ. The experimental results are plotted as points with error bars corresponding to single standard deviations. The solid line through the points represents the quantum mechanical predictions, suitably modified to account for instrumental inefficiencies. The dashed line, which does not appear in the original, represents the predictions of simple local hidden variable theories.

for $\theta = 90°$, perfect correlation would give $R(\theta)/R_o = 0$. Why 0.5 and not 1.0? The pile of plates polarization analysers transmit only one type of polarization, so even in a perfect world half the photons are lost. In measuring R_o without the analysers the apparatus detects *all* photons and so perfect correlation corresponds to 0.5.

The ratio $R(\theta)/R_o$ refers not to a single probability but rather to the correlation $P_{++}(\theta) - P_{+-}(\theta) - P_{-+}(\theta) + P_{--}(\theta)$, where $+$ and $-$ now refer to vertical (V) and horizontal (H) polarization, respectively.[39]

The quantum-mechanical predictions are $P_{++}(\theta) = P_{--}(\theta) = \frac{1}{2}\cos^2\theta$ and $P_{+-}(\theta) = P_{-+}(\theta) = \frac{1}{2}\sin^2\theta$. These combine to give the result $\cos^2\theta - \sin^2\theta = \cos 2\theta$, graphed as the bold curve drawn through the data points in Fig. 14. Simple local-hidden-variable theories predict a straight-line relationship (the dashed line) and deviate most from the expectations of quantum mechanics at the 'magic' angles $\theta = 22\frac{1}{2}°$ and $67\frac{1}{2}°$. Freedman and Clauser used the results for these angles in a version of the CHSH inequality and found $0.050 \pm 0.008 \leq 0$, a violation of the inequality by more than six standard deviations and a firm justification of their conclusion: 'We consider these results to be strong evidence against local-hidden-variable theories'.[40]

A cloud immediately came over this clear result. Holt and Pipkin at Harvard found $-0.034 \pm 0.013 \leq 0$, consistent with Bell's inequality and favourable to local hidden variables. Since few people expected this outcome, Holt and Pipkin decided not to publish it; it survives, however, in Holt's PhD thesis of 1973. They were right to be wary. Clauser repeated their experiments at Berkeley in 1976 and found $0.0385 \pm 0.0093 \leq 0$ (page 208).[41] In 1975, Kasday, Ullman, and Wu confirmed Freedman and Clauser's results; in 1976, Edward Fry and Randall Thompson at Texas A&M University added $0.046 \pm 0.014 \leq 0$ to the symphony. Physicists surprised or alarmed by the violation of Bell's inequalities were to devote 20 years to 'strenuous efforts' to find 'loopholes' in the experiments.[42] But the experimental evidence pointed firmly in the direction of quantum mechanics.

The properties of the screws in our parabell are determined by their process of manufacture. Extrapolated to pairs of correlated photons, the parabell suggests that hidden variables fix the photon polarizations at their factory settings the instant they are produced and emitted. Scientists unfamiliar with the infelicities of entanglement would assume that the photons maintain their pre-determined properties as they make their journeys to their respective detectors. From this consideration they might be tempted to conclude that it is the kind of classical determinism representable by factory settings that conflicts with quantum mechanics and results in the violations of Bell's inequality witnessed in the laboratory.

Clauser (pictured) repeated Holt and Pipkin's experiments at Berkeley in 1976
and found in favour of quantum mechanics.

Not so. At issue was still the main point that had been buried in the
EPR paper: separability and local realism. Perhaps, Bell speculated, the
instruments themselves contain hidden variables that affect the results.[43]
Perhaps the evolution of the two-particle system between emission and
detection suffers random fluctuations that render the particle properties
unknowable in flight but still correlated on measurement. Clauser and
Horne defined a class of *objective local theories* that required the outcome
of a measurement performed on one part of a composite system to be
independent of whatever aspects of the other component the experimenter

might choose to measure. They showed that, with the adoption of a weaker 'no-enhancement' assumption (the intrinsic probability of photon counting is unaffected by the presence or absence of a polarizer), these locally realistic theories would still be expected to violate Bell's inequality.[44] The emphasis began to shift, from tests of local hidden variables to tests of locally realistic theories, no matter how constituted.

The Freedman–Clauser experiment secured Freedman's PhD but not Clauser's future. Neither this success nor his subsequent contributions to the Bell-test literature had shaken the world. The quantum-mechanical predictions had been confirmed. Clauser recalled the situation years later with a bitterness that had likely grown with time. He reminisced that a 'nearly furious' Feynman had once dismissed him with the admonition, 'if you find something wrong with quantum mechanics, come back then and we can talk'; this must have been the Feynman who remarked that because physics had not solved all its problems it was safe against 'the encroachment of philosophers and other fools on the subject of knowledge'.[45] As for Commins, Clauser remembered that his benefactor had deprecated his experiment with Freedman as 'a pointless waste of time'. Was this the same Commins who advised his students to take up only meaningful problems?[46]

Clauser's neglect of radio astronomy while at Berkeley may have hurt his career and tainted his recollections. He thought that Thaddeus advised potential employers to avoid him if they caught a whiff of quantum foundations about him. 'Don't hire this guy if there's any chance that's what he's going to do, because it's all junk science'.[47] Shimony questioned a friend to whose university Clauser had applied for a position. Why hadn't Clauser got the job? 'Oh, my colleagues thought that the whole field was controversial'.[48]

Clauser ascribed the stigma associated with research on foundations to stubborn adherence to the 'religion' fostered by Bohr, von Neumann, and the Copenhagen school. 'The net impact of this stigma was that any physicist who openly criticized or even seriously questioned these foundations (or predictions) was immediately branded as a "quack". Quacks naturally found it difficult to find decent jobs within the profession'.[49] This cannot be the whole story. In his own specialty, experimental Bell tests, Clauser could count nine varieties, including two of his own, performed up to 1977.[50] Not, perhaps, a thriving area of research, but vigorous for a subject still in its infancy.

In the wider physics community Clauser may have been met with indifference or, more actively, dismissal as an adept of 'junk' or fringe science. Nevertheless, as an interloper in Wu's group at Columbia he immediately secured a colleague in Kasday and as a fresh PhD quickly found others in his collaboration with Shimony. At Berkeley he had the powerful support of Townes, the help of Commins, and the encouragement of Feyerabend, then at Berkeley, who offered seminars on the Freedman–Clauser experiment as it ran.[51] We do not doubt that most physicists, then pursuing their careers within the bounds of Kuhn's 'normal science', were content to assume that Bohr and his associates had settled quantum foundations once and for all. Few could afford to take time from the competitive pursuit of their specialty to question foundations, and many of those who believed that they followed Bohr nevertheless satisfied their intuitions by thinking like Einstein. But we cannot credit the persecution of heretics with the inquisitorial zeal that Clauser later recalled. Who had the interest and energy to do it? Copenhagen's Grand Inquisitor, Rosenfeld, had died in 1974.

Certainly, Clauser shouldered a big burden in forcing Bell tests on his colleagues. Established science does not encourage upstarts pushing unfashionable research that would consume the resources and might injure the reputation of a laboratory. That Bell's fundamental paper of 1964 was ignored for many years involved no quasi-religious opposition. Peter Higgs' now famous paper introducing his mass-giving boson, also published in 1964, suffered a similar fate although, since it appeared in the main-line American journal, *Physical Review Letters*, rather than in the short-lived fringe journal, *Physics, Physique, Fizika*, it had a better chance.[52] Higgs had wanted to deliver his epistle in Europe, but the European journal *Physics Letters* rejected it, because, a colleague told him, particle theorists did not see the point of his work. 'In retrospect [Higgs wrote], this is not surprising: in 1964 . . . quantum field theory was out of fashion'.[53]

Clauser worked as a postdoc at Berkeley and at the Lawrence Berkeley National Laboratory until 1975. He then became a research physicist at the Lawrence Livermore National Laboratory, set up to offer Los Alamos some competition in designing nuclear bombs. The largely apolitical Clauser did not identify with his employer. Valuing his freedom of speech, he refused to engage in classified research.[54] After reviewing the field with Shimony in 1977, Clauser stayed away from foundational questions until the demand for quantum reminiscences began to grow in the early 1990s.

Clauser rejoined the debate in later life. As an experimentalist he had necessarily worked in the Euclidean space of the laboratory, and the scientific
realist in him perceived a direct conflict: ' . . . lab-space formulated standard
quantum mechanics provides an "improper" description for matter waves,
where "propriety" instead requires the formulation in configuration space
only'. Despite his role in demonstrating the veracity of quantum mechanics,
his brain still 'kind of refused to do it'. The realist continued to resist the
embrace of complementarity and Bohr's symbolism. 'Whatever quantum
mechanics does indeed describe is sadly very difficult to visualize'.[55]

While the Photons Are Dancing

The Danish Order of the Elephant enrols only exalted persons and royalty. A coat of arms is required. Bohr chose as his charge the Chinese symbol of Yin and Yang, representing the union of opposites, placed under a scroll reading *contraria sunt complementa* (opposites are complementary). Pauli had pointed out that complementarity bore a close resemblance to the philosophy of Lao Tsu, who taught that 'true words seem contradictory'. Like the Copenhagen sage, the Old Master's teachings were 'nebulous and blurred . . . cavernous and dark', replete with truths hard to grasp but fundamental to wisdom. And Bohr might have said with Lao Tsu that 'Those who understand me are few', especially in the United States, where, as Pauli reported, ignorance and hypocrisy prevailed, especially among physicists. To their ignorance of the foundations of their science many of them added the hypocrisy of piously obeying the sixth commandment (against adultery) while vigorously violating the fifth (against killing) by working for the military.[1] As we know, opposition to capitalistic and military applications of physics often went hand in hand with advocacy for the investigation of foundations.

The founders had perceived connections between quantum foundations and Eastern philosophy and had indulged themselves in occasional attempts at spiritual profundity. In the 1970s the awakening interest in foundations sparked by Bell ran headlong into a burgeoning New Age spiritualism that nurtured the hippie counterculture of the 1960s. Like many religious or spiritual groupies, the New Agers rejected the rationality and reductionism of scientific practice, but the 'new physics' of quantum entanglement and non-locality suited them perfectly. In ever-creative Berkeley this collision gave rise to an ill-assorted company, which called itself the Fundamental Fysiks Group (FFG). Formed in 1975 at the Lawrence Berkeley Laboratory, the FFG's early recruits turned the joblessness they suffered owing to a dramatic decline in funding academic physics during the Vietnam war into the productive probing of foundational questions.

Perhaps the best-known work associated with the FFG was the best-seller, *The Tao of physics* (1975). Its author, the Austrian physicist Fritjof Capra, was a student in Paris during the riots in 1968, protested the war, and took refuge in the woody, hippie campus of the University of California at Santa Cruz. There he split his time between quantum physics, politics, Buddhist classics, psychedelics, and advanced lifestyles. A powerful trance-like experience on the beach in Santa Cruz prompted him to draw parallels between the physical energy driving change in the material world around him and the Dancing Shiva in Hinduism, reflecting creation and destruction. In the early 1970s, he commenced parallel projects to write a physics textbook and absorb the teachings of Eastern texts when, on advice from his former thesis adviser at Santa Cruz, he decided to weave these projects together. He also profited from the advice and support of Heisenberg. He reached the Berkeley Mecca in 1973 and returned in 1975, *Tao* in hand, in time to join the newly formed FFG.[2]

There he met a wondrous world of serious physicists like Clauser ('They would routinely drag me down there once a year'[3]) and Henry Stapp, a theorist at the Berkeley laboratory interested in Bell's work. The group included unemployed PhDs seeking to couple quantum physics with consciousness, like Nick Herbert, who delighted in Bell's inequalities and designed faster-than-light machines, and eccentrics like Jack Sarfatti, who had a PhD from the University of California's branch at Riverside and post-doc training in Munich. Sarfatti busied himself with parapsychology and the mind- and spoon-bending tricks of the 'super psychic' Uri Geller, whom he had watched in Bristol when Bohm was investigating Geller's psychokinetic abilities as a possible key to the implicate order.[4]

Geller consented to bend a real key by psychic power alone and, as an encore, to set off a Geiger counter by pure thought. Bohm's group accepted these phenomena at face value and so informed the world of science. Sarfatti observed that the Princeton school of quantum mechanics left 'ample room for the possibility of psychokinetic and telepathic effects'. The Princeton gurus Wheeler and Wigner took an interest in the FFG. So did Bohm, d'Espagnat, and Kuhn.[5] No doubt the attractions of the Esalen Institute on the California coast at Big Sur, where Herbert, Sarfatti, and other members of the FFG held month-long conferences on the foundations of physics and relaxed their minds and bodies in hot tubs, helped to lure fellow travellers.[6]

The success of *Tao* had followed that of Robert M. Pirsig's *Zen and the art of motorcycle maintenance* (1974). Commercial publishers took to feeding the new and substantial appetite for New Age literature and brought out an inferior in-house imitation of *Tao*, *The dancing Wu Li masters* (1979) by Gary Zukav, a non-physicist drawn to the FFG by Sarfatti's mix of physics and consciousness. After soaking in the atmosphere at Esalen, Zukav wrote a layman's account of what he called the 'essence of quantum mechanics'. Wu Li, 'patterns of organic energy', is a Chinese term for physics; 'master' means a teacher of essences; and 'dancing', or perhaps dancing around, signifies the master's pedagogy.[7]

With the help of Sarfatti and Herbert and a lift from Bohm, Zukav ran through physics from Planck to Bell with pointers to Eastern parallels to Western concepts. The distinction between virtual and real disappears, and the quantum turns up in the sutras. 'The appearance of physical reality, according to Mahayana Buddhism, is based upon the interdependence of all things'. The truly enlightened, physicists and Buddhists alike, while acknowledging the transiency of the parts have experienced the unity of the enduring whole. Like electrons, we go through life entangled. Like Bohm, Buddhists lived in a world of 'unbroken wholeness'.[8] Returning West, Zukav stopped in ancient Greece to forge the most acute analogy in his book. '[EPR's] thought experiment is the Pandora's box of modern physics'. Bell found the treasure hidden in its deepest recess. A very great treasure: Stapp rated it 'the most profound theorem of science'.[9]

Wu Li came out to warm applause for its intent and biting criticism for its physics. In the book's first reprinting, Zukav downplayed or excised the more exotic material and the book went on to become a best-seller. Still, it gave a rosier picture of foundational research than its practitioners had experienced. Bell: 'It gives the wrong impression of what is happening in physics institutes. People are not all desperately discussing Buddhism and Bell's theorem and the like. A precious few are doing that, whereas his book gives the impression that that's what we're doing all the time'.[10]

The FFG championed Bell. Three-quarters of the papers published about his theorem and inequalities in the US during the 1970s came from members of the group. Although their historian, the physicist-historian David Kaiser, overreaches in declaring that they saved Bell from oblivion and physics from itself, his account of the connections between anti-war protests, fringe culture, and fascination with fundamental quantum questions fits our drama comfortably.[11] Some of the grand masters had been captivated by Eastern philosophy. Schrödinger had studied Sanskrit and scrutinized Hindu scriptures, the Upanishads and Vedanta, in the years before he invented wave mechanics and Bohm submitted himself to a real guru, Jiddu Krishnamurti, with whom he began conversations in the early 1960s that spanned two decades.[12] Many of their exchanges were recorded and some of them published.

Krishnamurti was a creation of Theosophy, a late nineteenth century Western mash-up of European and Eastern philosophies. Eventually he rejected its tenets, along with all organized belief. He attracted Bohm with quasi-quantum mantras like 'the thinker is the thought, the observer the observed', and 'insight is the perception of the whole'. Bohm also liked Krishnamurti's anti-scientific programme of freeing the mind by emptying it. Consciousness disturbs contemplation of the universal unity, learning leads to pride and fosters the illusions of individuality, only properly conducted moderation produces wisdom. Beginning in 1976, Bohm made several trips to Krishnamurti's headquarters in rustic Ojai, California. His visit in 1978 gave him the opportunity to lecture to the FFG in Berkeley and, perhaps, to take his turn in the hot tubs of Esalen.[13]

Bohm's new obsessions had led him to neglect the hidden variable theory for which he was most widely known. His interest revived in 1978 when his Birkbeck graduate student Christopher Philippidis, who had learned of the theory not from Bohm himself but from reading the correspondence between Einstein and Born, developed a computer program to build what looked like a topological map of Bohm's quantum potential, replete with ridges and valleys.[14] Philippidis joined forces with another research student at Birkbeck, Christopher Dewdney, who also had learned about Bohm's theory in a roundabout way, from Frederik Belinfante's *A survey of hidden variable theories* (1973). They asked Dewdney's professor and Bohm's colleague Basil Hiley to reopen Bohm's old work. 'Why don't you and David Bohm talk about this stuff?' Hiley replied that it was all wrong. It then occurred to him that he had never read beyond the introduction to Bohm's infamous papers of 1952. He studied them the following weekend and they survived his scrutiny. 'It seems perfectly all right. Whether that's the way nature behaves is another matter'. He went back to the 'two Chris's': 'Okay, let's now work out what the trajectories are, work out what the quantum potential looks like in various situations'.[15]

Once again quantum mechanics turned to Young's arrangement. The computer-generated graphs programmed by the two Chris's showed the interference pattern of alternating bright and dark fringes resulting from constraints imposed on the particle trajectories by the quantum potential (Fig. 15).[16] A roller-coaster potential makes a nice icon of the scarier aspects of quantum mechanics. The dark places were dark because the quantum potential disfavoured trajectories leading to them. In this picture a single particle passes through one slit or the other, steered by the quantum potential towards regions of space that will show bright fringes. Blocking one of the slits changes the shape of the quantum potential and enforces a new set of trajectories that generates a diffraction pattern instead of the fringes typical of two-slit interference.

With this inspiration, Bohm resurrected his old theory, which he now called the *causal interpretation* of quantum mechanics and, from the mid-1980s, the *ontological interpretation*. With Hiley he sought to fuse this approach with notions of wholeness and the 'implicate order', a deeper and more fundamental layer of reality than standard physics postulated.

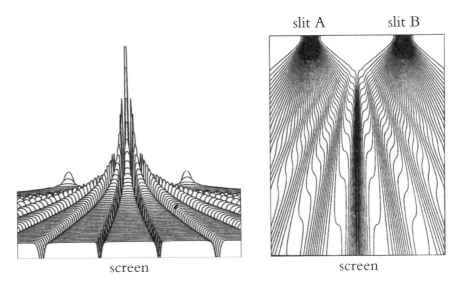

Fig. 15 Computer simulations of the quantum potential as viewed from the screen (left) show peaks, ridges, and valleys that guide the motions of the particulate electrons as though 'rolling' along on them. The electrons are more likely to be trapped in the valleys than not, leading to particle trajectories (right) that mimic two-slit interference.

In April 1976, the international circle of foundationalists returned to the attractions of Italy. Antonio Zichichi, whom Bell had met at CERN, had founded the Ettore Majorana Foundation and Centre for Scientific Culture, based in Erice, Sicily, in 1962. (The Foundation was named for a physicist whom his Italian colleagues ranked with Galileo and Newton; Majorana deprived the world of the full fruits of his genius by disappearing in 1938, at the age of 32.) Bell had co-signed the Centre's charter. Neither an academy nor a university, the Centre promotes 'science without secrets and without frontiers' (though in a strongly Catholic setting) where leading scientists and promising students come together for mutual instruction. Bell and d'Espagnat selected the attendees for the Erice 'thinkshop' for 1976. The subject was 'Experimental quantum mechanics'. Although far from the heady attractions of the quantum mysticism of California, Erice had its allure. It was founded by a son of Venus and Neptune, or maybe by Trojans fleeing the fall of Troy. It boasted ancient and Medieval architecture for sightseeing, former monasteries for accommodation, and nearby beaches instead of hot tubs.

Although advertised as a 'thinkshop', the Erice meeting was firmly about experiments, the genuine article, as Bell observed; 'we do a service to future generations by replacing gedanken experiments with real experiments'.[17] Many who had already done their service were there—Clauser, Horne, Shimony, Fry, Pipkin, and Ullman, among others—and two newcomers who would contribute as much and more—Alain Aspect (then 28), invited by his countryman d'Espagnat, and Austrian physicist Anton Zeilinger (30).

Aspect had completed a degree in physics at the Université d'Orsay in 1969. He became an assistant lecturer and completed a master's thesis at the university's Institut d'Optique in 1971. He recalls his undergraduate courses in classical physics as 'fantastic', those on quantum mechanics (delivered by an excellent spectroscopist who, alas, did not appear to understand the subject) as 'absolutely terrible'. He suffered from quantum malaise much as Clauser had done and for the same reasons. He could not discern a 'physical world behind the mathematics'. 'It did not mean anything to me'.[18] Partly in frustration, he decided to go abroad. His wife's family had historical associations with Africa and with the help of the international development organization Voluntary Service Overseas, he joined a college of higher

education in Yaoundé, Cameroon, as an employee of the French government and in lieu of mandatory military service. He taught Cameroon's
future teachers of science. His wife Annie taught them chemistry.

He found time to attack the quantum again. A new textbook, *Mecanique
quantique* (1973), by Claude Cohen-Tannoudji, Bernard Diu, and Franck
Laloë came to hand. It privileges problems and solutions, anticipates and
corrects misunderstandings, and disposes of foundational questions by references to the literature. It gave Aspect access to the connection he had
missed between the mathematics of the quantum and the behaviour of the
physical world. He now re-taught himself quantum mechanics, working
through the enlightening text, all 889 pages of it. Anglophones able to stay
the course through the three volumes of the latest English edition (2020) can
try to repeat Aspect's feat. The burden may be lightened if the ambitious
Anglophone does as Aspect did and pays little attention to the extensive
bibliography on foundational questions.

Aspect returned to France in 1974 to teach physics at the Ecole Nor-
male Supérieure at Cachan, south of Paris, and to pursue a doctorate in
any relevant field that captured his interest. He visited several laboratories in France seeking inspiration. Back at the Institut d'Optique in Orsay
he learned about experiments by Christian Imbert that demonstrated an
extraordinary feature of light. *Individual* photons appeared to show interference phenomena. Aspect asked Imbert for 'a [research] subject equivalent to
that'. It happened that Shimony had recently visited Orsay at the invitation
of d'Espagnat and had given lectures on Bell's inequality. Imbert told Aspect
about the Bell tests conducted at Berkeley and at Harvard and handed him
a file. The first document in the file was Bell's breakthrough paper of 1964.

Now well versed in the quantum formalism, sympathetic to Einstein's
classical conception of physical reality, and perplexed by complementarity
('Bohr is impossible to understand'), Aspect realized that Bell had pin-
pointed the core of his conflict. Are there 'quantum mechanical predictions
[that] cannot be mimicked by any local realistic model in the spirit of Einstein's ideas?' Aspect would be equally content with an answer yes or no, but
not with no answer at all. Bell also had worried over the conflict between
Bohr and Einstein and, what was more, knew how to resolve it. 'John Bell
was showing [the way out of] the conflict! This is crazy! This is the most
interesting programme in the world!' 'I decided that my doctoral thesis
would deal with experimental tests of Bell's inequalities and Imbert accepted
to take me under his wing'.[19] Aspect would take on the hardest form of the

test. He would attempt what Bell had called for ten years earlier. He would perform 'experiments of the type proposed by Bohm and Aharonov, in which the settings are changed during the flight of the particles'.[20]

Bell tests involve two kinds of locality condition. The first assumes the reasonable criterion that there can be no connection (no 'mutual rapport') between the distant polarization analysers and detectors, or between them and the source of photons. The second, more stringent requirement, prohibits interactions occurring at speeds faster than light. This was the sort of locality that Aspect wanted to test by fast switching of the analysers. With a change in terminology, the test could be made more stringent still. 'D'Espagnat suggested to use the word "separable" to distinguish the relativistic locality from "reasonable" locality'.[21] If 'separable', systems cannot interact at faster-than-light speeds under any circumstances, whereas 'reasonable' locality only prohibits sending *useful information* at speeds faster than light.

On Imbert's recommendation, Aspect visited Bell at CERN in 1975 to discuss the proposed experiments. Bell listened in silence. Then he asked one of his penetrating questions. 'Are you tenured?' Aspect explained that he was still only a graduate student. He had yet to secure his doctorate, although he held a permanent teaching position. 'You must be a very courageous graduate student . . . ' But Bell was friendly and encouraging, and helpfully explained what he meant by 'crackpot'. It is the way the physics community is likely to regard a junior member of the profession who pursued such willow-the-wisps as you do.[22] But Aspect's institutional connections were much stronger than Clauser's had been and influential colleagues agreed that the experiment he proposed was necessary.[23]

In June 1975, on advice from Bell, Aspect wrote up an experiment 'to test separable hidden variable theories'. A more detailed version, written a few months later, bore a new title that expressed a decisive change in perspective: 'Proposed experiment to test the nonseparability of quantum mechanics'.[24] Aspect had shifted from testing hidden variables and awarding the palm to Einstein or Bohr. He now proposed to use Bell's theorem not necessarily to advance the *interpretation* of quantum mechanics but to explore an unsuspected *consequence* of quantum physics: entanglement. He had reconceived his project as a response to the core challenge of EPR, the challenge Einstein first threw down at Solvay V. If ψ describes each individual entangled pair, and if it is taken to be a real function capable of exerting real physical effects, then the particles in the pair are not separable and quantum mechanics cannot be completed in a way that would have satisfied Einstein.

The Erice thinkshop of 1976 offered a stage for raising entanglement and non-locality to the main action, or, in Clauser's metaphor, for the quantum sub-culture's coming 'out of the closet'.[25] Aspect presented his experimental arrangement (Fig. 16). He met Clauser and Fry ('Pipkin was hiding in his corner'), and Valentin Telegdi, invited to advocate for the devil and argue for quantum orthodoxy. Telegdi told Aspect that he would never have suggested the proposed change-in-flight experiments but, impressed with Aspect's preparations and the passion of his arguments, he would have been willing to take the risk. '[I]f you would come to me with your enthusiasm and clarity, I would certainly give you space in my lab'.[26] Aspect was very persuasive. A co-author of the magical textbook that had taught him quantum mechanics, Laloë, heard him at Erice and rated his proposals 'quite interesting'. Aspect's wife took this encouragement as welcome reassurance that her husband was not crazy.[27]

For most of the attendees, Clauser's repeat of Holt–Pipkin and the results of the Fry–Thompson experiments put the matter beyond reasonable doubt, as Laloë observed: 'no-one could anymore still think seriously that

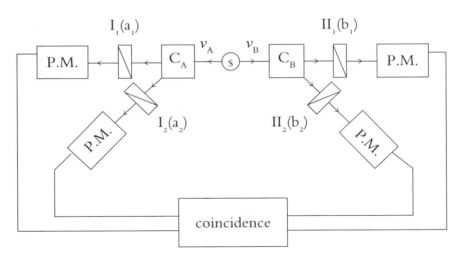

Fig. 16 Aspect's original experiment, proposed at Erice in 1976. The source s of cascade emission produces two correlated photons, labelled ν_A and ν_B. The 'commutators' C_A and C_B randomly switch each photon to one of two differently oriented polarization analysers I_1 (angle a_1) or I_2 (a_2) on the left, and II_1 (b_1) or II_2 (b_2) on the right. The photons pass through the analysers to the photomultipliers (PM) whose output runs to a coincidence counter.

the Bell inequalities were obeyed by physics'.[28] Perhaps. But the drama had not ended for subtle thinkers. As Bell observed, they could maintain that 'the experiments so far described have nothing whatever to do with Einstein locality ... It is therefore of the very highest interest that an atomic cascade experiment is now under way, presented here by Aspect, in which the polarization analysers are in effect *re-set while the photons are in flight*'.[29]

Imbert set aside laboratory space in the basement of the optical institute and Aspect began to beg, borrow, and build the necessary apparatus. He had to develop expertise in atomic beams, optics, and (the lesson from Fry and Thompson) lasers to prepare the atoms in the state required for cascade emission. Fry: 'Alain, if you can use a laser, use a laser'.[30] He fixed on the calcium atom source that Freedman and Clauser had used. He borrowed a krypton–ion laser from Imbert and photon coincidence-counting equipment from the workshop at the Commissariat à l'Energie Atomique (CEA) branch in Saclay. On Clauser's suggestion he wrote to Berkeley to borrow the custom-made optical filters that Clauser had used to isolate the blue and green photons, left and right. He yielded to persuasion to apply for funding, which was granted, to purchase a commercial tuneable (variable wavelength) dye laser. Slowly, as the months extended to years, the apparatus came together with the help of Gérard Roger, a gifted laboratory technician loaned by Imbert,[31] and André Villing, an outstanding young electronics engineer.[32]

Aspect's respectability was greatly enhanced in 1979 when Cohen-Tannoudji, the main author of the transformational textbook and France's leading experimental physicist, paid him a visit. Laloë had introduced them and Cohen-Tannoudji had expressed interest in the photon correlation apparatus. Aspect was at liberty. His laser had broken, and he could do little until it was fixed. He and Cohen-Tannoudji agreed to work together on experiments unrelated to Bell's inequality. The experiments succeeded and Aspect wrote up his first scientific research paper under the guidance of a master.[33]

The great day came for the first experimental runs. Over lunch Aspect told Annie that he had checked absolutely everything, 'I don't know what else I can do'. That afternoon the new source delivered its first pairs of entangled photons. It enabled Aspect to develop three tests. The first, published in August 1981, was a variation of the Freedman–Clauser experiments.

Aspect prepared the excited state of atomic calcium through simultaneous absorption of two photons, one from the krypton–ion laser and the other from the dye laser, passed the entangled photons from the cascade through 'pile of plates' polarizers, and violated Bell's inequality by more than 13 standard deviations.[34] Placing the polarizers 6.5 metres from the source of the cascade emission (and so 13 metres apart from each other) produced no change in the results. Aspect's apparatus could count coincidences within a 19-nanosecond window, 19 billionths of a second, half the time it took the light to travel 13 metres across the laboratory. Aspect thus demonstrated that entanglement persists over distances longer than the so-called 'coherence length' of the cascade photons, the distance over which the photons could be expected to remain in phase, about 1.5 metres.[35]

Aspect included a graduate student, Philippe Grangier, along with Roger as co-authors of this first Bell-test paper. Grangier would upgrade the experiment to include two-channel polarization analysers. Whereas the 'pile of plates' analysers passed only one type of polarization, rejecting (and wasting) the other type, polarizing cubes formed from two prisms stuck together would pass both types by sending them in different directions. The cubes act on photons much like Stern–Gerlach magnets act on spinning atoms or electrons, but they were not easy to construct. Aspect appealed to a friend working in the Philips laboratory in the Netherlands, which built them for free. '[T]hey were custom-made, but I was a customer who could not pay'.[36] The group now expanded the electronics to study four-fold coincidences: vertical and horizontal polarization on the left and on the right. A run that had taken several hundred hours for Freedman and Clauser could now be completed in just 100 seconds.

Aspect's group tested a variation of the CHSH inequality, which Bell's theorem demands should be less than or equal to 2 but for which, for the magic difference angles θ of $22\frac{1}{2}°$ and $67\frac{1}{2}°$, quantum mechanics predicts $2\sqrt{2} = 2.828$. The group found $2.697 \pm 0.015 \leq 2$, a violation by almost 50 standard deviations. The results, published in July 1982, encouraged another student, Jean Dalibard, then working under Cohen-Tannoudji, to ask to join the group when Aspect was setting up the apparatus for his ultimate test.[37] Why should such a student not prefer a much more prestigious laboratory for his mandatory military period? 'At a time when people were still smiling at me, Dalibard said that he wanted to turn the knobs of an experiment that will be in the books'.[38]

In 1976 Aspect had planned to use acoustic-optical switches as the 'commutators' of Fig. 15, diverting the paths of the flying entangled photons. Passing a standing ultrasonic wave through a crystal will change its diffraction properties and turning the sound wave on and off alternates between different diffraction paths. In Aspect's experiment, the different paths would lead to differently orientated analysers, defeating any conspiracy requiring foreknowledge of their settings and frustrating any tendency to 'mutual rapport' between them. He placed an order for a set of commercial switches, but the supplier could not make them. Aspect ransacked the literature and found that he could build his own by replacing the crystal with a small volume of water. The water cure worked. Even then Aspect did not presume the outcome of his experiment. '[F]or me there was a real possibility that when we would switch the polarizers something [other than quantum mechanics] would happen'.[39] It did not.

With the photons switched between analysers while in flight and the analysers 13 metres apart, the experiment yielded a violation of Bell's inequality by five standard deviations. The use of additional optical devices had inevitably increased inefficiencies and reduced the accuracy and precision of the results, but they were not ambiguous. They spilled from the apparatus at two in the morning. Aspect called Annie. She arrived at the laboratory with bread, duck paté, and champagne. It remained to write up the results. They were published in December 1982.[40]

Aspect's third paper confirmed the recognition that had dawned slowly on the international circle of foundationalists over the previous decade: quantum entanglement and non-locality are real phenomena. What could be the use of so grand a discovery? The FFG's Herbert had proposed schemes to utilize the faster-than-light interactions he supposed sanctioned by Bell's theorem for practical communication. His schemes had fatal flaws, which, when identified, led William Wootters and Wojciech Zurek (and, independently, Dennis Dieks) in 1982 to declare an important rule, which in Wheeler's colourful language became the 'no cloning' principle. It states that the information encoded in the state of a quantum object cannot be copied to another object without destroying the original.[41] Aspect's first experimental paper appears in their list of authorities.

Although entanglement did not support faster-than-light signals, it might have some use in ordinary communication. At an international conference on computers, systems, and signal processing held in Bangalore, India, in December 1984, Charles Bennett (IBM) and Giles Brassard (University of

'[F]or me there was a real possibility that when we would switch the polarizers something [other than quantum mechanics] would happen'. Above: Aspect explains his experimental test of Bell's inequality. Below: With the photons switched between analysers while in flight and the analysers 13 metres apart, the experiment yielded a violation of Bell's inequality by 5 standard deviations.

Montreal) suggested entanglement as the basis for a new system of *quantum cryptography*. The 'no cloning' principle would be essential to its security. Among their authorities was Aspect's second experimental paper.[42]

Feynman had preceded them. In a keynote speech in 1981 he referred to two-photon correlation experiments, deduced his own version of Bell's theorem and inequality, and speculated about building a *quantum computer*.[43] After a little more thought he concluded that 'the laws of physics present no barrier to reducing the size of computers until bits are the size of atoms, and quantum behavior holds dominant sway'.[44] Binary information coded in the physical properties of quantum objects (such as electron spin or photon polarization) would become known as *qubits*.

Despite his rigid insistence on double-checking every aspect of his experiments before he ran them, Aspect seems to have associated himself with Feynman's breezy way with foundations. He concluded a seminar he gave in 1984 at Caltech, Feynman's base, with a quote from Feynman himself, revealed one line at a time: 'It has not yet become obvious to me that there's no real problem. I cannot define the real problem, therefore I suspect there's no real problem, but I'm [not] sure there's no real problem. So that's why I like to investigate things'.[45] Nobody laughed until Feynman did. Feynman later acknowledged to Aspect that the 'real problem' did indeed refer to the subject of Aspect's seminar, and 'was quite in context'.[46] Apparently, Aspect did not approach foundations with the visceral philosophical commitments of Bohr or Einstein.

The 'Aspect experiments' became iconic, penetrating a wider public consciousness with a speed more remarkable for being driven by interest in the science of quantum 'weirdness' rather than New Age quantum mysticism. One of John Gribbin's duties as a consultant to the popular British weekly magazine *New Scientist* was to scan the latest physics periodicals for stories that might interest the magazine's readers. Encountering Aspect's second experimental paper revived his interest in quantum foundations (he had trained as a physicist) and he wrote a short unattributed piece for the magazine's issue of 19 August 1982. He was further inspired to write an accessible introductory text on the subject. *In search of Schrödinger's cat* was published two years later after eight or nine publishers turned it down as 'too scientific and not hippie enough'. Undaunted by its scientific tone, its publisher, Bantam Press, still placed it in its New Age list.[47]

Herbert's *Quantum reality* (1985) followed. Bell admired it. 'Repeatedly, I came upon striking and novel foundations of important points'. Among them are explanations of the connection between hidden variables, the indistinguishability of electrons, and the claimed omniscience of the wavefunction; the characterization of a measurement as a miracle; the obligation of realists to admit superluminal interactions; a clear accessible presentation of Bell's inequalities; and a machine for reviving dead Schrödinger cats.[48] Gribbin too thought something of Herbert's *Quantum reality*. 'It seems to me (trying to be objective) that my book and Nick Herbert's book . . . were the two that started the ball rolling, reaching a "market" that already existed but had not been catered to'. But their books did not signal the end of New Age quantum 'woo'. Deepak Chopra's *Quantum healing* was published in 1989.

On 11 May 1984, BBC Radio 3 broadcast *The ghost in the atom*, a documentary inspired by Aspect's experiments and based on discussions with Aspect, Bell, Wheeler, Peierls, Oxford physicist David Deutsch, John Taylor at King's College London, Bohm, and Hiley. The British physicist and science popularizer Paul Davies led the discussion. A book containing enlarged versions of the interview transcripts appeared in 1986. Among other informed physicists who helped to give prominence to Aspect's experiments was Mermin, whose 'Is the Moon there when nobody looks? Reality and the quantum theory', gained a wide readership when published in *Physics Today*. The story was then retold in Alastair Rae's *Quantum physics: Illusion or reality?* (1986) and Euan Squires' *The mystery of the quantum world* (1986). A co-author of our drama (JB) was relatively late to the party with *The meaning of quantum theory* (1992). The flow persists as each generation discovers 'quantum weirdness' and patronizes writers and publishers who have found the market indefinitely elastic.

Bell regarded Aspect's experiment as very important, even decisive. 'Perhaps it marks the point where one should stop and think for a time', Bell told the BBC, 'but I certainly hope it is not the end. I think that the probing of what quantum mechanics means must continue, and in fact it will continue, whether we agree or not that it is worth while, because many people are sufficiently fascinated and perturbed by this that it will go on'.[49] It was indeed not the end, but rather the beginning of a boom in foundational research that would draw the attentions of many more physicists and philosophers.

Aspect was not among them. He had no desire to build a career as a professional Bell ringer. In a parting shot, he showed that interference effects in strongly attenuated light could *not* be described as interference by individual photons, contradicting the conclusions of the research programme that had first drawn him to Imbert more than ten years earlier. Aspect and Grangier used the calcium cascade emission system to produce genuine single photon pulses (this was Grangier's PhD thesis project). These states do not (indeed cannot) split apart and 'interfere with themselves' as they pass through an apparatus. As Einstein had deduced for electrons at Solvay V in 1927, the description of individual events has to be particulate: only single particles are detected. Interference results from the observation of many such events or from arrangements involving large numbers of particles not individually identified. Aspect and Grangier echoed Bohr: 'Of course, the two complementary descriptions correspond to mutually exclusive experimental set-ups'.[50]

In 1985, Aspect joined Cohen-Tannoudji at the Collège de France and embarked on a research programme to develop new techniques using lasers to cool and trap atoms, work that would lead eventually to the Nobel physics prize of 1997, which Cohen-Tannoudji shared with Steven Chu and William Phillips.

Adventures in Quantum Information

The elaborate experiments involving entangled pairs of particles conducted in the 1970s and 1980s concerned the quantum world of photons and charged particles. Would chargeless massive particles reveal more oddities? Interferometry provided the answer. An interferometer splits a beam of light, sends the resultant parts along two different paths, and subsequently reunites them. The lengths of these paths determine the manner of their meeting, best understood in terms of waves. If the waves meet in phase—peak to peak or trough to trough—the light persists, equivalent to a bright fringe. But if the waves meet out of phase—peak to trough—the light fails, equivalent to a dark fringe. Distances of billionths of a metre can make the difference between light and darkness. So sensitive is optical interferometry that its failure to detect motion through the suppositious luminiferous ether helped to persuade many to accept relativity theory. In 2015 interferometry of extraordinary sensitivity confirmed a prediction of general relativity by detecting the ripples in space-time we call gravitational waves.

Participants in the Erice thinkfest learned from Zeilinger about the behaviour of a beam of neutrons split into two secondary beams by reflection from a single crystal and subsequently recombined. Did the recombined beam show interference effects such as light would show if run through an interferometer? It seemed so. Zeilinger reported that interference accumulated by passing one neutron at a time through the apparatus. Just as Aspect and Grangier would later demonstrate for photons, a single neutron does not 'split' in the interferometer; interference arises only when many neutrons have passed through the apparatus. But the experimenter cannot say which path each neutron follows in its journey.[1]

That set up another conundrum. Like electrons, neutrons have spins that can orient either up or down in a magnetic field. Quantum mechanics predicts that rotating the spin of such a particle through 360° changes the *sign* of its wavefunction, from + to − or *vice versa*. To bring the wavefunction back to its original sign requires a second rotation through 360°. This seems very odd but can be visualized if we assume the equivalent of 'rotating' the neutron is to move it along the surface of a Möbius strip. Zeilinger and his colleagues applied a magnetic field to achieve this change in the wavefunction in one of the secondary beams. On recombination, the partial wavefunctions no longer had the same sign, + met −, and the two cancelled out: a dark fringe. To appreciate the craziness of the situation, recall that Zeilinger had operated on the wavefunction of one of the secondary beams: of the whereabouts of each neutron before the recombination of the partial beams he had no inkling.

The co-author of this magic, Anton Zeilinger, had received his PhD from the University of Vienna in 1971 under Helmut Rauch without feeling the quantum-mechanical malaise that Bell, Clauser, and Aspect had suffered. Perhaps he owed this relative robustness to the Austrian mix of science and positivist philosophy. The teachings of Mach and the Vienna Circle did not encourage a search for an instinctively appealing, visualizable physics. Against this background, Bohr's arguments appeared sound. Zeilinger reckoned them 'the most interesting in the field of interpretation of quantum mechanics'.[2]

With his degree in hand, Zeilinger joined Rauch's research group to work on the experiment on neutron spin rotation, whose results they published in 1975.[3] The Erice thinkshop was then being planned. Zeilinger informed his boss: 'Do you see there is a conference which is interesting for us? Maybe you should go'. Rauch was too busy: there were many pressing experiments still to be done with neutrons.[4] Zeilinger went instead, contributed as reported, understood little, and came away inspired. The meeting proved 'immediately crucial' as it taught him about entanglement and put him in touch with Horne and Telegdi.[5]

After completing his PhD with Shimony, Horne had joined the faculty at Stonehill, a new Catholic teaching college near Boston. Neither the teaching load nor the conviction that 'Bell's theorem was finished' had blunted his interest in research and he was intrigued by Zeilinger's single-particle interference experiments. At Erice they 'found out that [they] very much liked each other's ideas about what's interesting'.[6] Working with Shimony, Horne had anticipated the experiments involving rotating the spin of the neutrons, unaware that Rauch's group was already performing them.[7]

At Erice, Zeilinger told Telegdi that he wanted to study in America but did not much care where. Telegdi: 'You are talking that way because you are a European. But if an American asks you, you have to give him the best possible institution. So let's start again. Where do you want to go?' Zeilinger wanted to go to MIT to work with '*the* pioneer in neutron diffraction', Clifford Shull. Telegdi made the necessary contact.[8] Horne sealed the deal. For a year he had visited Shull's group on his days off from Stonehill 'and weekends and summers and Christmas'. One day Shull waved a letter at him. 'Do you know someone named Anton Zeilinger?' Indeed, he did. The old-boy network worked. Zeilinger joined Schull's group as a Fulbright Fellow in 1977.[9]

Horne and Zeilinger played with a neutron interferometer that Shull had recently built. 'The idea was to find interesting things to do with this new interferometer ... Disturb it in some way, you know, rotate it, turn it on its side, subject it to various fields—things that would definitely cause a phase shift'. They did not expect surprise departures from the predictions of quantum mechanics and were rewarded by finding none. When Shull and his laboratory retired in 1985, Zeilinger split his time between Cambridge, Massachusetts, and Vienna, where he rose to a professorship at the Technical University. Nine years later the Nobel prize rewarded Shull for his work in neutron diffraction.

'Have you ever been to Finland?' Zeilinger had spotted another interesting conference, on '50 years of the Einstein-Podolsky-Rosen gedanken experiment', to be held in June 1985 in Joensuu, Finland. He wanted to go but he and Horne did not have anything suitable to report. They had been working on single-particle interference. EPR experiments required two particles. They needed 'to find a connection between interferometry and entanglement'. They did: a novel proposal for a Bell-type experiment that did not rely on spin states. Differences in polarizer orientations would be replaced by differences in the *phases* of particles following different paths in a two-particle interferometer. This 'mode entanglement', a novelty in 1985, is now routinely achieved in the laboratory.

Horne and Zeilinger presented their ideas again at a conference on 'New Techniques and Ideas in Quantum Measurement Theory' held at the New York Academy of Sciences in January 1986. The conference hosted a broad church spanning the generations (it was dedicated to Wigner) and reflected the now burgeoning interest in quantum foundations. Many physicists already active in our drama were there (Aharonov,

Aspect, Bell, Clauser, Horne, Mermin, Selleri, Shimony, Stapp, Vigier, Wheeler, Wigner, Zeilinger, Zurek) and some newcomers who have parts yet to play (James Franson, Tony Leggett, Michael Redhead, and Marlan Scully).

In June 1978, at a conference on neutron interferometry at the Institut Laue-Langevin in Grenoble, Horne, Shull, and Zeilinger met Daniel Greenberger, formerly an army cryptologist and a Berkeley high-energy particle theorist, then cultivating general relativity as a professor at the City College of New York (CCNY). Greenberger was soon working with Horne and Zeilinger on extending entanglement theory to systems of more than two particles. During a sabbatical in Vienna in 1986, he deduced the inequalities that could be derived for a four-particle system and refined them until they overpowered him. He derived the astonishing result that, for a certain type of experiment, quantum mechanics forbids *all* possible outcomes predicted using local hidden variables. Greenberger and his friends were nonplussed: '[it] had to hit us over the head. Like everyone else we found our result totally unexpected'.[10]

The trick can be done with three photons, *A*, *B*, and *C*. Let the linear polarization states of these photons be entangled in a superposition such that they are either all vertically polarized or all horizontally polarized relative to the laboratory z-direction. The entangled photons set off in different directions and are passed through separate polarization analysers. Our first experiment (1A), measures the *linear* polarization (vertical/horizontal, V/H) of photon *A* with the analyser set at 45° to the z-direction, and the *circular* polarization (left/right circular, L/R) of photons *B* and *C*. There are eight possible outcomes: VLL, VLR, VRL, VRR, HLL, HLR, HRL, and HRR, each triplet designating the outcomes for photons *A*, *B*, and *C*, respectively.

The quantum formalism demands a transformation of the description of the entangled photons just after their production to a description representative of their possible states after they have passed through the analysers. The little quantum algebra required kills half of these possibilities by reducing their probabilities to zero. The four spared, VLL, HLR, HRL, and VRR, indicate a striking pattern. A measurement that places photon *A* in a V-state implies that photons B and C must occupy *identical* circular polarization states, either both L or both R (VLL, VRR), whereas detection of photon A in an H-state implies that its companions must be found in *opposing* circular polarization states, either L/R or R/L (HLR, HRL).

Similar measurements can be made on the linear polarization state of photon B (or C) and the circular polarization states of A and C (or A and B)—experiments 1B and 1C. Algebra again cuts down the eight possible permutations to four in each case and the same pattern recurs. For experiment 1B the possibilities are LVL, LHR, RHL, and RVR, and for 1C, LLV, LRH, RLH, and RRV: Vs go with LL or RR, Hs with LR or RL.

With a little ingenuity we can reproduce this pattern with a system of local hidden variables. We need only suppose that each photon possesses two factory settings, one pre-determining the outcome of the V/H measurement and the other that of the L/R measurement. Let these variables have the values $+1$ for V and L, -1 for H and R. A period of quiet reflection reveals that we need eight combinations of these hidden variables to reproduce the 12 outcomes predicted by quantum mechanics in experiments 1A, 1B, and 1C. These are summarized in Fig. 17. We draw a veil over the awkward question why these eight combinations and no others occur.

This impressive bit of bookkeeping should be interpreted as follows. The left-hand column under each photon gives values of the local hidden variable corresponding to V/H measurements, the right-hand column values corresponding to L/R measurements. For example, if we perform the V/H procedure on photon A and find V ($+1$), the entries in the right-hand columns for photons B and C predict that both will yield either $+1$ (both L) or -1 (both R), corresponding to VLL and VRR (each occurring twice). Alternatively, if we find that photon A is H (-1) the entries in the right-hand columns for photons B and C predict $+1, -1$ (L/R) or $-1, +1$ (R/L), corresponding to HLR and HRL (again, each occurring twice). Despite the double counting, all the outcomes agree with the quantum-mechanical predictions.

Let us now perform V/H measurements (at $45°$) *on all three photons*. Call this experiment 2. We can see immediately from the factory settings in Fig. 17 that if one of the photons is detected in a V-state, the other two must have identical linear polarizations, either V/V or H/H. If one of the photons shows up in an H-state, the other two must be either V/H or H/V.

Photon A		Photon B		Photon C		Expt. 1			Expt. 2
V/H	L/R	V/H	L/R	V/H	L/R	1A	1B	1C	
+1	+1	+1	+1	+1	+1	VLL	LVL	LLV	VVV
+1	−1	+1	−1	+1	−1	VRR	RVR	RRV	
+1	−1	−1	+1	−1	+1	VLL	RHL	RLH	VHH
+1	+1	−1	−1	−1	−1	VRR	LHR	LRH	
−1	+1	+1	−1	−1	+1	HRL	LVL	LRH	HVH
−1	−1	+1	+1	−1	−1	HLR	RVR	RLH	
−1	+1	−1	+1	+1	−1	HLR	LHR	LLV	HHV
−1	−1	−1	−1	+1	+1	HRL	RHL	RRV	

Fig. 17 Eight combinations of the local hidden variables for the three entangled photons (left-hand table) are required to reproduce the 12 outcomes (VLL, etc.) predicted by quantum mechanics for experiments 1A, 1B, and 1C (centre table). If all three photons are subjected to V/H measurements (experiment 2), this hidden-variable scheme predicts the four equally probable outcomes given in the right-hand table.

This system of hidden variables predicts that the only possible outcomes in experiment 2 are VVV, VHH, HVH, and HHV. They can be obtained immediately by reading across the rows of V/H entries under A, B, C in fertile Fig. 17. These results are all equally probable.

We know that quantum mechanics does not respect factory settings. Astonishingly, the quantum algebra predicts that in experiment 2 we can expect the outcomes VVH, VHV, HVV, and HHH, all equally probable, and *all different* from the outcomes predicted by our system of local hidden variables. We could of course devise another set of factory settings that would reproduce the quantum-mechanical predictions for experiment 2, but then we could not recover the predictions for experiment 1. With three entangled photons the predictions of local hidden variables and quantum mechanics are mutually exclusive. A truly extraordinary result. We should note that a different initial superposition would yield a different set of predictions but would not change the fundamental fact of mutual exclusivity.

After Greenberger had discussed his work at conferences but before he published it, he received a preprint in which Michael Redhead, then professor of history and philosophy of science at the University of Cambridge, claimed to provide a more rigorous derivation. That prompted a short paper by GHZ (Greenberger, Horne, and Zeilinger).[11] A fuller version required the help of Shimony and took two weeks of drafting, 'fighting over every detail (the way good friends fight)'.[12] Mermin suggested some improvements before the paper, 'Bell's theorem without inequalities', saw daylight in July 1990.[13] Zeilinger's group managed to produce three-particle 'GHZ states' in 1998,[14] and reported the results of experiments equivalent to our 1 and 2 in 2000. The results agreed perfectly with quantum mechanics (Fig. 18).[15]

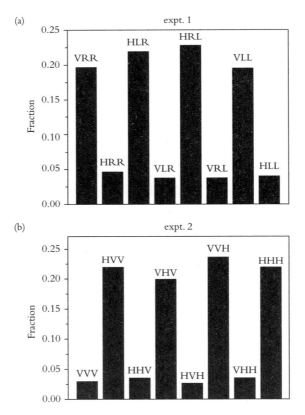

Fig. 18 Experiments on three-photon 'GHZ states' performed by Zeilinger and his colleagues demonstrated a clear preference for all 12 of the measurement outcomes predicted by quantum mechanics as detailed in experiments 1A, 1B, and 1C. The results for experiment 1A are shown in (a). Experiment 2 demonstrated a clear preference for quantum mechanics compared with the predictions of local hidden variable theories, (b).

Although the case against locality would seem unanswerable, theorists who would have made good lawyers discerned that defenders of Einstein still had escape routes. Aspect, Dalibard, and Roger had sought to close the 'locality loophole' by switching between different polarizer orientations while the photons flew. Their experiment succeeded FAPP but was open to the objection that the devices used did not provide completely random switching. Zeilinger's group came to the rescue in 1998 with switching driven by a random number generator.[16]

Another possible escape, the 'efficiency loophole', arises because the number of particle pairs detected falls far short of the number generated. Do the detected pairs represent a true statistical sample (a 'fair sample') of the total? Suppose the sub-ensemble of detected pairs violates Bell's inequality, but the total ensemble does not. In a more devilish conspiracy, the hidden variables themselves might determine the efficiency of the detectors.[17] Physicists closed this loophole in 2001 with experiments on entangled pairs of large, positively charged ions. They detected all the entangled ion pairs and obtained a violation of Bell's inequality by eight standard deviations.[18]

The final refuge of refuseniks, the 'freedom-of-choice loophole', supposed an influence of the instrument settings on the hidden variables. Zeilinger obtained the resources to plug this hole after the Austrian Academy of Sciences established a new Institute for Quantum Optics and Quantum Information (IQOQI) in Vienna under him as director. In 2010 a team from IQOQI and the University of Vienna reported results of closure experiments they conducted in the Canary Islands between La Palma and Tenerife, a distance of 144 km. They placed the random chooser of polarizer settings 1.2 km from the source of entangled photons. An electronic delay on Tenerife ensured that the choice of setting took place before a signal could be received from the source. Their results violated Bell's inequality by more than 18 standard deviations.[19]

That would seem dispositive. But in 2017, another team from IQOQI and institutions in China, Germany, and the US rose to the heavens to obtain light from Milky Way stars to trigger the selection of the instrument settings, which increased the interval between emission of the trigger photons and the selection of settings to at least 600 years.[20] That should have ruled out conspiracies between detectors, experimenters, and plausible third parties. Still, following the imperative that all practicable experiments that find funding should be tried, the group devised a cosmic circuit stunt to exploit

In 2003, Zeilinger (pictured) was appointed director of the Institute for Quantum Optics and Quantum Information (IQOQI) in Vienna.

light from distant quasars. That pushed the time delay to about 7.9 billion years, long before the creation of physicists.[21]

Industry did not wait for purists to close the loopholes before entangling foundational research with commercial development. During the 1980s the Hitachi laboratories in Japan organized two conferences on possible applications. The second, in 1986, brought 178 participants together from 15 countries to hear 44 papers dealing mainly with the latest technologies for testing foundational questions. Anatomists of quantum mechanics were rushing hand in hand with inventors of microdevices towards the region where irremovable quantum fluctuations would either arrest their progress or open new prospects. The industrialist hosts hoped that as new technologies helped to convert gedanken experiments into real ones, so inquiry into foundations might uncover ways to convert entangled quanta into hard cash. Already speculative physicists thought to metamorphose a CAT (like Schrödinger's) into a SQUID (a superconducting quantum interference device), which can support two superposed macroscopic states. The perceptive speculator might see in the superposed states, consisting of two oppositely directed but separate super currents, an analogy to the suspended animation of the cat.[22] We will return to this possibility momentarily.

If the contribution of a pair of particle physicists to the Tokyo proceedings represented the consensus, much, or almost everything, remained to be done on the foundation side of the collaboration. '[EPR] is today just as far as it was 50 years ago from having a solution consistent with all aspects of quantum mechanics'. Perhaps it is not the sort of argument that has a solution.[23] Tony Leggett, who will be a principal actor in our final scene, found the foundationalists wanting just where they were most needed. They had not solved the measurement problem, the immanent, demanding, perplexing join between the micro and the macro worlds. He had been trying, not yet successfully, to develop macroscopic superposition in SQUIDs. That the problem remained open was 'a glaring indication of the inadequacy of quantum mechanics as a total world view'. It claims to be a 'totalitarian' theory. Treat it so! Open your eyes! '[A]s a consequence of ingrained reductionist prejudices and perhaps to some extent sociological factors, we may have been looking in precisely the wrong direction'.[24]

Looking where they pleased, some semi-practical people saw that they did not have to probe the depths of non-locality to use it to confound nefarious others. And so modern cryptography, which relies on a mathematical algorithm to transmit information that can be read only by its intended recipients, came to be. In earlier systems the cipher used to encrypt and decrypt information had to be held secret. The encrypting algorithms in modern use can be made public without compromising security provided that only the necessary few know the key.

Suppose Alice wants to send a secret message to her friend Bob. At her request, Bob chooses two extremely large prime numbers (at least 100 digits each), multiplies them together, performs a series of arithmetical manipulations on the product, and chooses an 'encryption exponent' in accordance with these manipulations and other restrictions. This concocted number in turn determines a 'decryption exponent'.[25] Bob now openly shares the product of the primes and the encryption exponent with Alice (this is the *public key*) and keeps the decryption exponent secret (the *private key*). Alice uses the public key to encrypt her message, which she sends to Bob, who then decrypts it.[26]

This is the RSA algorithm, named for mathematicians Ronald Rivest, Adi Shamir, and Leonard Adelman who developed it at MIT in 1977; the exchange of the public key, the Diffie–Hellman protocol, is named for American cryptologists Whitfield Diffie and Martin Hellman. An English mathematician, Clifford Cocks, developed an equivalent algorithm in 1973 and, with John Ellis and David Williamson, also developed a version of the Diffie–Hellman protocol. The British signals intelligence agency classified their work, which remained secret until 1997.

The protocol is secure because the encrypted information cannot be decoded without knowledge of the huge prime numbers used to compute their product. These remain hidden even from a fast computer. Computing speeds are measured in 'CPU time', the time a server-grade central processing unit spends processing instructions. The (real) elapsed time is always greater than or at best equal to the CPU time, since the CPU occasionally idles or waits for inputs and outputs. New world records for finding the prime factors of large numbers are set every few years, but it typically takes a few thousand CPU years to factor a number with a few hundred digits.[27] An RSA key consisting of 2,048 bits has 617 decimal digits, and is thought to be impossible to factor on a conventional computer. The RSA system now figures in Virtual Private Networks, email services, web browsers, digital signing, and some banking and online shopping services.

Suppose Bob has an answer for Alice. If he wants to employ the same factors he must send her the decryption key via a communication channel that might be vulnerable to eavesdropping. The eavesdropper will learn nothing, however, if Bob sends particles he has entangled with ones he retains. He does not have to invent the procedure. Bennett and Brassard had devised a protocol for quantum cryptography in 1984.[28] A year later, Oxford theorist David Deutsch suggested making use of entangled states for key distribution and the Polish-born Oxford mathematician and physicist Artur Ekert proposed a formal procedure based on Bell's theorem in 1991.[29]

Using Ekert's procedure, Bob and Alice each make Bell-type measurements on their particles and share such information as the sequence of polarizer orientations they use (the public key). But each will know the outcomes only for the measurements he or she makes. The correlation established between these outcomes can be used to generate a private key, now known only to both Alice and Bob. Any attempt to eavesdrop or substitute false information will destroy the entanglement, which they can quickly detect through regular tests of Bell's inequality.

For every defence there is an offence. The quantum computers first imagined by Feynman promise an exponential increase in the efficiency of computation. They are still in their infancy, however, and face a challenging childhood. A capital problem concerns the reliable transfer of quantum data within a single computer or across a network of quantum computers without error, attenuation, or loss. A potential solution, identified in 1993 by Bennett and his IBM colleagues, involves *quantum teleportation*, in which Alice makes use of non-locality to send Bob a copy of a photon in an unknown state (destroying the original in the process, so escaping the 'no cloning' principle). There is no theoretical constraint on the distance over which this teleportation can take place.

With Richard Jozsa, Deutsch devised the first quantum computer algorithm in 1992.[30] Two years later Peter Shor at AT&T (now Nokia) Bell Laboratories in New Jersey found a counter algorithm that threatens all conventional computer-based cryptographic systems.[31] Its power is terrifying: it is estimated that a quantum computer consisting of 4,099 qubits could break the RSA encryption in just 10 seconds.[32] Although such computers lie in the future, the pace of progress is such that the widely-used RSA system could soon be under threat. Quantum cryptography would provide the only secure defence.

Any practical quantum cryptographic system must meet certain criteria relating to the source of entangled particles, their transmission over long distances, their detection and counting. Bennett and Brassard's prototype tabletop experiment reached only 30 cm. Going to greater distances raised severe challenges. The most eligible particles for entanglement at a distance are photons. Unfortunately, the optimal wavelengths for the detectors used for photon counting in Bell tests did not match the optimal wavelengths of light transmitted along the optical fibres in telecommunications networks. Transmission through free space is feasible but limited since the transmitter and receiver must 'see' one another and the system and the photons must brave weather hazards. Specialized optical fibres could be produced to handle different wavelengths, but that would have required costly duplications of existing networks. The obvious course was also pursued, but photon-counting detectors optimized for use with installed optical fibres did not exist.

Into this complex of foundational and applied physics entered Nicolas
Gisin, who, when still a PhD student at the University of Geneva, had
met Aspect and Bell, and had heard Bell lecture. Their programme gave
him a sabbath occupation. 'I am a quantum engineer, but on Sundays I
have principles'. He could not get Bell's teaching out of his thoughts. After
obtaining his doctorate in 1981 and completing post-doctoral studies in the
US, he joined a small Swiss start-up specializing in fibre-optic instrumen-
tation. Working with the national telecommunications operator Swisscom,
he contributed to the introduction of fibre-optic networks in Switzerland.
He returned to academia in 1988 as head of the optics section in the Group
of Applied Physics at the University of Geneva. In 1992, he caught up with
developments in quantum cryptography and realized that he had all the
ingredients for an experiment. At a conference in Davos he asked Ekert's
opinion: 'Yes, no one has done these experiments. But if you know how
to do it, you should do them'.[33]

In experiments performed in 1997, Gisin used an elegant laser source of
high-quality entangled infra-red photons. Following a suggestion by James
Franson of Johns Hopkins University, Gisin created energy–time entangled
pairs by slightly delaying one of the members. Measuring the energy of one
photon instantaneously determined the energy of the other or, alternatively,
measuring the time elapsed since creation of one (its 'age') gave the age of
the other. The energy–time uncertainty relation ensures that the energies
and the ages of the photons cannot be known precisely and simultaneously.
The EPR analogy is complete.

Infra-red photons have the advantage that they can be transmitted along
optical fibres. Gisin and his colleagues gained access to a telecommunications centre near the railway station in Geneva, where they set up the
source. The output fibres, 8.1 km and 9.3 km long, ran to photon counting systems and interferometers at Bellevue north of Geneva and at Bernex
to the south. Mary Bell came by to cheer them on.[34] The results were
most gratifying. One experiment yielded a violation of Bell's inequality by
nearly 25 standard deviations.[35] No impairment of the correlation between
the photons resulting from transmission over such long lengths of optical
fibre occurred. Practical quantum cryptography based on entangled pairs
was feasible. Later that same year, Zeilinger's group in Innsbruck used the
same source of entangled photons to demonstrate quantum teleportation in
the laboratory.[36]

The ultimate demonstration of the practical feasibility of quantum communications channels was the transmission of entangled pairs of photons via satellite. In 2017 a Chinese team bounced such pairs from a dedicated experimental satellite named *Micius* separately to ground stations in Delingha (Qinghai province) and Nanshan (Xinjiang province), 1,120 km apart, and to Delingha and Lijiang (Yunnan province), 1,203 km apart. The team completed 1,167 Bell tests in just 1,059 seconds. Their whirlwind average result violated Bell's inequality by four standard deviations.[37] The experimentalists were unstoppable. A few months later the Chinese group published a demonstration of ground-to-satellite teleportation over 1,400 km.[38]

As the technologists prepared for quantum communication with the heavens, the pioneers whose work brought it all about, Bohm and Bell, were returning to the earth. Bell went first, like a prophet of old. The year before his death, at a workshop to celebrate '62 years of Uncertainty' held in Erice in August 1989, he attacked the orthodox interpretation of quantum measurement in a fiery polemic, 'Against measurement'. He railed against physicists who, like 'why bother' Dirac, expressed indifference, if not outright hostility, to questions about the meaning of a theory that demonstrably works, FAPP. He exposed the confused and often contradictory descriptions of quantum measurement proffered in the 'good books' that served as student texts and pushed alternative interpretations to Copenhagen, such as de Broglie–Bohm's hidden variables and modified quantum mechanics with spontaneous collapse mechanisms.[39]

Mermin listened to Bell's talk (page 242). It 'came close to being the most spell-binding lecture I have ever heard', made memorable by Bell's 'brilliance and his wit [and] ... the music of his voice'. A year later, in August 1990, Mermin heard Bell again, at a small invitation-only conference at Amherst College in Massachusetts. Among the participants were Mary Bell, Kurt Gottfried (whose *Quantum mechanics* (1966) Bell had criticized in 'Against measurement'), Greenberger, Horne, Leggett, Shimony, and Zeilinger. There were 'no prepared talks, no schedule, no proceedings. Just wonderful conversations'.[40] Bell managed some 'passionate' exchanges with Gottfried, but his fire was going out. On 1 October he had a stroke

In August 1990, Mermin heard Bell lecture again, at a small invitation-only
conference at Amhurst College in Massachusetts. L–R, 1st row: Leggett and Bell
are first and fourth. 2nd row: Philip Pearle, Mermin, and Weisskopf are first,
fourth and fifth. 3rd row: Horne and Kurt Gottfried are first and fourth. 4th row:
Greenberger is fourth. 5th row: Shimony and Zeilinger are third and fifth.

that killed him. He was 62. He had convinced and inspired many physicists, but by no means all. 'His articles [on the interpretation of quantum
mechanics] were always provocative, and even those who did not agree with
his point of view found them thought-provoking and instructive'. Thus ran
the cool tribute by Bell's old thesis advisor Peierls.[41]

Peierl's contemporary, Nevil Mott, who had a Nobel prize in physics,
wrote to him expressing solidarity. 'I see you do not agree with Bell or with
hidden variables. Nor do I—but I've never thought about it very much'.
Bohr did not 'brainwash' a generation of physicists, as Murray Gell-Mann,
another Nobel theorist, had claimed; if they were brainwashed, then, like
Mott, their indifference, not Bohr's proselytizing, was the likely cause.[42]

Bohm soon followed Bell, and with him went the last of the *Prinzipienfuchser* most like Bohr. We see something of Bohr in Bohm's habit of

developing his ideas in conversation, making quantum jumps in discourse, and tolerating vague and even incomprehensible questions.[43] And so Bohm could consider with Krishnamurti whether 'the brain can be aware of its own structure' or 'silence is a part of emptiness' or 'thought has entangled the brain in time'. Bohm grew addicted to such questions and descended into deep depression when Krishnamurti broke off their conversations in 1984. Nonetheless, Bohm acted as a spokesman for his guru during the remaining two years of Krishnamurti's life. Bohm's health began to deteriorate alarmingly in 1991. He suffered bouts of severe depression and underwent electric shock therapy. In October 1992, two months before his 75th birthday, he had a heart attack in a taxi on his way home from Birkbeck. He was rushed to hospital but could not be revived.[44]

What had given Bohm the energy and confidence to challenge the grand masters of the Copenhagen school and the Western ideal of a physics built up from independent elements? He had pondered the question. The spring of his creativity, of creativity in general, he wrote in 1980, is 'an act, permeated by intense passion, that makes possible great clarity in the sense that it preserves and dissolves subtle but strong emotional, social, linguistic, and intellectual pressures tending to hold the mind in rigid grooves and fixed compartments'. Bohm's associates recognized quiet but intense passion in pursuit of original ideas as his distinctive trait.[45]

The passion could not free him entirely from inhaling the same sorts of social whiffs that other physicists breathed. Even Bohr was not exempt, for he had admitted 'privatively' to Bohm to being attracted by the writings of the existentialist philosopher Søren Kierkegaard. And, Bohm continued, Bohr's philosophy, however constructed, had propelled him to recognize the radical inability of the past to determine the future and to express this openness in his concept of complementarity. It is a pity he did not stop there, before ruling out, in his eagerness to declare the domain of quantum mechanics closed, the possibility of 'new categories of thought concerning space, time, [and] existence'. These last propositions come from the first letters in an extensive correspondence that Bohm maintained with an unusually assertive artist, Charles Biedermann.[46] Bohm advised Biedermann that everything is in flux, forming and dissolving, ceaselessly opening opportunity to creativity; 'we are liberated from the dominion of the past'.[47] That is the way forward. Free yourself from the past, empty your mind, amalgamate the I and the not-I, have insights, create.[48]

Bohm settled on the problem of 'the unity of inner and outer' as the key to science, philosophy, art, and human relations. 'The true reality is in the individual . . . but a whole world is in him'. The separation of 'I' from 'me', which also fascinated Bohr, is no more than a bad habit that we learn when young from our competitive society.[49] The Marxist lingered in Bohm. Action is essential not only to insight, but also to the material universe. 'Things are to be referred to as sides of actions . . . "Subject" and "object" are non-exclusively comprehended as sides or aspects of the act or process as a whole'. This may be a little confused. No matter; 'our minds are generally confused . . . mainly because of our training and conditioning'.[50]

Bohm thought that most of his colleagues lived and worked contentedly caught up and blinkered in what Kuhn called their paradigms. They had not assimilated quantum theory but some ill-digested mixture of Popper's philosophy, Heisenberg's hagiography of mathematics, and Einstein's requirement of locality. '[I]n fact, [the theory] has gone into greater and greater confusion as to what its meaning and interpretation is'. Bohm admitted that some little progress had been made since his intervention in 1952, 'but it's still not anything fundamental'.[51] The reception of his later efforts was mixed. Enthusiasm from non-scientists ready to embrace his blend of physics and mysticism; warmth from the small community of 'Bohmian' physicists; and disdain for his metaphysical vagueness from the rest of the physics community.

Where to Cut? Which Way to Go?

Experimental research on Bell's theorem and inequality over many decades confirmed that no system of local hidden variables could describe the quantum world with the accuracy that modern laser and optical technology had made accessible. Pilot-wave theories, like those Einstein and de Broglie entertained and discarded in 1927, proved untenable and Bohm-type non-local versions, though still viable, wanted development. The alternative that Einstein had recommended at Solvay V, the interpretation of the wavefunction as the statistics of an ensemble, had also struck out. All evidence confirmed that quantum theory is a complete theory of individual processes. The domain of quantum mechanics appears firmly closed.

Quantum non-locality, though an established fact, played havoc with the 'ordinary view of a scientist'.[1] It inhibited reliance on intuition. Bohr's intricate interpretation could be made to paper over the difficulty, but it too left holes in understanding. The 'measurement problem' remained. Bohr had dismissed it because he did not regard the ψ wave as a physical thing able to produce physical effects. Nicolaas van Kampen, a no-nonsense student of Kramer's, made this dismissal the fourth of his ten theorems of measurement in quantum theory: 'Whoever endows ψ with more meaning than is needed for computing observable phenomena is responsible for the consequences'.[2]

Should we then limit the damage and, with Bohr, take quantum mechanics merely as an abstract theory that for reasons unstatable or unknowable 'makes use of a mathematical formalism in which the variables of classical physical theory are replaced by symbols subject to a non-commutable algorism involving Planck's constant'?[3] The medicine seems to invite the disease it was intended to cure. If the classical concepts are unavoidable because human physicists cannot describe and interpret phenomena without them, if they are indispensable and irreplaceable because they are the product of an evolutionary alignment of mind and nature, why do they not offer a firm basis for understanding the world that gave them birth?

Still limiting damage, we can join with Bohr in seeing no paradox in EPR and no threat from Bell's theorem. We can freely admit that knowledge about unentangled particles is both necessary and sufficient to predict the correlations exhibited by entangled ones. And we can agree that the Schrödinger equation does not supply a physical mechanism by which particles become entangled or probabilities become actualities. It is an abstract, purely symbolic creation wrought from our understanding of the microworld as measured by our classical apparatus. No doubt we are puzzled to learn that the square of the solution to the Schrödinger equation, $|\psi|^2$, provides all the information possible about the outcomes of experiments on the microsystem that ψ describes. But that is the way it is.

The Copenhagen way advised its adherents not to confuse themselves by talking about 'projection' or 'collapse' (these are 'wrongly put'). Let us just accept that we have extraordinarily successful recipes for predicting the classical manifestations of an unknowable quantum physics, FAPP. There is no need to worry about 'non-locality'. It does not affect what counts, that is, measurements. No doubt correlations exist between distant particles. Not to worry. Do as Bohr's tranquillizing philosophy advises and ask not how these correlations arise. Indeed, complementarity does not allow you to make the inquiry; it offers 'no way even to raise the question of what is their origin'.[4] Tests of Bell's inequality did not, could not, challenge the Copenhagen interpretation.

But the more the Copenhagen position is hedged around, the more exposed its shortcomings become. That is a prize example of complementarity! It bites, however. Bohr's clear distinction between a quantum world knowable only in some of its potential aspects and the classical world of fully specifiable polarization analysers and coincidence counters raises the fundamental question, where does the quantum world end and the classical begin? At the level of atoms and molecules? Cats? Human observers? Even if we accept that the quantum formalism is a purely symbolic calculus, at some point the symbols common to quantum and classical physics must connect with the same world and, in the sense of the correspondence principle, predict the same measurement outcomes.

Bohr placed the 'cut' or, perhaps better, the junction between micro and macro in or at the irreversible interaction of the quantum system with the classical measuring device. 'In the analysis of single atomic particles [physical experience] is made possible by irreversible amplification effects—such as a spot on a photographic plate left by the impact of an electron, or an electric discharge created in a counter device'. 'The amplification of atomic effects, which makes it possible to base the [experimenter's] account on measurable quantities and which gives the phenomena a closed character, only emphasizes the irreversibility characteristic of the very concept of observation'.[5] Since abstract symbols do not interact with measuring devices, the atomic particle to which Bohr refers must be shorthand for what the unknowable particle-in-itself 'really is', and whatever it is 'really doing'. The irreversible act of amplification gives us what we accept as the particle-as-observed, with the properties assigned to it by our choice of experimental procedure.

As we witnessed earlier in our drama, Heisenberg was more flexible in placing the cut.[6] Why does it matter? It makes no difference to the measurement outcomes or our understanding of them. Why not use the symbolic quantum formalism to summarize our knowledge of the whole, of the quantum system, classical apparatus, cat, and—yes—the experimenter? We can then allow the symbols to give way to real physics only when the mind of the experimenter receives knowledge of the measurement outcome. Where else can the regress find a natural stopping place? In this end game, mind does not 'collapse the wavefunction'; that would be physically meaningless, since the wavefunction is not real.

There was too much immoral ambiguity in Heisenberg's lackadaisical placing of the 'shifty split' for the moral guardian of the 'great enterprise'. Bell: 'No explicit guidance is given as to how in practice this shifty division [into "system" and "apparatus"] is to be made . . . some authors of "quantum measurement" theories seem to be trying to [get rid of it]. It is like a snake trying to swallow itself by the tail. It can be done—up to a point. But it becomes embarrassing for the spectators even before it becomes uncomfortable for the snake'.[7]

We must not omit the question that Wigner had sought to address. In a theory exclusively concerned with the 'results of measurement', what makes the mind of the experimenter unique? Why not extend the cut to include her friend? Or the entire community of physicists who will read about the result in a published paper? Or the entire universe? 'When the "system" in question is the whole world where is the "measurer" to be found? . . . What exactly qualifies some subsystems to play this role? Was the world wave function waiting to jump for thousands of millions of years until a single-celled living creature appeared? Or did it have to wait a little longer for some more highly qualified measurer—with a Ph.D.?'[8]

The reductionist, for whom large things are composed of smaller things, would probably prefer Bohr's physical analysis to Heisenberg's epistemological one. Bohr's 'irreversible act of amplification' mobilizes an intuitive physical mechanism for the transition from quantum to classical behaviour. In the inaugural issue of *Foundations of Physics* in 1970, Zeh acknowledged (citing Bohm's account of Bohr) 'the importance of the amplification of the result of a measurement up to the macroscopic scale, thus leading to a natural position of "Heisenberg's cut"'.[9]

The first steps in a measurement interaction occur at the level of the microscopic components of the macroscopic device. The interactions multiply prolifically in the complex environment of the device and the *coherence* of the wavefunction rapidly degrades whether judged as a real physical thing or as a symbolic summary of the information we assume it to hold about the quantum system. Any tendency to interference contained in it is lost as the interaction cascades irreversibly towards the final measurement outcome. The dials that give readouts of experiments would not have to point in two or more directions at the same time; the cat need not be dead and alive simultaneously. Zeh wrote of a 'dynamical decoupling' or *decoherence*. The larger the object with which the quantum system interacts (the greater the number of quantum particles it contains), the shorter the 'decoherence time'. For a cat, or an object designed to detect correlated photons and count coincidences, decoherence is essentially instantaneous.

Decoherence suppresses interference terms that appear in the quantum algebra by diluting them over the vast number of states in the measuring device and its environment. But as a physical mechanism decoherence cannot force the choice between measurement possibilities. That we randomly get either vertical or horizontal in a polarization measurement remains mysterious. Decoherence cannot account for quantum probability or convert a superposition of vertical *and* horizontal into vertical *or* horizontal. Bell again: 'The idea that elimination of coherence, in one way or another, implies the replacement of "and" by "or" is a very common one among solvers of the "measurement problem". It has always puzzled me'.[10]

Roger Penrose agreed: '[Decoherence] does not help us to determine that the cat is actually either alive or dead . . . we need more . . . What we do not have is a thing which I call OR standing for *Objective Reduction*'.[11] To secure it he needed the help of general relativity, from which he derived a measure of the endurance of superpositions of macroscopic quantum-gravitational states, an approach anticipated nearly ten years earlier by Lajos Diósi. On this scheme, a water droplet with a radius of a thousandth of a centimetre might enjoy the macroscopic distinction of belonging in two places at the same time before the gravitational energy difference between the two states thrusts it into one of them.[12] Gravity does the work that Bohr assigned to the measuring apparatus and others vaguely to the 'environment'.

One way or another, the physicist had to make peace with the measurement problem and the puzzles of entanglement. In none of the standard forms of pacification did Einstein appear to have a part. Entanglement had made his natural philosophy untenable, no? No. A notable holdout or holdover was (and is) Leggett. He struggled with realist interpretations based on decoherence: 'We all agree, myself included, that by the time we reach the level of cats and counters and so forth, decoherence means that the phenomenon of interference goes away. But the formalism of [quantum mechanics] has not changed one bit between the atom and the cat'. As long as the divide between the classical and the quantum mechanical remains open, it would not be prudent to drop Einstein's philosophy. 'We cannot actually prove it [quantum mechanics] wrong, but somehow we cannot quite believe it is the end of the story'.[13]

We might suppose, as Einstein did, that quantum mechanics is incomplete; the domain left open. Avoiding macroscopic superpositions of cats or human observers requires a new physical mechanism to collapse the (real) wavefunction, such as the Diósi–Penrose gravitational scheme. Leggett dubbed such theories 'macro-realistic': they proffer some mechanism that rules out superpositions of macroscopically distinct states or quickly kills them off. He argued that in this context the completeness question can be answered experimentally. We should strive to create macroscopically distinct superposed states and to detect their interference. Reluctantly, we must spare the cat as too ambitious and try for an inanimate preparation of tiny macroscopic dimensions. All macro-realistic theories predict that the attempt should fail.

Leggett's journey to physics helps to explain his independence. He grew up an 'embattled Catholic' in the suburbs of London. His father taught physics and chemistry at Beaumont College, a Jesuit school near Windsor, which offered free tuition to his three sons. Leggett had a choice between 'classics' and modern languages, mathematics, or science and, since he was academically gifted, his teachers directed him towards what they understood to be the more prestigious slog through Latin and Greek. Precocious and aloof, Leggett did not socialize well. 'As a result, the five years I spent [at Beaumont], while not unhappy, do not stand out in retrospect as a particularly joyful period of my life'.[14] He went on long, solitary hikes, took a mountaineering course, and fell in love—with the craggy and sometimes dangerous mountains of Snowdonia.

In 1955 Leggett matriculated at Balliol College, Oxford, intending to continue his ordained academic path by reading for a degree in Literae Humaniores, a mix of classics, ancient history, and philosophy. It covered all the classical philosophers from Plato to Bertrand Russell and Ludwig Wittgenstein but did not make much of the philosophy of science: 'to most of the philosophy department in Oxford, the philosophy of science quite generally really didn't exist'.[15] Nevertheless, the diet of analytical philosophy would later shape the way Leggett looked at the problems of physics.

Leggett soon discovered that the philosophers he met with in Oxford were obsessed with the nuances of language and lacked objective criteria for deciding between right and wrong, good and bad, in philosophy. He did not suffer this regime for long. During his final months at Beaumont, he had studied advanced mathematics, which opened a path to physics and, unlike Oxford philosophy, offered a chance for being wrong 'for interesting and non-trivial reasons'.[16] He applied to switch his degree to physics. The timing was good. In reaction to the Soviet Union's launch of *Sputnik* in October 1957, British politicians had attacked the culture that led many of its most able young students to 'waste' their talents on the humanities instead of devoting themselves to useful subjects like physics.

Perhaps with such encouragement, Leggett crammed the three-year physics course into two. He studied quantum mechanics from a textbook that presented the subject axiomatically and did not bother with questions of interpretation. He earned a DPhil in solid state physics at Oxford, did postdoctoral work at the University of Illinois and Kyoto University in Japan, and, in 1967, settled into a lectureship at the then new University of Sussex in Brighton. The liberal and collegial *Geist* at Sussex sanctioned Leggett's inroads into philosophy and quantum foundations. An early fruit from this trespass was a review of *Criticism and the growth of knowledge*, the edited proceedings of an international colloquium in the philosophy of science held in London in July 1965. Leggett became acquainted with the views of Kuhn, Popper, and Feyerabend, and the programme of Bohm.[17] His Sussex colleague Brian Easlea, a theoretical physicist who had done time in Copenhagen and turned historian, introduced him to the work of Bohm and Bell and convinced him that he had been naïve about the foundations of quantum mechanics. Luckily none of this deflected Leggett from his work on superfluid helium (^{3}He), which won him a share of the Nobel prize in 2003.

Secure in a professorship at the University of Illinois, to which he advanced in 1982 ('out of the blue'), Leggett had the freedom to continue the exploration of foundational questions that he had begun at Sussex. That meant, for him, locating the join, or fixing a continuity, between the micro and the macro, working through successive approximations towards 'building Schrödinger's cat in the laboratory'. He was well prepared, toughened, he thought, by earlier skirmishes he had had in defence of his Catholicism. 'I did have to fight back consciously against that and it did make me perhaps, more prepared than I might otherwise have been, keen to embrace opinions that might not be the current orthodoxy'.[18]

To investigate whether a macroscopic body can occupy distinct states capable of interfering with one another evidently demands a definition of 'macroscopic'. Leggett wanted to answer the question without being unduly distracted by the circular declaration: 'The very fact that you have seen interference shows that [the system] cannot be truly macroscopic', which he found 'tedious'.[19] He invented a semi-quantitative measure, 'disconnectivity' D, 'a sort of rather crude and preliminary version of what is now called multiparticle entanglement'. In the case of two photons, a simple statistical mixture of states gives $D = 1$; but for two entangled particles, as in Bell tests, $D = 2$. Superpositions of macroscopic states involving large numbers of particles imply a high D. 'For the case of Schrödinger's cat, if indeed it were possible to show the interference of the states of the living and dead cat, then in those two states, something like 10^{23} particles are behaving differently. So D is something like 10^{23}'.[20]

Remaining in the macro world, Leggett collaborated with one of his colleagues in Illinois, Anupam Garg, to formulate a Bell-style inequality.[21] They made the comfortable assumption that an object with at least two macroscopically distinct states is always definitely in one of them and that an adroit experimenter can determine which without disturbing it in any essential way. The assumption of a definite determinable state contradicts the quantum mechanics of micro-objects. But that did not necessarily put Leggett and Garg in opposition to Bell, for he had derived his inequality to

test the separability of entangled microsystems, while they derived theirs to
test the limits of quantum coherence in macroscopic objects. What they had
in mind is easily paraphrased. Suppose a macroscopic object has an observable Q that can take values of $+1$ or -1. Measure Q at three consecutive
times, t_1, t_2, and t_3, yielding values Q_1, Q_2, and Q_3. Since by basic assumption the act of measurement does not change the value of Q, the different
times play a role like the polarizer settings in the usual Bell experiments;
we have $Q_1 Q_2 + Q_1 Q_3 + Q_2 Q_3 \geq -1$. But if Q is subject to interference
effects or alters under measurement, this inequality may be violated.

Let us advance to measurement. In 1999 Zeilinger's team at the University of Vienna poured some new wine into an old bottle by sending
a beam of buckminsterfullerene molecules, each consisting of 60 carbon
atoms formed in a soccer-ball structure (in later experiments, fullerene-70),
through a variation of the classic two-slit experiment.[22] Each of these structures contains around 1,000 nuclear particles and a corresponding value of
D. They came from an oven heated to 900 degrees and so were a little
excited. On average, each emitted or absorbed a few infrared photons on
its journey though the slits. Perhaps then we might determine which slit
it passed through by an adaptation of Heisenberg's microscope? No. The
wavelength of the infrared photons is large compared with the distance
between the slits and therefore cannot be used to determine which slit a
given particle had chosen. The quantum mechanical situation held and the
bucky-ball beam formed an interference pattern.

We know that quantum mechanics wipes out the interference pattern if
the experimenter tries to determine the trajectory of photons or electrons
sent through a Young's device. As Bohr put the point in his contribution
to the Schilpp volume on Einstein, 'we are presented with a choice of
either tracing the path of a particle *or* observing interference effects, which
allows us to escape from the paradoxical necessity of concluding that the
behaviour of an electron or a photon should depend on the presence of a
slit . . . through which it could be proved not to pass'.[23] That might be good
enough for an electron or a photon; how is it with a beam of bucky-balls
prepared in an oven so hot that the photons they emit have wavelengths
short enough to locate them? Zeilinger et al. gave it a try and wiped out
the interference pattern.[24] Experiments with hot C_{70} molecules confirmed
the uncertainty principle.[25] Quantum mechanics may be unintelligible, but
it is consistent.

Zeilinger's experiments came between ones performed in 1989 showing interference between beams of particles much larger than buckminsterfullerene. The particles were as odd as the results: uniform magnetic biomolecules (obtained without the donors' consent) from a horse's spleen (1992). Similar experiments with molecules with 430 atoms succeeded in 2011 and, in 2018, with living sulphur bacteria (anaerobic bacteria found in microbial mats that can metabolize sulphur).[26] It is still a long way, however, from a bacterium measuring a few millionths of a metre to a kitten. Experimentalists sought to close the gap with the SQUIDs we mentioned earlier.[27]

The phenomenon of superconductivity derives from the co-operative motions of pairs of electrons, which meet no resistance as they pass through the body of a conducting material. These 'Cooper pairs' are named for Leon Cooper who, with John Bardeen and Robert Schrieffer, received a Nobel prize in 1972 for explaining superconductivity. The collective behaviour of large numbers of Cooper pairs can be described by a single wavefunction. Interference among them can be facilitated by a thin barrier of insulating material inserted into the superconductor. If the barrier is thin enough, the wavefunctions of Cooper pairs from both sides can leak through it without breaking up. The British physicist Brian David Josephson predicted this *quantum tunnelling* effect in 1962 and received a Nobel physics prize in 1973 following its demonstration. The barrier is called a Josephson junction.

Form the superconductor into a ring, insert two Josephson junctions, and the result is a direct-current SQUID. The two junctions act much like the two slits in Young's experiment. The wavefunctions of the Cooper pairs at the two junctions interfere. Instead of bright and dark fringes, the interference shows up as changes to the superconducting current and voltage. As these quantities also depend on the strength of a magnetic field flowing through the centre of the ring, SQUIDs can be used as extraordinarily sensitive magnetometers. They can detect changes in magnetic flux (the product of magnetic field strength and the area enclosed by the ring) corresponding roughly to the energy required to raise a single electron one millimetre against Earth's gravity. More sensitive devices approach the limits imposed by the uncertainty principle.[28]

Leggett had in mind a different type of SQUID, one consisting of a superconducting ring hosting a single Josephson junction. Here a small external magnetic field induces a supercurrent in the ring by forcing the Cooper pairs to tunnel their way through the junction. The current may run clock-wise or anticlockwise around the ring. The magnetic flux 'trapped' in the ring is quantized, and when it reaches close to half-integer values of the magnetic flux quantum ($h/2e$), the ring may support multiple stable states, including states in which the current runs clockwise and anticlockwise *at the same time*. These states may then interfere at the junction.[29]

In 2000, two groups of experimentalists, one a collaboration between the State University of New York, Stony Brook and the Delft Institute for Micro Electronics and Submicron Technology, and the other at MIT detected interference between states involving about ten billion Cooper pairs ($D \sim 10^7\text{-}10^{10}$) travelling in opposite directions around a SQUID ring large enough to be visible to the naked eye.[30] This achievement carried Leggett '40% of the way between the atom and a real-life Schrödinger's cat'—on a logarithmic scale.[31] Many subsequent experimental tests have confirmed the phenomenon.[32] In another approach, Leggett and his col-leagues observed a superposition of magnetic moments corresponding to 'several hundred thousand static electron spins pointing in opposite directions simultaneously'.[33]

No limit is yet in sight that excludes the possibility of macroscopic quan-tum interference in environments that minimize decoherence. A project still at the proposal stage, the 'medium-sized' MAQRO satellite mission, could prove decisive. It will carry macroscopic quantum resonators (hence MAQRO) to test the predictions of quantum mechanics for superposi-tions of objects with billions of atoms 'in principle, visible by eye'.[34] First proposed in 2010, the mission has made slow but steady progress. It was judged to be 'challenging yet feasible' in a study organized by the European Space Agency (ESA) in 2018.[35] NASA and the ESA are now considering the mission.

The trade-off between wave-like interference fringes and particle-like tra-jectories can be expressed in a simple 'complementarity inequality' between the 'distinguishability' D of the particle paths (not to be confused with Leggett's D) and the 'visibility' V of interference fringes: $D^2 + V^2 \leq 1$.[36] If both paths through the slits or both arms of an interferometer are open, D vanishes and $V^2 \approx 1$. If one slit or arm is blocked, the fringes disappear but the path is definite: $D^2 \approx 1$. Like the uncertainty principle, the complementarity inequality admits a significant penumbra. Fringe visibility can be traded for which-way information.

In 1982 Marlan Scully and Kai Drühl, working at the Max Planck Institute for Quantum Optics in Munich, conceived an extraordinary thought experiment based on the interference of photons emitted from two atoms trapped in an interferometer.[37] The atoms act as distinct sources of coherent light like the two slits in Young's experiment. Let a pulse from a laser simultaneously excite an electron in each atom to a higher-energy state. The atoms subsequently emit photons as the electrons return directly to their ground states. The photons move along different paths in the interferometer. Because they are identical, no experiment can tell which photon has come from which atom. In these circumstances quantum mechanics predicts that the emitted photons should exhibit two-particle interference effects.

Now suppose that the excited atoms can decay to ground via an intermediate state and arrange the experiment to detect only photons arising from decays via this state. A chink in the armour of quantum mechanics? The experimenter has an opportunity to learn whether after emission an atom is in the ground or intermediate state and distinguish which atom the detected photon came from. This which-way information comes by interrogating the *atoms* without jostling the photons, which continue undisturbed through the apparatus until brought to interfere. The experiment could determine the wave-like (interference) and particle-like (which-way) behaviour of the photons simultaneously! If this feat were feasible, it would go badly for quantum mechanics. But it cannot be done. Even though the photons propagate undisturbed through the interferometer, they are inexorably entangled with the quantum states of the atoms they left behind. Scully and Drühl showed that in granting access to which-way information the quantum formalism eliminates the possibility of interference.

What if we set up the experiment to give which-way information, but choose not to look at it? Does our indifference restore the interference pattern? What if we wait until the photons have passed through the apparatus and have been detected and *then* decide whether we want to inquire which way they went? Can we switch the interference pattern on and off by choosing whether to look or not *after* the photons have been detected? According to Scully and Drühl's analysis, the interference pattern does indeed come back if we decide not to look. These thought experiments are 'quantum-erasers', so called because, as in Bohr's complementarity, the delayed choice of one possibility cancels the other.

The ingenuity of the new breed of experimental philosophers turned these speculations into physical tests. In 1995 Zeilinger and his colleagues reported on quantum-eraser experiments using pairs of entangled photons placed inside 'micromaser' cavities. The results bore out Scully and Drühl's analysis. 'The use of mutually exclusive settings of the experimental apparatus implies the complementarity between complete path information and the occurrence of interference . . . our results corroborate Bohr's view that the whole experimental setup determines the possible experimental predictions'.[38] Complementarity, deduced from deep reflection and articulated in tortured Bohrish, could now be understood as a direct consequence of the formalism itself. Bohr's prophetic words have been rendered into mathematics and corroborated by experiment.

Quantum-eraser experiments work with single particles, pairs of particles, and 'entanglement swapping' (a generalization of teleportation) with causally disconnected delayed choice.[39] Experiments reported in 2013 determined the complementarity inequality across a continuous range between 'particle measurements' and 'wave measurements'. A pair of linear polarization-entangled photons is converted into states corresponding to two different paths in an interferometer. The experimenter can then choose to measure the polarization of one of the pair, revealing the path taken by the second and so providing 'which-way' information. Or the experimenter may choose to pass one of the entangled pair through a device called an electro-optic modulator that projects its (unknown) linear polarization state into a superposition of left/right circular polarization states. There is no possibility of discovering the path of the second photon in these circumstances: the experiment erases 'which-way' information. But the result, L or R, conveys information about the relative phase of the photons in the pair. One result corresponds to a phase difference equivalent to an interference fringe, another to a phase difference equivalent to an anti-fringe. Here 'which-way' information is traded for interference.

In these experiments, the voltage applied to the electro-optic modulator could be continuously varied, allowing a smooth transition between the extremes of full interference visibility and full 'which-way' distinguishability. Fig. 19 shows the results.[40] The dotted curve in this diagram is the

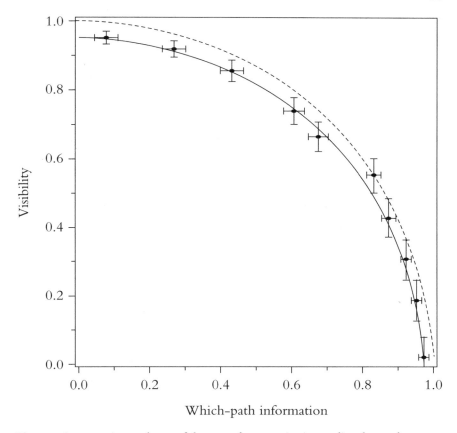

Fig. 19 An experimental test of the complementarity inequality shows the trade-off between fringe visibility and which-way information.

expectation assuming no experimental error. The solid curve takes these into account and corresponds to $(V/0.95)^2 + (D/0.97)^2 \leq 1$.

In 1976 Leggett took a temporary one-semester teaching exchange at the University of Science and Technology in Kumasi, Ghana, and applied himself to a problem that had attracted his attention some years earlier. All local hidden variable theories are characterized by two further assumptions. The first supposes (as did EPR) that the outcome of the measurement on one particle cannot be affected by the outcome of the measurement on the second, and *vice versa*. The second supposes that the outcome of the measurement on one particle cannot be affected by the *setting* of the instrument used to perform the measurement on the second, no matter how long the choice of setting is delayed. Experiments that violate Bell's inequality invalidate one or the other of these assumptions.

Leggett examined the predictions of a general class of non-local hidden variable theories that drop the setting assumption. By retaining the outcome assumption, he defined a class of non-local hidden variable theories in which the individual particles possess defined properties *before* the act of measurement. What is measured will of course depend on the settings, changes of which *do* affect what we can expect to measure for distant particles. Leggett showed that all the predictions of quantum mechanics cannot be recovered in theories that drop the setting assumption. Just as Bell had done in 1964, Leggett now derived an inequality (not the Leggett–Garg inequality!) obeyed by such hidden variable theories but violated for certain combinations of the settings by quantum mechanics. At stake was whether quantum particles have the properties we assign to them *before the act of measurement*.

On his return to England, Leggett drafted a paper presenting his ideas but did not submit it for publication. He put the draft in a drawer until the advent of reliable sources of entangled photons of the kind that have since been used extensively to test Bell's inequality. He dusted off the old manuscript and published it in October 2003.[41] His tardiness allowed rediscovery of the result by Jon Jarrett, then at Harvard University, who published it in 1984. Jarrett, too, decomposed Bell's locality condition into two weaker conditions, which he called 'locality' and the 'completeness' of the description of the quantum state afforded by quantum mechanics.[42] Shimony suggested renaming these conditions 'parameter independence' and 'outcome independence', similar to Leggett's setting and outcome assumptions.[43]

At a scientific conference in Minnesota in May 2004, Leggett shared his results with Markus Aspelmeyer, who worked alongside Zeilinger's group at the University of Vienna and the IQOQI. Aspelmeyer returned to Vienna and, with Leggett's help, collaborated with a local theorist to rederive the inequality for a specific experimental arrangement that could be realized in the laboratory. Aspelmeyer's student Simon Gröblacher carried out the experiments over one weekend.

They took their results to the ubiquitous Zeilinger, who agreed to repeat them. The two groups submitted their joint results for publication in December 2006. For a specific combination of settings, Leggett's inequality demanded a result no larger than 3.779. Quantum mechanics predicted 3.879, and experiment gave 3.8521 ± 0.0227. The conclusion: 'We believe that our results lend strong support to the view that any future extension of quantum theory that is in agreement with experiments must abandon certain features of realistic descriptions'.[44]

The experimental result was never in any real doubt. A year later, Roger Colbeck and Renato Renner at the ETH proved that *any* general hidden variable theory that admits both local and non-local parts (and of which Leggett's model is a special case) will predict correlations incompatible with the predictions of quantum mechanics.[45] This result could have been established by 2003, when Leggett's paper was published, though not in 1976 when he did the original work.[46] A further theorem published in 2012 claims that any statistical interpretation of ψ based on the assumption of pre-existing, real underlying states (in essence, Einstein's conception I of 1927) is unable to reproduce all the predictions of quantum mechanics.[47]

Continually meeting with the incomprehensible has so numbed us that the conclusions just quoted no longer seem startling. We have no problem with the premise that we cannot know the state of a quantum object before we look. But must we now abandon on the altar of experiment the common-sense notion that an object exists in *some* state, in *any* state, known or unknown, before measurement? This was a sacrifice Bohr was willing to make already in 1936, long before experiment would demand it.[48] The struggle to apply natural language to increasingly sophisticated experiments on the microworld seems to have wiped the universe clean of things. That should disquiet even the most jaded student of quantum mechanics.

Epilogue

The drama of our story lies not only in the clash of ideas, but also in the human passions and social contexts in which the ideas were conceived, debated, refined, accepted, or rejected. The cultivation of physics at the highest and deepest level is an art. Planck and Einstein described one another as artists.[1] Bohr's physics attained the 'highest form of musicality', according to the violinist Einstein.[2] Like artists and musicians, the creators of quantum mechanics approached their work with the full range of human emotions. We have the controlled, overpowering commitment of Bohr and Einstein, the jealous passion of Heisenberg, the breakdown of Pauli, the self-doubt of Born, the suicidal despair of Ehrenfest, the assertiveness of Bohm and Bell, the triumph and disappointment of Clauser, the confident virtuosity of Aspect, Zeilinger, and Leggett.

These characteristics became fruitful in accordance with social and cultural circumstances and the state of science. Bohr and Einstein were bred on classical physics and the old culture of Europe but also recognized during the agitated decade before World War I that problems in their science presaged fundamental change. After the war a new generation of physicists, thrown in upon themselves by Germany's ignominious defeat, rampant inflation, and political turmoil, rushed to master, remake, and replace the revolutionary ideas of Bohr and Einstein. Heisenberg, Pauli, and Jordan possessed the perfect combination of brilliance, energy, ambition, and submission to struggle in the hothouses of Göttingen and Copenhagen for answers to the quantum riddle. Exhausted by their accomplishment and dismayed by the appearance of the rival approaches of Schrödinger and Dirac, Heisenberg, Pauli, and Jordan broke down. Heisenberg soon recovered his spirits, but Pauli and Jordan went to the psychiatrist. Soon they were trying to annex psychology to physics.

Marxism powered many objections to Bohr's complementarity. Podolsky and Rosen, who provided the shaft if not the point of EPR, admired the

Soviet Union. Bohm cleaved to his Marxism throughout his challenge to standard quantum mechanics and obtained his first effective following among French communists. Soviet physicist-philosophers lent their support by finding positivist tendencies in Bohr's teaching at odds with dialectical materialism. In this the dictators on the right agreed with their enemies on the left, for the Nazis supported a Deutsche Physik that condemned complementarity and relativity as poisonous Jewish abstractions.

Jews are conspicuous in our drama. Although not religious, indeed, in Bohr's case, atheistic, our protagonists owed much to their Jewish heritage. Bohr's style of reasoning, which Einstein once called 'Talmudic', combined elements singled out by Jews as peculiar to their coreligionists. These elements run as follows in the catalogue of Henri Nathansen, a popular Danish-Jewish playwright of the time. Jews are fierce competitors, sometimes over-critical, domineering, and arrogant. But they are also champions of truth, justice, freedom, and human rights. With family and friends, their competitiveness turns to humour, irony, satire, wordplay, banter, 'wily, equivocal, ambiguous, double-edge wit combined with irony and self-irony'. Their world is the world of the mind. 'From the special exclusivity of this life of the mind Jewish "chutzpa", boldness, something between courage and insolence, has developed, and also the Jewish "chain", the artistic, sensitive union of grace and taste, something between enchantment and enticement'.[3] Much of this also applies to Einstein.

Many of the founding theorists of quantum mechanics, de Broglie, Heisenberg, Jordan, Dirac, and Schrödinger, were not Jewish. By the time they came on the scene, however, Bohr, Einstein, Ehrenfest, and Born had established ways of thought (the correspondence principle, the translation programme, the interpretation problem) that would dominate the early development of quantum mechanics. In this development and its challenges, Jews were disproportionally represented by, among others, Pauli, Wigner, von Neumann, Rosen, and Bohm. After Bell's intervention enticed in experimentalists, the proportion declined significantly.

During the last 20 years experimental physicists have learned to live intimately with entangled particles. They have demonstrated the possibility of quantum banking by sending messages safely through the sewers of Vienna via entangled photons.[4] Ammunition for double-slit experiments now goes well beyond electrons, photons, and buckyballs and may have arrived at bacteria. Ways to stave off decoherence give promise of pushing interference fully into the macroworld. Perhaps live-dead cats with some scale up

will become a feasible and attractive alternative to cryogenics for time travellers. You need only have yourself made catatonic in Schrödinger's sense
and direct (before undergoing the procedure) that someone look at you in
a century or two. You have a 50–50 chance of being observed alive.

The promoter of searches for macro interference, Leggett, still hopes that
his programme will fail and realism be salvaged from the wreckage of Einstein's scientific principles. As we know, results of tests that he inspired 'lend
strong support to the view that any future extension of quantum theory that
is in agreement with experiments must abandon certain features of realistic descriptions'.[5] But neither these results nor later confirmations of them
have converted Leggett. He still does not accept that quantum mechanics
is complete. 'I'm in a small minority with that point of view', he admitted,
'And I wouldn't stake my life on it'.[6]

The debate continues. The modern *Prinzipienfuchser* have been busy por-
ing over the formalism and the evidence wrought by the experimentalists,
looking for points of weakness and elaborating a variety of alternative interpretations. Some agree with Bohr that the ψ function is purely symbolic
but seek to eliminate the special place that Copenhagen appears to reserve
for measurement. Those sharing Einstein's discomfort insist that there must
be more to physics than 'piddling laboratory operations' and seek to sup-
plement the theory with new ingredients. Yet others are willing to escape
into a multiverse of possibilities, stretching the definition of 'scientific' to
its limits. A more sober approach, the search for a deep union of quantum mechanics and general relativity, has not yet succeeded. Despite much
speculation, the leading candidates for a theory of quantum gravity build on
quantum mechanics, rather than seek to rebuild on different foundations.

In April 1992, Columbia University in New York hosted a symposium
on 'The Interpretation of Quantum Theory: Where Do We Stand?' The
science journalist John Horgan attended. What he heard might have been
a discussion among the builders of Babel. 'For the most part each speaker
seemed to have arrived at a private understanding of quantum mechanics, couched in an idiosyncratic language; no one seemed to understand,
let alone agree with, anyone else'.[7]

The babble of Babel. We recall that the multiplication of natural languages and the confusion the multiplicity occasioned were a chastisement
of humans who thought to climb to heaven. The Lord did not tolerate this
attempt to invade his privacy. 'Behold', he said, 'the people is one, and

they have all one language; and this they begin to do: and now nothing will be restrained from them, which they have imagined to do'.[8] Perhaps that original universal language contained the concepts needed for correct intuitions about the microworld. But it has been lost, no doubt irrevocably, and physicists must do their best with their cacophony of tongues. That they have been able to reach so deeply and so quickly into a region they do not have words to describe may be as incomprehensible as the microworld itself.

It took two millennia to expand the domain of Euclidean geometry with bizarre regular polygons, two centuries to displace Newton's 'occult' gravitational force with Einstein's curved space-time, and two years to produce, in Bohr's complementarity, a method of using ordinary language where it does not suffice. This method in one form or another has dominated interpretations of quantum mechanics since the showdown at Solvay V in 1927. The centenary of that formative meeting will soon be upon us. Physicists will celebrate the beginnings of the debate that led a half-century later to the experimental demonstration of quantum entanglement and non-locality and the origins of what might yet prove to be a quantum information revolution.

We began our drama with an account of the work that brought Bohr and Einstein their Nobel prizes in 1922. By then they had developed opinions about the nature and purpose of physics that would shape their debate about the essence of quantum mechanics. The debate did not progress much beyond the stalemate reached by EPR in 1935 until experiments prompted by Bohm and Bell made the phenomenon of entanglement irrefutable. No Nobel prizes went to Bohm or Bell. The experimenters Clauser, Aspect, and Zeilinger also were passed over by the Royal Swedish Academy of Sciences, though they did have the satisfaction of receiving a lesser award, the Wolf prize, in 2010.

The endower of this distinction, Richard Wolf, was a German-Jewish inventor who emigrated to Cuba before the first world war and, after the second, backed Fidel Castro's revolution. At his request, Castro made him ambassador to Israel, where he remained after Cuba broke off diplomatic ties in 1973 and established his foundation for honouring scientists and artists. Its physics prize has been a good indicator of decisions later made in Stockholm: between 1978, the first year of the prize, and 2010 more than half the Wolf winners in physics went on to receive a Nobel prize.

Clauser, Aspect, and Zeilinger have been favourites among handicappers of the Nobel physics prize every year since their winning of the Wolf prize.[9] In October 2022, as we were finishing the final draft of our book, the news came that the three would at last be recognized by the Swedish Academy, 'for experiments with entangled photons, establishing the violation of Bell inequalities and pioneering quantum information science'.[10] By what we take to be pure chance—no hidden variables in the recognition— the prize came exactly 100 years after those of Einstein and Bohr. That a century elapsed between the awarding of the highest honour in physics to the founders of quantum physics and formulators of its deepest riddles, and the attribution of the same award to those who did most to expose the true nature of these riddles is a pleasing addition to the symbols of our drama. Foundational work took a long time; recognizing its value took longer.

The Nobel laudation for Clauser acknowledges his development of Bell's ideas into a plan for a practicable experiment. 'When he took the measurements, they supported quantum mechanics by clearly violating a Bell inequality'. Aspect (the laudation continues) 'closed an important loophole. He was able to switch the measurement settings after an entangled pair had left its source, so the setting that existed when they were emitted could not affect the result'. Clauser and Aspect left the field soon after completing their prize-worthy experiments, but Zeilinger, a late-comer to them, stayed on and, '[u]sing refined tools [in a] long series of experiments', helped to refashion questions of interpretation into the nascent science of quantum information.

It would be unsafe to infer that the Nobel institution has become more friendly to foundational research than it was a century ago. The prize comes 50 years after Clauser's first experiments. The chair of the Nobel Committee for Physics, Anders Irbäck, referred to a new estimate of possible practical applications of the award-winning work in an implicit explanation of the late timing of the prize. 'It has become increasingly clear that a new kind of quantum technology is emerging. We can see that the laureates' work with entangled states is of great importance, even beyond the fundamental questions about the interpretation of quantum mechanics'. Or, rather, their work is of great importance because *Prinzipienfuchser* have proved themselves useful.

This work began with impassioned, healthy debate among the founders of quantum mechanics and their disciples. Science owes a large debt to those who kept the discussions going with thought experiments against the apathy and indifference of most physicists before definitive experimental inquiries became possible. Although experiment moved the Bohr-Einstein debate to a new level and drew many into foundational research, it has by no means removed or resolved the fundamental question underlying the initial debate. Is the primary goal of science the closure of domains of experience through ever more detailed description and control of phenomena, or is it rather a search for deeper insights into the nature of physical reality? Can progress in control continue indefinitely without support from the search for deeper insights? There will be no Nobel prize for an answer. That will not shut off discussion. Our drama will continue beyond our telling of it and is unlikely to reach its final scene before science ceases or the world ends.

Acknowledgements

Ours has been a long and thoroughly enjoyable collaboration. Along the way it has provided both of us with many opportunities for learning, the sharing of insights and what might pass for wisdom, and good humour. Inevitably, we find ourselves indebted to many who have played a part in the drama of creating our drama, and we will here attempt some small measure of repayment.

It is, perhaps, unlikely that our collaboration would ever have begun without the foresight of Joanna Ashbourn, Director of the St Cross Centre for the History and Philosophy of Physics in Oxford, who seated us together at dinner following a day's discussion on paradigm shifts in June 2019. Our dinner conversation resulted in the book you now hold. In our endeavours we have benefitted from personal accounts and comments on our draft provided by our interviewees, Alain Aspect at the Institut d'Optique, John Clauser of J.F. Clauser & Associates in California, Tony Leggett at the University of Illinois, and Anton Zeilinger at the University of Vienna and the IQOQI. We thank Olival Freire Jr. (Bahia) and Don Howard (Notre Dame) for their thorough and insightful reviews of our manuscript. Substantial improvements ensued.

We have enjoyed timely access to published and unpublished documents provided by Alain Aspect, Guido Bacciagaluppi (Utrecht), Angelo Baracca (Florence), Elise Crull (New York), Olival Freire Jr., Don Howard, Alexei Kojevnikov (British Columbia), and Andrew Whitaker (Belfast). JB especially thanks N. David Mermin (Cornell University) for the article he read in 1987 which set him on a 36-year quest to understand why he could not understand quantum mechanics. JH especially thanks Diana Buchwald, Director and General Editor at the Einstein Papers Project at Caltech, and the staff of the DocuServe information delivery service at Caltech Library, for their enduring support.

A book is more than its words and illustrations. Careful assembly is required, demanding the attentions of a knowledgeable, uncompromising, yet sympathetic editor. We have both been fortunate in having several of our previous books, as well as this drama, guided to publication by Latha Menon at Oxford University Press. Here our debt runs so deep as to be—like quantum mechanics—unfathomable. Our collaboration has involved much to-and-fro, but there was no need for lengthy discussion about our dedication. Latha, this is for you.

Jim Baggott & John Heilbron
July 2023

Note: It is with considerable sadness that we acknowledge the death of John Heilbron, on 5 November 2023. John's death marks the passing of much more than a distinguished mind. He was one of a rare breed now fast disappearing: a natural polymath, endlessly curious, equally at home in science, technology, and the humanities. We will greatly miss his scholarship and his friendship.

Jim Baggott & Latha Menon
December 2023

Figure and Photo Credits

Fig. 1 (a) From Thomas Young, *Course of lectures on natural philosophy and the mechanical arts* [1807], Kelland, ed. (1845), **2**, fig. 267.

Fig. 2 From C. Christiansen, *Lærebog i physic*, Copenhagen: Gyldendal, 1910.

Fig. 3 Adapted from Kramers and Holst, *The atom and the Bohr theory of its structure: An elementary presentation*, London: Gyldendal, 1923.

Photo p. 31 Photograph by Paul Ehrenfest, courtesy AIP Emilio Segrè Visual Archives.

Fig. 6 Adapted from Filk, in Bacciagaluppi and Crull, *Grete Hermann between physics and philosophy*, Dordrecht: Springer, 2017, 74 (Figs. 5.1 and 5.2).

Photo p. 68 Photograph by Benjamin Couprie, Institut International de Physique Solvay, courtesy AIP Emilio Segrè Visual Archives.

Fig. 7 Adapted from *ECP*, **16**, 254.

Fig. 8 From Schilpp, *Einstein: Philosopher-scientist,* Evanston, IL: Library of Living Philosophers, 1949, 227.

Photo p. 104 Max-Planck-Institute, courtesy of AIP Emilio Segrè Visual Archives.

Fig. 9 From Gamow, *Thirty years that shook physics,* New York: Doubleday, 1966; New York: Dover, 1985, pp. 178, 188, 190, 192 and 208.

Photos p. 116 Above: photograph by Paul Ehrenfest, Jr., courtesy AIP Emilio Segrè Visual Archives, Weisskopf Collection. Below: Niels Bohr Institute, courtesy AIP Emilio Segrè Visual Archives.

Photo p. 132 Photo by Keystone France/Gamma Keystone via Getty Images.

Photo p. 146 Library of Congress, New York World – Telegram and Sun Collection, courtesy AIP Emilio Segrè Visual Archives.

Photo p. 161 Photograph by Alan Richards, courtesy AIP Emilio Segrè Visual Archives.

Photo p. 171 AIP Emilio Segrè Visual Archives, Physics Today Collection. This image has been cropped from the original.

Photo p. 182 Photo courtesy CERN. © CERN.

Fig. 12 From Bell, *J. de Physique*, **42** C2 Suppl. 3 (1981), in *Speakable and unspeakable in quantum mechanics,* Cambridge: Cambridge University Press, 1987, 139.

Fig. 13 Adapted from Freedman and Clauser, *Phys. Rev. Lett.*, **28**:14 (1972), 939.

Fig. 14 Adapted from Freedman and Clauser, *Phys. Rev. Lett.*, **28**:14 (1972), 941.

Photo p. 208 Photo by Lawrence Berkeley Laboratory x BB 7511 8634. Courtesy John Clauser.

Fig. 15 Adapted from Philippidis, Dewdney and Hiley, *Nuovo Cimento B*, **52** (1979), 22 (Fig. 2), 23 (Fig. 3).

Fig. 16 From Aspect, *Prog. Sci. Culture*, **1**:4 (1977), 450.

Photos p. 224 Photos courtesy of Alain Aspect.

Fig. 18 Adapted from Pan et al., *Nature*, **403** (3 February 2000), 517 (Figure 3c), 518 (Figure 4c).

Photo p. 236 Photo by HANS KLAUS TECHT/APA/AFP via Getty Images.

Photo p. 242 Photo courtesy AIP Emilio Segrè Visual Archives, Gift of Abner Shimony.

Fig. 19 Adapted from Xiao-song Ma, et al., *PNAS*, **110** (2013), 1224 (Figure 4).

Endnotes

The following abbreviations are used:

AIP   Niels Bohr Library, American Institute of Physics, College Park, MD

BCW   Niels Bohr, *Collected works*, Léon Rosenfeld et al., eds. (1972–2007)

BSC   Bohr Scientific Correspondence, Niels Bohr Archive, Copenhagen

ECP   Albert Einstein, *Collected papers*, John Stachel et al., eds. (1987+)

EPP   Einstein Papers Project, California Institute of Technology, Pasadena

HGW:C Werner Heisenberg, *Gesammelte Werke*, Abt C.

HSPS   *Historical Studies in the Physical Sciences* (1969–1985) and *Historical Studies in the Physical and Biological Sciences* (1986–2007)

MIT   Massachusetts Institute of Technology

PNAS   *Proceedings of the [US] National Academy of Sciences*

PWB   Wolfgang Pauli, *Wissenschaftlicher Briefwechsel* (1979–1999)

QT&M   John Wheeler and Wojciech Zurek, eds., *Quantum theory and measurement* (1983)

Q(Un)S Reinhold Bertlmann and Anton Zeilinger, eds., *Quantum (Un)speakables* (2 vols, 2002, 2017)

SHPMP *Studies in History and Philosophy of Science Part B: Studies in History and Philosophy of Modern Physics*

SHPS   *Studies in History and Philosophy of Science Part A*

Solvay V Institut International de Physique Solvay, *Rapports et discussions du cinquième Conseil . . . 1927* (1928)

Solvay VI Institut International de Physique Solvay, *Rapports et discussions du sixième Conseil . . . 1930* (1931)

Prologue

1. Weinberg, *Dreams* (1993), 13.
2. Newton to Richard Bentley, 25 February 1692/3, in Newton, *Correspondence* (1959–1977), **3**, 254.
3. Newton, *Principles* (1934), 547.
4. Cornu, in *Congrès* (1900), **1**, 5.
5. Forman, *HSPS*, **3** (1971), 8–37.
6. For documentation for this and the following paragraphs, see Heilbron, in Bernhard, Crawford, Sörbom, eds., *Science, technology* (1983), 52–6.
7. Rosenfeld, *Z. Phys.*, **171**:1 (1963), 243.
8. Galilei, *The Assayer* (1623), in Drake and O'Malley, *Controversy* (1960), 184.

ACT I: CORRESPONDENCE TO COMPLEMENTARITY

Mutual Admiration

1. Young, *Lectures* (1845), Kelland, ed., **1**, 364–5, 382, and **2**, fig. 267 (text of 1807).
2. Peacock, *Life* (1855), 162.
3. Ibid., 382.
4. Ibid., 370.
5. Kirsten and Körber, *Wahlvorschläge* (1975), 202.
6. Newton, *Principles* (1934), 398 (text of 1712).
7. Georg Hevesy, reporting on Einstein to Bohr, 23 September 1913, in *BCW*, **2**, 532.
8. Arnold Sommerfeld, *Atombau* [1919] (1921), 400 ('Zauberstab').
9. *ECP*, **6**, 367–8 (parallel to radioactivity), 383 ('verblüffend einfacher').
10. Einstein to Max and Hedi Born, 19 April 1924, *ECP*, **14**, 371.
11. *ECP*, **6**, 386–97.
12. Kojevnikov, *Copenhagen network* (2020), 3–4, 23–5, 69, 75–6, 103–4.
13. Enz, *No time* (2002), 49, 55; Cassidy, *Uncertainty* (1992), 109; Sommerfeld, *Atombau* (1921), vii.
14. Enz, *No time* (2002), 7, 11, 13.
15. Enz, *No time* (2002), 49–52; Cassidy, *Uncertainty* (1992), 97–8, 109, 151–3.
16. Enz, *No time* (2002), 147, 196–7; Cassidy, *Uncertainty* (1992), 139.
17. Einstein to Jun Ishiwara, 27 March 22, in *ECP*, **13**, 213, and to Planck and others, May–July 1922, in ibid., 321–2, 388, 398–9, 400, 404, 408–9. Einstein rejoined the ICIC in 1924.
18. Einstein to Ehrenfest, 23 March 1922, in *ECP*, **13**, 202–3.
19. Bohr to Einstein, 11 November 1922, and reply, 10 January 1923, in *ECP*, **13**, 593, 697–8.

An Honourable Funeral

1. Einstein to Nobelkommitté för Fysik, 26 October 1923, in *ECP*, **14**, 220, and ibid., 765–6 (text of 7 May 1925).
2. Einstein, in de Haas-Lorentz, *Lorentz* (1957), 8.

3. Ehrenfest, *Coll. sci. papers* (1959), 478.
4. Einstein to Ehrenfest, 20 June 1912, in *ECP*, **5**, 484 (quote); ibid., 473, 486 (re: habilitation).
5. Lorentz to Einstein, 13 November 1921, in *ECP*, **12**, 347–51.
6. Einstein to Max and Hedi Born, 19 April 1924, in *ECP*, **14**, 271.
7. Einstein to Sommerfeld, 4 January 1921, in *ECP*, **12**, 27; *ECP*, **7**, 484–6 (the experiment); Einstein to Max and Hedi Born, 30 December 1921, in *ECP*, **12**, 399 (quote); Ehrenfest to Einstein, 19 January 1922, in *ECP*, **13**, 106–8.
8. Einstein to Sommerfeld, 28 June 1922, in *ECP*, **13**, 120 (quote).
9. Van Dongen, in Aaserud and Kragh, *Hundred years* (2015), 318–26.
10. Ehrenfest to Einstein, 16 September 1925, in *ECP*, **15**, 122–4.
11. Ibid., and Ehrenfest to Einstein, 29 May 1928, in *ECP*, **16**, 334.
12. Ehrenfest to Einstein, 16 September, and reply, 18 September 1925, in *ECP*, **15**, 123, 126–7; cf. Seth, *Crafting* (2010), 186.
13. Heisenberg to Pauli, 26 March 1923, in *PBW*, **1**, 86.
14. Pauli to Sommerfeld, 6 June 1923, in *PBW*, **1**, 95.
15. Born, *Naturwiss.*, **11**:27 (1923), 542.
16. Pauli to Bohr, 11 March 1924, in *PBW*, **1**, 143–4; Heisenberg to his father, 29 November 1923, quoted in Cassidy, *Uncertainty* (1992), 171.
17. Heisenberg to his parents, 16 June 1922, 29 November 1923, 23 March 1924, and 19 November 1924, in Heisenberg, *Liebe Eltern* (2003), 34, 56, 72, 82, resp.
18. Pauli to Sommerfeld, 6 June 1923, *PBW*, **1**, 97.
19. Pauli to Kronig, 21 May 1925, in *PBW*, **1**, 214.
20. Pauli to Heisenberg, 12 December, and reply, 15 December 1924, in *PBW*, **1**, 190, 192–3.
21. Bohr to Pauli, 22 December, and reply, 31 December 1924, in *PBW*, **1**, 194–5, 197–8.
22. Ehrenfest, *Coll. sci. papers* (1959), 618.
23. Heilbron, *HSPS*, **13**:2 (1983), 309–10.
24. Enz, *No time* (2002), 110–17; Jammer, *Development* (1966), 149–50.
25. Pais, *Inward bound* (1986), 278, 279.
26. Bohr, letter to Kronig, 26 March 1926, quoted in Pais, *Niels Bohr's times* (1991), 243.
27. Enz, *No time* (2002), 117.
28. Pauli to Kramers, 8 March 1926, in *PBW*, **1**, 304, 307, and Ehrenfest to Pauli, 26 November 1928, ibid., 477.
29. Enz, *No time* (2002), 112–15. Llewellyn Hilleth Thomas, a Cambridge graduate working in Copenhagen, noticed the relativity error.
30. Schrödinger to Bohr, 24 May 1924, in *BCW*, **5**, 29–30.
31. Report on Einstein's seminar of 28 May 1924, in *ECP*, **14**, 393.
32. Enz, *No time* (2002), 158; Cassidy, *Uncertainty* (1992), 190.
33. Bohr to Darwin, 21 April 1925, quoted in Pais, *Niels Bohr's times* (1991), 238.
34. Einstein, *Berliner Tageblatt*, 20 April 1924, in *ECP*, **14**, 395–6.

New Ways to Calculate

1. Kramers, *Nature*, **114** (30 August 1924), 311; cf. Konno, *Centaurus*, **36**:2 (1993), 117–66.

2. Born to Einstein, 15 July 1925, in Born, *Born–Einstein letters* (1971), 83–4. 'Prof. Born and Prof. Frank . . . inform me [a representative of the International Education Board] that Dr Jordan is perhaps the most outstanding man of his age in Germany today'; letter of 1 Mar 1927, in Kojevnikov, *Copenhagen network* (2020), 116.

3. Einstein to Bohr, 20 October 1925, in *BCW*, **5**, 312.

4. Bohr, *Nature*, **116** Suppl. (5 December 1925), 852, = *BCW*, **5**, 280; cf. ibid., 265.

5. Bohr, late 1925, in *BCW*, **5**, 265–6; repeated in Bohr, *Nature*, **116** Suppl. (5 December 1925), 852, = *BCW*, **5**, 280.

6. Bohr to Born, 1 May 1925, in *BSC*, **5**, 311, and in *Nature*, **116** Suppl. (5 December 1925), 848, = *BCW*, **5**, 276.

7. Einstein to Gustav Mie, 22 May 1926, in *ECP*, **15**, 484.

8. Louis de Broglie, from the 1963 version of his PhD thesis, quoted in Pais, *Subtle is the Lord* (1982), 436.

9. Einstein to Paul Langevin, 16 December 1924, in *ECP*, **14**, 608, and to Lorentz, same date, ibid., 610.

10. The kinetic energy $T_n = \frac{1}{2}nh\omega_n = \frac{1}{2}mv_n^2$ or $\frac{1}{2}p_nv_n$, where $p_n = mv_n$ is the instantaneous *linear* momentum of the electron in the nth orbit. The orbital frequency $\omega_n = v_n/2\pi r_n$, where r_n is the orbital radius. It follows that $\frac{1}{2}nhv_n/2\pi r_n = \frac{1}{2}p_nv_n$, or $nh/2\pi = p_nr_n$, which is the *angular* momentum of the electron in the nth orbit.

11. Jammer, *Development* (1966), 244–5.

12. Ibid., 247.

13. Schrödinger to Einstein, 3 November 1925, in *ECP*, **15**, 181–2, and 23 April 1926, ibid., 442.

14. Bloch, *Phys. Today*, **29**:12 (1976), 23.

15. Schrödinger, *Ann. Phys.*, **79**:4 (1926), 361, in Schrödinger, *Papers* (1982), 1.

16. Schrödinger, *Ann. Phys.*, **79**:4 (1926), 375, in Schrödinger, *Papers* (1982), 10–11.

17. Jammer, *Development* (1966), 271–5.

18. Planck to Schrödinger, 2 April, 24 May, and 4 June 1926, in Schrödinger, et al., *Briefe* (1963), 3, 6, 12.

19. Einstein to Ehrenfest, 12 April 1926, in *ECP*, **15**, 429–30 (geistreich, Höllenmachine), and 20 November 1925, ibid., 202 (egg).

20. Einstein to Ehrenfest, 28 July 1926, in *ECP*, **15**, 575.

21. Bloch, *Phys. Today*, **29**:12 (1976), 24, for the German verse, attributed to Erich Hückel, one of Debye's collaborators.

22. Quoted in Moore, *Schrödinger* (1989), 208, from a text of 1926.

23. Born, *Z. Phys.*, **37**:12 (1926), 863–4.

24. Born, *Z. Phys.*, **37**:12 (1926), 866.

25. Pauli to Heisenberg, 19 October, and reply, 28 October 1926, in *PWB*, **1**, 347, 349.
26. Born, *Z. Phys.*, **38**:11/12 (1926), 803–4. Cf. Jammer, *Philosophy* (1974), 41–3.
27. Born, *Z. Phys.*, **38**:11/12 (1926), 826–7.
28. Jammer, *Philosophy* (1974), 63.
29. Born, *Z. Phys.*, **37**:12 (1926), 866 (quotes); ibid., **38**:11/12 (1926), 803–4 (connection with Einstein).
30. Born, *Z. Phys.*, **38**:11/12 (1926), 826–7.
31. Born, *Z. Phys.*, **37**:12 (1926), 864; Schrödinger, quoted in Meyenn, *Gesnerus*, **44**:1/{2} (1987), 109.
32. Heisenberg to Pauli, 8 June and 28 July 1926, in *PWB*, **1**, 328, 338.
33. Born to Einstein, 30 November 1926, in *ECP*, **15**, 647–8.
34. Schrödinger to Willy Wien, 26 August 1926, in Meyenn, *Gesnerus*, **44**:1/{2} (1987), 108.
35. Heisenberg, *Physics and beyond* (1971), 73–6.
36. Bohr to Fowler, 26 October 1926, in *BCW*, **6**, 423–4.
37. Schrödinger to Wien, quoted without date in Moore, *Schrödinger* (1989), 228.
38. Schrödinger to Bohr, 23 October 1926, in *BCW*, **6**, 459–60.
39. Schrödinger to Planck, 4 July 1927, in Przibram, *Letters* (1967), 19–20.
40. Quoted in Moore, *Schrödinger* (1989), 251, from a text of 1930.
41. Schrödinger to Bohr, 23 October 1926, in *BCW*, **6**, 459–60; ' . . . ' is in the original text.
42. Ibid., 460.

New Ways to Think

1. Heisenberg, *Z. Phys.*, **43**:3/4 (1927), 174–5.
2. Ibid., 197–8.
3. Letter of 30 May 1927, in Heisenberg, *Liebe Eltern* (2003), 122.
4. Heisenberg, *Physical principles* (1930), 21.
5. Camilleri, *Heisenberg* (2009), 104–5.
6. Heisenberg, *Physical principles* (1930), 14.
7. Margenau, *The Monist*, **42**:2 (1932), 183–4.
8. Weizsäcker, *Z. Phys.*, **70**:1/2 (1931), 114–30.
9. Filk, in Bacciagaluppi and Crull, *Hermann* (2017), 71–83.
10. Bohr to Einstein, 19 April 1927, in *BCW*, **6**, 21–4, 418–21.
11. Notes, 10 July 1927, in *BCW*, **6**, 59–62.
12. *BCW*, **6**, 28.
13. Rosenfeld, interview by Thomas Kuhn and John Heilbron, 1 July 1963, quoted in Pais, *Niels Bohr's times* (1991), 315.
14. Discussion, reprinted in *BCW*, **6**, 137–41; Heisenberg, *Z. Phys.*, **43**:3/4 (1927), 179–80.
15. Pauli to Bohr, 13 March 1928, in *BCW*, **6**, 43–4.

16. The following paragraphs summarize Bohr, *Nature*, **121** Suppl. (14 April 1928), 580–90, in *BCW*, **6**, 148–58.

17. Schrödinger to Bohr, 5 May 1928, and reply, 23 May 1928, in *BCW*, **6**, 47–9, 463–7.

18. Howard, in Faye and Folse, *Niels Bohr* (1994), 203–4, 213–7, 223–5, and in Freire, *Oxford companion* (2022), 527, 534–5.

ACT II: UNCERTAINTY TO ORTHODOXY

Incompatible Conceptions

1. Einstein to Elsa Einstein, 23 October 1927, in *ECP*, **16**, 133–4.

2. Bacciagaluppi and Valentini, *Quantum theory* (2009), 10.

3. Heilbron, *Dilemmas* (2000), 106–12.

4. Lorentz to Einstein, 6 April 1926, in *ECP*, **15**, 418; Heilbron, *Dilemmas* (2000), 107–8.

5. Lorentz to Einstein, 6 April 1926, and reply, 12 April 1926, in *ECP*, **15**, 418, 431.

6. Lorentz to Einstein, 28 April 1926, and reply, 1 May 1926, in *ECP*, **15**, 448–9, 452–3.

7. Einstein to Lorentz, 17 June 1927, in *ECP*, **16**, 57.

8. *ECP*, **15**, 813.

9. Einstein to Ehrenfest, 5 May 1927, and to Born, ca. 10 May 1927, and Heisenberg to Einstein, 19 May 1927; in *ECP*, **15**, 815, 817, 823–4.

10. *ECP*, **15**, 814; Belousek, *SHPMP*, **27**:4 (1996), 438–53.

11. Bragg, in Solvay V, 32–7; Compton, in ibid., 56–7, 75–6, 80, 85 (quote).

12. De Broglie, *Journal de physique*, **8** (1927), 225, 232, 241.

13. De Broglie, in Solvay V, 108–9 (singularity), 111, 114, 116 (quote), 127.

14. Ibid., 45–8, 135.

15. Lorentz, in ibid., 53, 86–7.

16. Bohr, in ibid., 91–2, 103.

17. Born and Heisenberg, in ibid., 143–5.

18. Ibid., 156–7, 160.

19. Ibid., 172, 178.

20. Ibid., 183, 184.

21. Ibid., 208–9, 211–12.

22. Bacciagaluppi and Valentini, *Quantum theory* (2009), 19–21.

23. *BSC*, **6**, 148–58; Bohr, *Naturwiss.*, **16**:15 (1928), 245–57; Solvay V, 215n.

24. Einstein, in Solvay V, 254–6, and *ECP*, **16**, 138–40; cf. Bacciagaluppi and Valentini, *Quantum theory* (2009), 486–8.

25. *BCW*, **6**, 103.

26. De Broglie, *Introduction* (1930), 142–5 (text of 1928).

27. Bacciagaluppi and Valentini, *Quantum theory* (2009), 512.

28. Lochak, in Barut et al., *Quantum* (1984), 23; Pauli to Bohr, 8 August 1929, in *PWB*, **1**, 404–5; Bacciagaluppi and Valentini, *Quantum theory* (2009), 55, 65–78.
29. Jammer, *Philosophy* (1974), 114.
30. Solvay V, 261–2, 264–5.
31. Bacciagaluppi and Valentini, *Quantum theory* (2009), 437, 462.
32. Solvay V, 285, 287–8.
33. Einstein to Sommerfeld, 9 November 1927, in *ECP*, **16**, 149.
34. Ehrenfest to Goudsmit, et al., 3 November 1927, in *BCW*, **6**, 37–40, slightly revised using the German text, ibid., 415–18.
35. *BCW*, **7**, 349–51.
36. *BCW*, **10**, 442–6.

Measurement and Impossibility

1. *HGW:C*, **1**, 23, 26–7; 'was geschieht . . . ', attributed to Bohr, ibid., 31.
2. *HGW:C*, **1**, 32–4, 36–7, 39, = *Erkenntnis*, **2**:1 (1931), 175–7, 179–80, 182.
3. Heisenberg, *Erkenntnis*, **2**:1 (1931), 183–8.
4. Heims, *John von Neumann* (1980), 48–53.
5. Van Hove, *Bull. Am. Math. Soc.*, **65** (1958), 95–99.
6. Von Neumann, *Math. foundations* (1955), 28.
7. Ibid., 31.
8. For this and the preceding paragraph, ibid., 311–14, 326–7, 305 (first quote), 328 (second), text of 1932.
9. Ibid., 420, n. 207, referring to Bohr, *Naturwiss.*, **17**:26 (1929), 483–6, = *BCW*, **6**, 203–6.
10. Ibid., 418–19.
11. Ibid., 351–3, 357, 420 (quote), 421.
12. Szilard, *Z. Phys.*, **53**:11/12 (1929), 840–56, in NASA Tech. Transl. F–16723 (1976), 14–18.
13. Jammer, *Philosophy* (1974), 480, italics added.
14. Pais, *Niels Bohr's times* (1991), 435.
15. Teller, *Proc. Biennial Meeting Phil. Sci. Assoc.*, **2** (1980), 201.
16. Cassidy, *Uncertainty* (1992), 208–11.
17. *BCW*, **6**, 208–11 (text of 1929).
18. *BCW*, **6**, 212–17.
19. Planck to Bohr, 14 July 1929, in *BCW*, **6**, 192, 456.
20. *BCW*, **6**, 244–5, 249–53.
21. Heisenberg, *Principles* (1930), 65.
22. Born and Jordan, *Quantenmechanik* (1930), v–vi, 324–5; Pauli, *Coll. sci. papers*, **2** (1964), 1397.

EPR, Faust, and the Cat

1. Solvay VI, 278–80.
2. Gerlach, *Metall–Wirtschaft*, **9** (1930), 939.
3. Bohr, in Schilpp, *Einstein* (1949), 224.

4. Bohr, in *BCW*, **7**, 364–8 (text of 1949).
5. Useful discussions in de la Torre, Daleo, and Garcia–Mata, *Eur. J. Phys.*, **21**:3 (2000), 253–60, and **23**:4 (2002), L15–16; Hnizdo, *Eur. J. Phys.*, **23**:4 (2002), L9–13.
6. Bohr, in *BCW*, **10**, 448, text of a talk given at Solvay XII (1961); Rosenfeld, *Phys. Today*, **16**:10 (1963), 54, the final sketch on the blackboard.
7. Quoted by Hendrik Casimir in Casimir to Pais, 31 December 1977, in Pais, *Subtle is the Lord* (1982), 449.
8. Ehrenfest to Bohr, 9 July 1931, in BSC, reel 18:583; Howard, *Iyyum*, **56** (January 2007), 75–9, and in Miller, *Sixty-two years* (1990), 98–9.
9. Robertson, *Phys. Rev.*, **34**:1 (1929), 163–4; Heisenberg, *Z. Phys.*, **43**:3/4 (1927), 180, hints at a connection with the commutator but does not derive one and arrives at a factor of $1/2\pi$ by another route.
10. Ehrenfest to Bohr and Einstein, 12 September 1931, in BSC, reel 18:586, and Einstein to Ehrenfest, 2 October 1931 ('im Bett geschrieben') (EPP).
11. Bohr, in *BCW*, **7**, 368–9, = Bohr, in Schilpp, *Einstein* (1949), 228–9.
12. *HGW:C*, **1**, 97; Cassidy, *Uncertainty* (1992), 379–81, 391–4.
13. Lenin, *Materialism* (1908), esp. chapters 1 and 6; 'gibberish', 93.
14. Denman, *Phys. Today*, **20**:3 (1967), 141; Haynes, et al., *Spies* (2009), 73–5.
15. Einstein to Molotov, 13 March and 4 July 1936 (EPP).
16. Rosen to Einstein, 26 February 1937 and 24 March 1938 (EPP).
17. Einstein to Rosen, 18 April 1938 and 19 November 1940 (EPP).
18. Einstein, Podolsky, and Rosen, *Phys. Rev.*, **47**:10 (1935), 777–80.
19. Rosenfeld, in Rozental, ed. *Niels Bohr* (1967), 128.
20. Jammer, *Philosophy* (1974), 171–4, summarizes the sallies.
21. Pauli to Heisenberg, 15 June 1935, in *PWB*, **2**, 403–4.
22. Heisenberg, unpublished manuscript, in *PWB*, **2**, 414. Trans. Bacciagaluppi and Crull, *Einstein's paradox* (in press), 263.
23. Heisenberg, *Neuere Fortschr. exak. Naturw.* (1936), 99–100, in Bacciagaluppi and Crull, *Einstein's paradox* (in press), 265.
24. Heisenberg, in Pauli, et al., *Niels Bohr* (1955), 26.
25. Calvino, *Classics* (1999), 135, concerning Stendahl's *Charterhouse of Parma*.
26. Heisenberg to Pauli, 2 July 35, in *PWB*, **2**, 407–8.
27. Bohr, *Phys. Rev.*, **48**:8 (1935), 696–703.
28. Heilbron, *Rev. Hist. Sci.*, **38**:3/4 (1985), 47n.
29. Einstein to Schrödinger, 19 June 1935 (EPP).
30. Fine, *Shaky game* (1996), 35–6.
31. *New York Times*, 4 May 1935.
32. Thomas, *Institute letter* (Fall 2013).
33. Einstein to Schrödinger, 19 June 1935 (EPP).
34. Einstein to Schrödinger, 8 August 1935 (EPP).
35. Schrödinger, *Naturwiss.*, **23**:48 (1935), 812.
36. Einstein to Schrödinger, 4 September 1935 (EPP).

37. Schrödinger, *Naturwiss.*, **23**:48 (1935), 807–12; **23**:49 (1935), 823–8; **23**:50 (1935), 844–9. Trans. Trimmer, *QT&M* (1983), 152–67 (161 quote).

38. Schrödinger, from Dublin, to Pauli, early July 1935, in *PBW*, **2**, 406–7.

39. Pauli to Schrödinger, 9 July 1935, in *PBW*, **2**, 419–21.

40. Schrödinger, *Naturwiss.*, **23**:48 (1935), 811–12; **23**:49 (1935), 823, 825–8.

41. Gavroglu, *London* (1995), 8–11, 170–5.

42. London and Bauer, *Théorie* (1939), 21–5, 38–43, 48–50 (quotes).

43. Crawford, *Nobel population* (2002), 110–31.

44. Max Delbrück, who would turn his talents to biology and finish his career as a professor at Caltech, composed most of the Copenhagen *Faust*; the version quoted here, with brilliant cartoons by George Gamow and a less brilliant translation by Barbara Gamow, appears in Gamow, *Thirty years* (1966), 70–218.

Missionaries of the Copenhagen Spirit

1. Delft, *Phys. Today*, **67**:1 (2014), 47.

2. Ehrenfest, letter to his students, 15 August 1932, in *PBW*, **2**, 142; and to Pauli, 28 November and 2 December 1932, in ibid., 142, 143.

3. Pauli to Heisenberg, 30 September 1933, and reply, 7 October 1933, in *PBW*, **2**, 216, 218.

4. Einstein, *Later years* (1950), 236–8.

5. Jordan, *Erkenntnis*, **4**:1 (1934), 247–8, *Naturwiss.*, **22**:29 (1934), 490, and *Quantentheorie* (1936), 313.

6. Jordan, *Verdrängen* (1948), 9–10.

7. Ibid., 45–7, 51–81.

8. Enz, *No time* (2002), 209–11; the marriage took place on 23 December 1929.

9. Jung, *Psychology and alchemy*, in *Works* (1953–79), **12**, 42, 46, 106.

10. Franz, *Number* (1974), 5.

11. Pauli, *Coll. sci. papers* (1964), **2**, 1108, 1156 (texts of 1948, 1950, resp.).

12. Ibid., 1158, 1215 (texts of 1950 and 1954, resp.).

13. Cassidy, *Uncertainty* (1992), 74, 368–73.

14. Heisenberg to Bohr, 1 October 1935 (BSC).

15. Einstein to Pauli, 24 December 1929, in *PBW*, **1**, 528; Bohr to Pauli, 25 January 1933, *PBW*, **2**, 154.

16. Pauli to Heisenberg, 13 May 1954, cited in Heilbron, *Rev. Hist. Sci.*, **38**:3/4 (1985), 220.

17. Pauli to Weisskopf, 23 February 1954, cited in Heilbron, *Rev. Hist. Sci.*, **38**:3/4 (1985), 221.

18. Einstein to Schrödinger, 9 August 1939, in Rosenfeld, *Selected papers* (1979), 520–1, 535; Bohr to Ehrenfest, 21 October 1925 (BSC). Rosenfeld had also studied with de Broglie and Born before meeting Bohr in 1929; Jacobsen, *HSPS*, **37** Suppl. (2007), 5–7.

19. Letters from Klein, 12 November 1934, and F.G. Donnan, 11 May 1934 and 1 June 1935 (BSC).

20. Pauli to Heisenberg, 13 May 1954, cited in Heilbron, *Rev. Hist. Sci.*, **38**:3/4 (1985), 220, and Pauli, *Coll. sci. papers* (1964), **2**, 738, resp.

21. Planck, *Wege* (1933), 258–9.

22. The heresiarch philosopher Karl Popper liked to play with the priestly metaphor, e.g., in his *Quantum theory* (1982), 99–100 (text of 1956/7).

23. Heilbron, *Rev. Hist. Sci.*, **38**:3/4 (1985), 205n28.

24. Pauli to Bohr, 16 June 1928, quoted in Heilbron, *Rev. Hist. Sci.*, **38**:3/4 (1985), 205.

25. Serber, *Phys. Today*, **20**:10 (1967), 35.

26. Bridgman, *Logic* (1927), 3.

27. Kemble, *Phys. Rev. Suppl.*, **1** (1929), 160.

28. Kemble, *J. Franklin Inst.*, **225**:3 (1938), 267, 269–270.

29. Heilbron, *Bohr* (2020), 62–3.

30. Schweber, *Einstein and Oppenheimer* (2008), 140–4; Pauli to Ehrenfest, 15 February 1929, in *PBW*, **1**, 486–7.

31. Compton, *Atomic quest* (1956), 125.

32. Pauli to Ehrenfest, 5 February 1929, in *PWB*, **1**, 486–7; Oppenheimer, interview with T.S. Kuhn, Session 2, 20 November 1963, 21.

33. Pauli, *Handbuch der Physik* (1933), 7, in *Coll. sci. papers* (1964), **1**, 777.

34. Cf. Serber, *Phys. Today*, **20**:10 (1967), 38.

35. Peters, *Notes* (1939), 8.

36. Jordan, *Quantentheorie* (1936), 286–9, 294–7, 310–12, and *Physik* (1936), 125–6, 127 (quote), = *Physics* (1944), 151–2, 153 (quote).

37. Bohr, *BCW*, **10**, 245, = *Atomic physics* (1958), 27 (quote).

38. Jordan, *Naturwiss.*, **20**:45 (1934), 819–20; *Quantentheorie* (1936), 300–1, 312; *Physik* (1936), 108–9, = *Physics* (1944), 130–1.

39. Einstein, as quoted in Shankland, *Am. J. Phys.*, **31**:1 (1963), 50, 53–4.

40. Bohr to Kramers, 14 March 1936, and Oseen to Bohr, 25 June 1935 (BSC).

41. Bohr, *Erkenntnis*, **6**:1 (1936), 279, 281, 283–4, and Lenzen, ibid., 286, 328–30, 334–5.

42. Bohr, *Erkenntnis*, **6**:1 (1936), 293, 297, 301–2, and in *Phil. Sci.*, **4**:3 (1937), 289, 295–7, = *BCW*, **10**, 39, 45–7 (46 quote).

43. Frank, *Erkenntnis*, **6**:1 (1936), 312, 315–16, and Schlick, ibid., 319, 321, 326.

44. Hansen-Shaberg, in Bacciagaluppi and Crull, *Hermann* (2017), 6–11; Hermann, in ibid., 257–69 (text of 1933); Heisenberg, as quoted in Gustav Heckmann to Hermann, 17 December 1933, in ibid., 221; von Weizsäcker, *NTM*, **1** (1993), 10–13; Heisenberg, *Physics and beyond* (1931), 17–24, a trivializing recollection of the discussions of 1934.

45. Soler, in Bacciagaluppi and Crull, *Hermann* (2017), 60–4; Seevinck, in ibid., 110–15; Hermann, in ibid., 232–4 (text of 1933), 251–4 (1935), and *Erkenntnis*, **6**:1 (1936), 342–3.

46. Filk, in Bacciagaluppi and Crull, *Hermann* (2017), 73–81; Frappier, in ibid., 100–1; Hermann, in ibid., 257–9 (1935); Jammer, *Philosophy* (1974), 207–9.

47. Vogel, *Wirken* (1961), 79; Jordan to Bohr, May 1945 (BSC); Jordan, *Physics* (1944), 144, 148, 155, 160, and letter to Bohr, May 1945 (BSC).

48. Heisenberg, *Physics and beyond* (1971), 143–54, 159–79, and *Physics and philosophy* (1958), 205 (quote); Elisabeth Heisenberg, *Leben* (1980), 53–4, 84, 91–101, 113–14.

49. Bavink, *Science and God* (1933), 71, 79, 167–8 (quotes).

50. Bavink, *Unsere Welt*, **25** (1933), 233, and *Science and God* (1933), 119, 135–6 (quote), 159 (last quote), 170.

51. Hentschel, *Sudhoffs Archiv*, **77** (1993), 5–11, 23–5.

52. Darwin, *Science*, **73** (19 June 1931), 660; Compton, *Science*, **74** (14 August 1931), 172–3.

53. Inge, *God and the astronomers* (1933, 1934), 58–9 (quotes); Means, *J. Phil.*, **33**:4 (1936), 88–9; Davidson, *Free will* (1937), 6, 23, 74.

54. Douglas, *Life* (1956), 131–7, 148–50.

55. Jordan, *Physics* (1944), 158–9.

56. Eddington, *Nature phys. world* (1928), viii, xv–xvii, 295, 344.

57. Ibid., 291, 292, 295 (quotes, resp.), 258–9, 299.

58. Ibid., 275, 277, 282, 311, 321–3, 327, 330–1.

59. Ibid., 350.

60. Hartshorne, *J. Phil.*, **29**:16 (1932), 421, and Tyrrell, *Phil.*, **7**:28 (1932), 410 ('wobble').

61. Samuel, *Phil.*, **11**:41 (1936), 3, 5, 15, 17 (the bishop's contribution).

62. Lenzen, *J. Phil.*, **30**:11 (1933), 287–8.

63. Aristotle, *Post. anal.*, 1.12 (77b 5–6, 14–15).

ACT III: ORTHODOXY TO UNCERTAINTY

Postwar Hostilities

1. Moore, *Schrödinger* (1989), 298; Pais, *Niels Bohr's times* (1991), 502.

2. McCrea, in Kilmister, *Schrödinger* (1987), 125–6, 128.

3. Schweber, *QED* (1994), 157–93.

4. Pauli, *Dialectica*, **2**:3/4 (1948), 307–9; Einstein, in ibid., 320–2; Pauli to Einstein, 21 April, and reply, 2 May 1948, in *PWB*, **3**, 520–1, 524–5.

5. Pauli to Bohr, 4 June and 16 October 1948, resp., in *PWB*, **3**, 530, 573.

6. Heisenberg, *Dialectica*, **2**:3/4 (1948), 333–6 (quote); de Broglie, ibid., 326–9; Pauli, ibid., 311.

7. Einstein, in Schilpp, *Einstein* (1949), 3.

8. Ibid., 83.

9. Ibid., 85.

10. Ibid., 87.

11. Margenau, in Schilpp, *Einstein* (1949), 265, 267.

12. Pauli, in Schilpp, *Einstein* (1949), 157.

13. Schilpp, *Einstein* (1949), 127 (de Broglie), 158 (Pauli), 163–4, 177 (Born).

14. Bridgman, in Schilpp, *Einstein* (1949), 354.

15. Lenzen, in Schilpp, *Einstein* (1949), 361–3, 371–3, 383–4 (quote).

16. Einstein, in Schilpp, *Einstein* (1949), 683.

17. Pauli to Bohr, 17 August and reply, 15 September 1948, in *PWB*, **3**, 565–9; Rozental, in Rozental, *Life* (1967), 180 (two-year gestation); Pais, in ibid., 225.

18. Bohr, *Dialectica*, **2**:3/4 (1948), 312–19, = *BCW*, **7**, 330–7, and in Schilpp, *Einstein* (1949), 210, 223, 237–40.

19. Bohr, in Schilpp, *Einstein* (1949), 239, 235.

20. Einstein, in Schilpp, *Einstein* (1949), 667, 669.

21. Ibid., 667, 681–2.

22. Gerard Bonnot, 'Oppenheimer parle d'Einstein', *L'Express*, 20–26 December 1965, in Schweber, *Einstein and Oppenheimer* (2008), 281–2.

23. Pauli to Einstein, 22 April 1948, in *PWB*, **3**, 522–3.

24. Schrödinger, *Acta Phys. Austriaca*, **1** (1948), 205–7, 211, 214–17 (quote).

25. Ibid., 229–30, 234–7, 242–4 (quote).

26. Schrödinger, Smith. Inst., *Report* (1950), 185–91.

27. Schrödinger, *Dublin seminars* (1995).

28. Schrödinger, *Brit. J. Phil. Sci.*, Part I, **3**:10 (1952), 109–17, 120–3; Part II, **3**:11 (1952), 233–4.

29. Pauli to Schrödinger, 26 June, and to Bohr, 16 September 1952, in *PWB*, **4**, 657–8, 730.

30. Quoted in Moore, *Schrödinger* (1989), 279 (text of 1959), 472.

31. Editor's comments, *PWB*, **4**, 656.

32. Born, *Brit. J. Phil. Sci.*, **4**:14 (1953), 95–6, 100–2.

33. Schrödinger, in George, *De Broglie* (1953), 16 (quote), 20, 24, 26.

34. Pauli, in George, *De Broglie* (1953), 35, 38, 40 (quote).

35. Rosenfeld, in George, *De Broglie* (1953), 52–3, 62–3 (quote).

36. Strauss to Rosenfeld, 30 May 1953, in *PWB*, **4**:1, 73; Pauli to Markus Fierz, 17 July 1948, in *PWB*, **3**, 545.

37. Camilleri, *Perspect. Sci.*, **17**:1 (2009), 29, 33–5, 37–40.

38. Camilleri, *SHPMP*, **38**:3 (2007), 516–19, 523–5; Heisenberg, *Physics and philosophy* (1958), 49.

39. Cf. Stapp, *Am. J. Phys.*, **40**:8 (1972), 1098–100, 1105–8, who identified these essential consequences as core beliefs and obtained Heisenberg and Rosenfeld's assent to his representation (ibid., 1113–15).

40. Camilleri, *Perspect Sci.*, **17**:1 (2009), 41; Heisenberg, in Pauli et al., *Niels Bohr* (1955), 12–13, 16, 22–5, 28, and *Physics and philosophy* (1958), 44–58.

41. Rosenfeld, *Nature*, **186** (11 June 1960), 831; Heisenberg, *Physics and philosophy* (1958), 177–82 (language), 47–8, 53–4, 58 (subjectivity); Pauli, as quoted in Camilleri, *Perspect. Sci.*, **17**:1 (2009), 43.

42. Pauli to Bohr, 3 October 1950, and to Fierz, 13 October 1951, in *PWB*, **4**:1, 169–72 and 387–8, resp.

43. de Toqueville, *Democracy* (1875), **2**, book 2, chap. 1.

44. Hofstadter, *Anti-intellectualism* (1966), 7.

45. Kaiser, *Hippies* (2011), 15–23, *HSPS*, **33** (2002), 131–59, and *Phys. World*, **20**:5 (2007), 28–33.

46. Serber, *Phys. Today*, **20**:10 (1967), 38.

47. Schiff, *Quantum mechanics* (1949), 8.

48. Whitaker, *Bell* (2016), 46, 54, 56.

49. Mermin, *Phys. Today*, **57**:5 (2004) 10–11.

50. Mermin, note to Baggott, 12 December 2019.

Skirmishes in Princeton

1. Peat, *Potential* (1997), 7–14, 21, 26–7, 3–7; Freire, *Bohm* (2019), 20–3.

2. Phillips, interview with Katherine Sopka, 5 December 1977, and with Alexei Kojevnikov, 2 December 1997 (AIP).

3. Peat, *Potential* (1997), 56–8; Freire, *Bohm* (2019), 35–8.

4. Peters, *Notes* (1939), 8 (quote); Peat, *Potential* (1997), 50–2; Oppenheimer, *Open mind* (1963), 82 (quote).

5. Bohm, *Quantum theory* (1951), iii (quote), v, 32; Freire, *Bohm* (2019), 49–51.

6. Bohm, *Quantum theory* (1951), 611.

7. Ibid., 29, 623.

8. Ibid., 614–15.

9. Ibid., 620–2 (quote).

10. Ibid., 623.

11. Taylor, in Klauder, *Magic* (1972), 475–6 (mannerisms); Wheeler, *Geons* (1998), 267 (quote).

12. *PWB*, **4**:1, 58–9.

13. Schweber, *Einstein and Oppenheimer* (2008), 206–7.

14. Pauli to Panofsky, 10 October 1950, 19 January 1952, and a dozen other letters, 1950–52; to Carl Alfred Meier (a psychologist), 26 February 1950; and editorial note, in *PWB*, **4**:1, 167–8, 507, *passim*, 36–7, 281–3, resp.

15. Thomas, *Institute letter* (Spring 2007).

16. Wigner, *Recollections* (1992), 38–9, 67, 75, 307–9.

17. Wigner, in Good, *Scientist speculates* (1961), = *QT&M* (1983), 177–8 (quotes).

18. Wigner, *Proc. Am. Phil. Soc.*, **94**:5 (1950), 427, and *Symmetries* (1970), 175–81.

19. Bohm, interview with Wilkins, Session III, 7 July 1986, 143 (AIP).

20. Bohm, *Zygon*, **20**:2 (1985), 113–14, quoted in Jammer, *Einstein* (2002), 227.

21. Bohm, quoted by Basil Hiley in a note to Baggott, 1 June 2009.

22. Peat, *Potential* (1997), 92.

23. Heilbron, in Carson and Hollinger, *Oppenheimer* (2005), 275–91; Pais, *Oppenheimer* (2006), 274; Schweber, *Einstein and Oppenheimer* (2008), 12–13, 22, 25–7, 199; Dyson, *Maker* (2018), 63, reporting Rudolf Peierls' opinion of Oppenheimer. Peierls would be John Bell's thesis advisor.

24. Wheeler, *Geons* (1998), 216.

25. Olwell, *Isis*, **90**:4 (1999), 744, 746; Wheeler, *Geons* (1998), 215–16; Peat, *Potential* (1997), 98–9, 104 (quote, from Einstein).

26. Sayen, *Einstein* (1985), 250–7.

27. Bohm to Miriam Yevick, 7 January 1952, in Peat, *Potential* (1997), 105.

28. Pais, *Oppenheimer* (2006), 107–8, 150; Bird and Sherwin, *Oppenheimer* (2006), 76, 374.

29. Peat, *Potential* (1997), 119–20, 148 (gifted and lovable), 160 (Christ and Judas).

30. Bohm to Hanna Loewy, 1950, in Talbot, *David Bohm* (2017), 106, 108, 110.

31. Graham, in French and Kennedy, *Bohr* (1985), 333–7; Jammer, *Philosophy* (1974), 248–50.

32. Freire, *Dissidents* (2015), 189.

33. Cross, *Soc. Stud. Sci.*, **21**:4 (1991), 736–40, 743.

34. Bohm to Hanna Loewy, early 1952, in Talbot, *David Bohm* (2017), 123–4.

35. Bohm, *Phys. Rev.*, **85**:2 (1952), 169–71.

36. Ibid., 172.

37. Ibid., 186.

38. Cushing, *SHPS*, **24**:5 (1993), 821–3.

39. Ibid., 831–3.

40. Bohm to a former girlfriend, in Peat, *Potential* (1997), 113 (1st quote); Bohm, in Bohm and Peat, *Science, order* (2000), 61 (2nd and 3rd quotes).

Juvenile Deviationism

1. Bohm, interview with Wilkins, Session IV, 25 September 1986, 10 (AIP).

2. Pauli to Fierz, 6 January 1952, in *PWB*, **4**:1, 499–500.

3. Bohm to Pauli, July and October 1951, in *PWB*, **4**:1, 343–4, 389–92.

4. Bohm to Pauli, 20 November 1951, and reply, 3 December 1951, in *PWB*, **4**:1, 430–2, 436–7; Fierz to Pauli, 17 October 1951, in ibid., 403.

5. Bohm to Pauli, mid-December 1951, in *PWB*, **4**:1, 441–2.

6. Ibid., 444–6.

7. Pauli to Fierz, 6 January 1952, in *PWB*, **4**:1, 499–501; Cross, *Soc. Stud. Sci.*, **21**:4 (1991), 747–50.

8. De Broglie, *New perspectives* (1962), 83, 102, 104; Peat, *Potential* (1997), 149–50.

9. Lochak, in Barut et al., *Quantum* (1984), 29–31 ('dazzling'); Pauli to Fierz, 6 January and 14 February 1952, in *PWB*, **4**:1, 499–501, 552; de Broglie to Pauli, 10 February 1952, and ed. note, ibid., 544–7, 514–15, resp.

10. Born to Einstein, 26 November 1953, in Born, *Born–Einstein letters* (1971), 206.

11. Pauli to Fierz, 25 and 26 January 1952, in *PWB*, **4**:1, 522–4, and in George, *De Broglie* (1953), 36–40; De Broglie, *New perspectives* (1962), 97–8 ('obscure clarity').

12. Pauli to Rosenfeld, 16 March 1952, and reply, 20 March 1952, in *PWB*, **4**:1, 582–3, 587–8.

13. Rosenfeld, in George, *De Broglie* (1953), 44, 51–2 (quote); Jacobsen, *HSPS*, **37** Suppl. (2007), 12, 20.

14. Pauli to Rosenfeld, 1 April 1952, in *PWB*, **4**:1, 591; Jacobsen, *HSPS*, **37** Suppl. (2007), 25 (quote).

15. Pauli to Rosenfeld, 1 and 16 April 1952, and response, 6 April 1952, in *PWB*, **4**:1, 592–3, 610–11, 598–600, resp.

16. Oppenheimer, as quoted by Max Dresden, in Peat, *Potential* (1997), 133.

17. Einstein to Born, 12 May 1952, in Born, *Born–Einstein letters* (1971), 192; Bell, *Speakable* (1987), 91, cited this opinion.

18. Peat, *Potential* (1997), 132–3 (juvenile and Pais); Einstein to Born, 12 May 1952, in Born, *Born–Einstein letters* (1971), 192–3. Cf. Freire, *Bohm* (2019), 71–3, 94, 97; Pais, *Oppenheimer* (2006), 101–2; Bohm, *Causality* (1961), 86.

19. Peat, *Potential* (1997), 136.

20. Bohm to Hanna Loewy, 6 October 1953, and to Melba Philipps, early 1954, in Talbot, *David Bohm* (2017), 125, 168.

21. Freire, *Bohm* (2019), 98–9.

22. Freire, *Bohm* (2019), 86–7, 96 (Bohr), 127; Holland, *Found. Phys.*, **25**:1 (1995), 1–4 (Vigier).

23. Bohm and Vigier, *Phys. Rev.*, **96**:1 (1954), 208–10, 214–15.

24. Bohm, *Causality* (1961), 82, 112–16.

25. Ibid., 84, 91 (quote), 96.

26. Einstein to S. Goldstein, 17 November 1951, and to Rosen, 11 March 1954 (EPP).

27. Einstein to Bohm, 10 February, 28 October, 24 November (quote) 1954, in Peat, *Potential* (1997), 158–60; ibid., 172 ('intellectual son').

28. Bohm to Melba Phillips, spring and fall 1956, in Talbot, *David Bohm* (2017), 182–97.

29. Peat, *Potential* (1997), 163, 175–8, 183–5; Freire, *Bohm* (2019), 100–1, 105–7, 110, 121–4, 128–9; Freire, *Dissidents* (2015), 49–53.

30. Bohr, in *BCW*, **12**, 165–6.

31. Kramers, *Quantum mech.* (1958), 4, = *Grundlagen* (1933), 6.

32. Einstein, in Edinburgh, *Sci. papers* (1953), 39–40, 38 ('astuteness'); and letter to Born, 12 October 1953, in *Born–Einstein letters* (1971), 199 ('nursery song'). Bohm successfully replied to Einstein's conclusion about immobility in Edinburgh, *Sci. papers* (1953), 14–19.

33. Born to Einstein, 26 November 1953, and reply 3 December 1953, in Born, *Born–Einstein letters* (1971), 205–9.

34. Born to Einstein, 22 December 1953, and reply, 1 January 1954, in ibid., 211–13.

35. Einstein to Born, 12 January 1954, and reply, 20 January 1954, in ibid., 214–17.

36. Pauli to Born, 3 March, 31 March, 15 April 1954, in ibid., 217–18, 221–6.

37. Born to Einstein, 28 November 1954, in ibid., 230 (Pauling), and Born's editorial comments (1965), ibid., 228 (Pauli), 229, 231.

38. Wigner, *Recollections* (1992), 281.

Passing the Torch

1. Einstein to Everett, 11 June 1943, in Shikhovtsev and Ford, 'Sketch'.

2. Everett, interview with Misner, May 1977, 2 (AIP); Byrne, in Saunders, *Many worlds* (2010), 521–6, 534.

3. Everett, interview with Misner, May 1977, 2, 5 (AIP).

4. Everett, in DeWitt and Graham, eds., *Many worlds* (1975), 8 (italics in original).

5. Ibid., 9.

6. Osnaghi, Freitas, and Freire, *SHPMP*, **40**:2 (2009), 110.

7. Freire, *Dissidents* (2015), 87–97, 108–14.

8. Byrne, in Saunders, *Many worlds* (2010), 527–9; Rosenfeld, Review of de Broglie, *La théorie de la mesure en mécanique ondulatoire* (1957), in Osnaghi, Freitas, and Freire, *SHPMP*, **40**:2 (2009), 117, likely aimed at Bohm.

9. Nancy G. Everett to Frank J. Tipler, 10 October 1983, in Shikhovtsev and Ford, 'Sketch'.

10. Rosenfeld to Belinfante, 22 June 1972, in Osnaghi, Freitas, and Freire, *SHPMP*, **40**:2 (2009), 113.

11. Everett, interview with Misner, May 1977, 19 (AIP).

12. Byrne, in Saunders, *Many worlds* (2010), 55.

13. Bohm, interview with Wilkins, Session XI, 3 April 1987, 34 (AIP).

14. Moore, *Schrödinger* (1989), citing texts of 1959 (thoughtless slogan) and 1961 (Braman), 473, 475–7, resp.

15. Bohr, in *BCW*, **12**, 357–9, on Kramers, and 364–5, on Pauli; Pauli, *Writings* (1994), 122–3, = *Physik* (1984), 89.

16. Heisenberg, in Pauli et al., *Niels Bohr* (1955), 17–23, 28, and *Physics and philosophy* (1958), 128–46, on 145–6.

17. Cassidy and Baker, *Heisenberg bibliog.* (1984), 40–75.

18. Bohr, *Atomic theory* (1961), 1–24.

19. Pauli, *Writings* (1994), 34–5, 42, 145–8, 155 (texts of the 1950s).

20. Ibid., 46–7 (text of 1954), 96–9 (1936).

21. Ibid., 218–79.

22. Jordan, *Science and the course of history* (1955), 3, 4 (1st quote), 33, 113 (2nd), 30–1 (3rd).

23. Ibid., 106, 32 (first quotes, resp.), 43, 84, 96, 112 (3rd quote), 119, 129.

24. Darwin to Bohr, 11 September 1961, and reply, 20 September 1961, in *BCW*, **10**, 463–4.

25. *BCW*, **10**, 436.

26. Dörries, *Frayn's Copenhagen* (2005), 119, 125, 135, 149, 169 (draft letters, Bohr to Heisenberg, November and December 1961 and March 1962, probably none of which was sent).

27. Wheeler, in Kuhn, et al., *Sources* (1967), vi.

28. Quoted with approval by Wheeler, *At home* (1994), 188.

29. Margenau, interview with Bruce Lindsay and W. James King, 6 May 1964 (AIP).

30. Körner, *Observation* (1957); Werner, *Foundations* (1962); Kožnjak, *SHPMP*, **62** (May 2018), 86.

31. Körner, *Observation* (1957), 42, 44.

32. Körner, *Observation* (1957), 34–6, 60–1 (Bohm), 43–5 ('definiteness'), 46 (Fierz), 49 (Vigier); Jacobsen, *HSPS*, **37** Suppl. (2007), 14–17, 28–30, 32, 52–7 ('obvious thing', on 53); Kožnjak, *SHPMP*, **62** (May 2018), 90–2.

33. Popper, in Körner, *Observation* (1957), 65–70, 98–9, *Logic* (1961), 223–46 (text of 1934), and *Unended quest* (1992), 90–1, 101–8 (quote on 104), 175–89; Shields, *Quanta*, **1**:1 (2012), 1–12.

34. Popper, *Quantum theory* (1982), 99–104 (text of 1956/7); Feyerabend, *Killing time* (1995), 77–8, 92–3, and *Physics* (2016), 9–73 (texts of 1954–58). Feyerabend later defended Bohr and complementarity before sliding into philosophical anarchy.

35. Körner, *Observation* (1957), 138–45 (quotes on 43).

36. Körner, *Observation* (1957), 144.

37. Werner, *Foundations* (1962), Mon AM, 3–4, 12–16; Mon PM, 16, 19 (quote, from Furry); Dirac, *Sci. Am.*, **54** (May 1963), 296–7, a development of his remarks at the symposium.

38. Werner, *Foundations* (1962), Tue AM, 12–15; Tue PM, 31–2; Mon PM, 26, 30 (quote).

39. Ibid., Tue AM, 23–7 (Rosen); Tue PM, 23 ('operational'), 25 ('Bohm'), 28 ('happening').

40. Ibid., Mon PM, 10–13; Tue PM, 2–4, 3 ('you don't'), 7–8 ('the collapse'), 10 ('piece of paper').

41. Ibid., Tue PM, 10, 11 (Heisenberg), 14, 17, 18 (Rosen and Wigner), 19 'no' etc.), 20 (Aharonov).

42. Ibid., Tue PM, 41 (Furry's fury); Mon AM, 12–13, and Tue AM, 17–18 (Rosen), 20 (Everett).

43. Ibid., Tue AM, 20, 22.

44. Ibid., Wed PM, 11; Dirac, *Sci. Am.*, **54** (May 1963), 300–1.

45. Werner, *Foundations* (1962), Fri AM, 7–8 (solipsism), 9–11.

46. Ibid., Fri AM, 15–21, 29 (Dirac, Podolsky, and Aharonov).

47. Dyson, *Sci. Am.*, **199** (September 1958), 78.

ACT IV: PRODUCTIVE INEQUALITIES

The Theorem of John S. Bell

1. Werner, *Foundations* (1962), Tue AM, 32–7, 43–5.

2. Kožnjak, *SHPMP*, **62** (May 2018), 93–4; Freire, *Found. Phys.*, **34**:11 (2004), 1746–7, and *Dissidents* (2018), 85.

3. Bohm and Aharonov, *Phys. Rev.*, **108**:4 (1957), 1071.

4. Wu turned the heads of several seasoned experimentalists; the appreciation quoted is in Alvarez, *Adventures* (1987), 55.

5. Bohm and Aharonov, *Phys. Rev.*, **108**:4 (1957), 1072–5; Wu and Shaknov, *Phys. Rev.*, **77**:1 (1950), 136.

6. Peat, *Potential* (1997), 168; Bell, *Found. Phys.*, **12**:10 (1982), 989, = *Speakable* (1987), 160.

7. Bell, *Speakable* (1987), 160 ('impossible'), 173 ('revelation'); Kerr, 'Recollections', §2 ('heated'), §9 ('steam rising'), §2 ('positivism'), §4 (women).

8. Bernstein, *Quantum profiles* (1991), 50–1.

9. Bell, *Phys. World*, **3**:8 (1990), in *Speakable* (2004), 215–16 (bad words), 217 ('piddling'), 214 ('FAPP'); Gottfried and Mermin, in Ellis and Amati, *Reflections* (2000), 191.

10. Dirac, *Directions* (1970), 20.

11. Bertlmann, in *Q(Un)S* **1**, (2002), 29.

12. Shimony, in *Q(Un)S* **1**, (2002), 51–2, 55, and d'Espagnat, in ibid., 21, 23 (thinking), 25 ('repellent').

13. Mary Bell, in *Q(Un)S* **1**, (2002), 5; Whitaker, *Bell* (2016), 6–18, 20, 28–31.

14. Whitaker, *Bell* (2016), 34, 47–8, 68.

15. Ibid., 98–9, 129–30, 151–2, 162–3; Gowing, *Britain* (1964), 40–3.

16. Bell to Peierls, 20 August 1980 and 21 February 1983; Peierls to Bell, 13 November 1980 and 27 April 1982, in Peierls, *Select. corr.* (2009), **2**, 802–3, 807 ('set of correlations'), 826, 846–7 (Einstein).

17. Landau and Lifschitz, *Quantum mechanics* (1958), 2, 3, 21, 22; Bell, *Speakable* (2004), 217–19 (quote, 219, text of 1989).

18. Whitaker, *Bell* (2016), 190–1.

19. Ibid., 188, 197–201; Bell, *Speakable* (1987), 1–5 (ideas germinated in 1952, written out in 1964, and published in 1966).

20. Bell, *Speakable* (1987), 5–9.

21. Whitaker, *Bell* (2016), 201.

22. Bell, *Rev. Mod. Phys.*, **38**:3 (1966), in *Speakable* (1987), 11 (quote).

23. Bell, in Davies and Brown, *Ghost* (1986), 57.

24. Einstein and Ehrenfest, *Z. Phys.*, **11**:1 (1922), 31–4, in *ECP*, **13**, 441–4, and editorial notes, 445–6; criticized, in Bohrish, in *BCW*, **3**, 484–5n. A silver atom contains 47 electrons arranged in 'shells' according to the configuration 2, 8, 18, 18, 1. The single outermost 'unpaired' electron accounts for the atoms' behaviour in a Stern–Gerlach apparatus.

25. Bell, *J. de Physique*, **42** C2 Suppl. 3 (1981), in *Speakable* (1987), 146–50.

26. D'Espagnat gives other parabells in *Sci. Am.*, **241** (November 1979), 158–81, and *Search* (1983), 26–50.

27. Bell, *J. de Physique*, **42** C2 Suppl. 3 (1981), in *Speakable* (1987), 146.

28. Ibid., 149.

29. Bell, *Epistemological Lett.*, November 1975, 2–6, = *Speakable* (1987), 65.

30. Bell, in Davies and Brown, *Ghost* (1986), 57.

31. Bell, *Physics*, **1** (1964), = *Speakable* (1987), 20.

32. Bohm and Aharonov, *Phys. Rev.*, **108**:4 (1957), 1072.

33. Whitaker, *Bell* (2016), 209; Bernstein, *Quantum profiles* (1991), 74.

34. Whitaker, *Bell* (2016), 209–10.

35. Bell, *Speakable* (1987), 139–43; Bertlmann, in *Q(Un)S* **1** (2017), 18, 38–40.

36. Heilbron, *Isis*, **67**:1 (1976), 7–20.

Bell Tests and Protests

1. Freire, *Dissidents* (2015), 161–2.
2. Shimony, interview with Bromberg, 9–10 September 2002, 31 ('From Brandeis'), 28 (Bell–Wigner theorem), 38 ('kooky paper', 'nonlocality') (AIP).
3. Ibid., 40–2.
4. Clauser, interview with Bromberg, 20, 21, and 23 May 2002, 6 ('model'), 8 ('kind of refused') (AIP).
5. Ibid., 11 (sat in a corner); 8 (Thaddeus advised), 71 ('once I got involved'), and interview with Baggott and Heilbron, 24 November 2021 ('angry').
6. Clauser, interview with Baggott and Heilbron, 24 November 2021.
7. Clauser, in Black, *Foundations* (1992), 169–70, a fuller version than Clauser's similar recollections in *Q(Un)S* **I** (2002), 79–81, which does not mention his visit to Wu; Clauser, interview with Baggott and Heilbron, 24 November 2021.
8. According to his student Steven Chu, *Art and symmetry* (2001), 25. The advice worked for Chu, who won a Nobel Prize.
9. Clauser, *Q(Un)S* **I** (2002), 79–80; Bell to Clauser, 5 March 1969 (courtesy of Clauser).
10. Clauser, *Q(Un)S* **I** (2002), 80; Clauser, interview with Baggott and Heilbron, 24 November 2021.
11. Clauser, *Bull. Am. Phys. Soc.*, **14** (1969), 578.
12. Clauser, et al., *Phys. Rev. Lett.*, **23**:15 (1969), 881–3.
13. Clauser, note to Baggott, 12 January 2023.
14. Editorial, *Found. Phys.*, **1**:1 (1970), 2–3.
15. De Broglie, *Found. Phys.*, **1**:1 (1970), 5–15; Wigner, ibid., 35–45; Zeh, ibid., 69–76.
16. 'Aims and scope', at the publisher's website for *Foundations of Physics*.
17. Freire, *Dissidents* (2015), 197 (quote), 200–3.
18. Romano, PhD thesis, 2020, 306–11.
19. Wigner, in d'Espagnat, *Foundations* (1971), 1, 4 ('riddle'), 14–16, 17 ('satisfactory'), 19 ('illusion').
20. Baracca, Bergia, and del Santo, *SHPMP*, **57** (February 2017), 65; Freire, *Dissidents* (2015), 159–60, 163–5.
21. Silva, in Freire, *Handbook* (2022), 744, 747–50.
22. Bell, in d'Espagnat, *Foundations* (1971), 172–3; Kasday, in ibid., 207–8.
23. Jauch, in d'Espagnat, *Foundations* (1971), 21 ('profound'), 22, 23 ('straightforward'), 36–8, 42–4. Cf. Zeh, in ibid., 263, 268.
24. Shimony, in d'Espagnat, *Foundations* (1971), 182–94, 471–6. Shimony added in a note to the proofs of his lecture that the Berkeley experiments had confirmed the predictions of quantum mechanics.
25. De Broglie, in d'Espagnat, *Foundations* (1971), 353–4, 358–61, 366 (quote); Bohm and Vigier, *Phys. Rev.*, **96**:1 (1954), 208.

26. DeWitt, in d'Espagnat, *Foundations* (1971), 211 ('bizarre', 'straightforward'), 212.

27. DeWitt, *Phys. Today*, **23**:9 (1970), 32, 30.

28. Wheeler, in Davies and Brown, *Ghost* (1986), 60.

29. Wheeler to Paul Benioff, 7 July 1977, in Shikhovtsev and Ford, 'Sketch'.

30. Leggett, *Problems* (2006), 144 ('damage').

31. Bohm, in d'Espagnat, *Foundations* (1971), 421, 429–34, 437 ('thinking'), 438–42, 445 ('total order'), 447 ('holomovement'), 451 ('a name'), 452, 461 ('whole-ness'), 464–5 ('negotiation'), 469 ('communition', 'relevates').

32. Prosperi, in d'Espagnat, *Foundations* (1971), 106; Wigner, in ibid., 12–5; Selleri, in ibid., 398 ('forced', 'social responsibility'), 402–3, 406; d'Espagnat, in ibid., 91–3.

33. Selleri to Geymonat, in Romano, PhD thesis, 2020, 144, and Selleri, interview with Freire, Sessions I (quote) and II, 25 June 2003 (AIP).

34. Selleri, interview with Freire, Session I, 24 June 2003; Baracca, Bergia, and del Santo, *SHPMP*, **57** (February 2017), 68–9.

35. Freire, *Dissidents* (2015), 210–11; Baracca, Bergia, and del Santo, *SHPMP*, **57** (February 2017), 70–1.

36. Romano, PhD thesis, 2020, 320–2, 985–7.

37. Selleri, *Crit. Marx. Quad.*, **6** (1972), 124.

38. Weiner, *History* (1977), xi; 'Statement on Vietnam', dated 12 August 1972 and signed by 58 participants. Typescript kindly furnished by Angelo Baracca, who attended the school. Heilbron, who lectured in it, recalls the incident of the Fermi film.

39. Clauser, et al., *Phys. Rev. Lett.*, **23**:15 (1969), 881–3.

40. Freedman and Clauser, *Phys. Rev. Lett.*, **28**:14 (1972), 939–41.

41. Clauser, *Phys. Rev. Lett.*, **36**:21 (1976), 1223–6.

42. Leggett, *Problems* (2006), 166 (surprise, alarm).

43. Bell, in d'Espagnat, *Foundations* (1971), 178, = *Speakable* (1987), 36–7.

44. Clauser and Horne, *Phys. Rev. D*, **10**:2 (1974), 526–34.

45. Feynman, quoted in Schweber, *Einstein and Oppenheimer* (2008), 250–1.

46. Clauser, interview with Baggott and Heilbron, 24 November 2021. This remark is not consistent with what we know of Commins' character.

47. Clauser, interview with Bromberg, 20, 21, and 23 May 2002 ('pointless waste of time'), 9; ('Don't hire this guy'), 8 (AIP).

48. Shimony, interview with Bromberg, 9–10 September 2002, 64 (AIP).

49. Clauser, *Q(Un)S* **I** (2002), 80 ('net impact of this stigma'), 72.

50. Clauser and Shimony, *Rep. Prog. Phys.*, **41**:12 (1978), 1903–17.

51. Feyerabend to Lakatos, 20 February 1970, in Motterlini, *For and against method* (1999), 191.

52. Peter Higgs, *Phys. Rev. Lett.*, **13**:16 (1964), 508–9. Two further papers on the same topic were published in the same volume: F. Englert and R. Brout, **13**:9 (1964), 321–3, and G.S. Guralnik, C.R. Hagan, and T.W.B. Kibble, **13**:20

(1964), 585–7. François Englert would go on to share the 2013 Nobel prize for physics with Higgs. Baggott, *Higgs* (2012), 85–6, 207.

53. Higgs, in Hoddeson et al., *Standard model* (1997), 508.
54. Clauser, interview with Baggott and Heilbron, 24 November 2021.
55. Clauser, in Jaeger et al., eds., *Quantum arrangements* (2021), 89.

While the Photons are Dancing

1. Pauli to Bohr, 3 October 1950, in *PWB*, **4**:1, 170–1; Lao Tsu, *Tao* (1998), 54, 85, 45 (first three quotes).
2. Kaiser, *Hippies* (2011), xv–xvi, 49–54, 178–82, 265–6.
3. Clauser, interview with Bromberg, 20, 21 and 23 May 2002, 68 (AIP).
4. Kaiser, *Hippies* (2011), 51–5, 64–5, 70–81 (quote on 73).
5. Ibid., 130–1, 134, 171, 174, 177–8.
6. Ibid., 109–12, 116, 119, 178–9, 185, 188.
7. Zukav, *Masters* (1979), 7–8, 24–5, 31 (2nd quote), 35 (1st).
8. Ibid., 254–5, 256 (quote), 312 (2nd quote, from a text of 1974).
9. Ibid., 302 (1st quote); Stapp, *Nuovo Cimento B*, **29**:2 (1975), 270–1 (2nd quote).
10. Bernstein, *Quantum profiles* (1991), 77.
11. Kaiser, *Hippies* (2011), xxiii, xxv, 40–1, 269.
12. Moore, *Schrödinger* (1989), 169–77, 252–3; Moody, *Collaboration* (2017), 2, 95.
13. Moody, *Collaboration* (2017), 30–3, 45, 50, 73 ('thinker'), 93 ('insight'), 163, 179; Kaiser, *Hippies* (2011), 102, 306n14.
14. Peat, *Potential* (1997), 266.
15. Hiley, interview with Freire, 11 January 2008 (AIP).
16. Philippidis, Dewdney, and Hiley, *Nuovo Cimento B*, **52**:1 (1979), 22, 23.
17. Bell, *Prog. Sci. Culture*, **1**:4 (1977), 440.
18. Aspect, interview with Baggott and Heilbron, 1 December 2021.
19. Aspect, in Ellis and Amati, *Reflections* (2008), 69 (text of 1991, 'quantum mechanical predictions'), and interview with Baggott and Heilbron, 1 December 2021 ('Einstein was clear', 'Bohr impossible', 'Bell showed the conflict'); Aspect, in Gisin, *Quantum chance* (2014), v ('I decided').
20. Bell, *Physics*, **1** (1964), 195–200.
21. Aspect, note to Baggott and Heilbron, 10 March 2022.
22. Phillips and Dalibard, *Eur. Phys. J. D*, **77**:8 (2023), 4.
23. Cf. Aczel, *Entanglement* (2002), 186; Phillips and Dalibard, in ibid., 4–5.
24. Aspect, *Phys. Rev. D*, **14**:8 (1976), 1944–51.
25. Clauser, in Black, *Foundations* (1992), 172.
26. Aspect, interview with Baggott and Heilbron, 1 December 2021 ('Pipkin hiding', 'enthusiasm and clarity').
27. Ibid.
28. Freire, *Dissidents* (2015), 273.
29. Bell, *Prog. Sci. Culture*, **1**:4 (1977), 442, bold highlights in the original.
30. Phillips and Dalibard, *Eur. Phys. J. D*, **77**:8 (2023), 6.
31. Aspect, note to Baggott, 13 November 2022.

32. Aspect, interview with Baggott and Heilbron, 1 December 2021.

33. Ibid.

34. Aspect, Grangier, and Roger, *Phys. Rev. Lett.*, **47**:7 (1981), 460–3.

35. Thus contradicting the hypothesis advanced by Wendell Furry in 1936, which suggested that entanglement might be restricted to short distances (Furry, *Phys. Rev.* **49**:5 (1936), 393–9). Aspect, note to Baggott, 13 November 2022; Phillips and Dalibard, *Eur. Phys. J. D*, **77**:8 (2023), 5.

36. Aspect, interview with Baggott and Heilbron, 1 December 2021; Aspect, note to Baggott, 26 December 2022.

37. Aspect, Grangier, and Roger, *Phys. Rev. Lett.*, **49**:2 (1982), 91–4.

38. Aspect, interview with Baggott and Heilbron, 1 December 2021.

39. Ibid.

40. Ibid; Aspect, Dalibard, and Roger, *Phys. Rev. Lett.*, **49**:25 (1982), 1804–7.

41. Kaiser, *Hippies* (2011), 205–17, 220–1, 230–1; Wootters and Zurek, *Nature*, **299** (28 October 1982), 802–3; Dieks, *Phys. Lett. A*, 92:6 (1982), 271–2; Peres, *Fort. der Physik*, 51:4/5 (2003), 458–61.

42. Bennett and Brassard, *ICCSSP*, **1** (1984), 175–9.

43. Feynman, *Int. J. Theor. Phys.*, **21**:6/7 (1982), 474–6.

44. Feynman, *Found. Phys.*, **16**:6 (1986), 530.

45. Freire, *Dissidents* (2015), 278. Aspect delivered his seminar where Clauser had spoken ten years earlier, and Clauser later claimed that nobody had thought to mention his earlier experiments, 'that kind of got left out'. Clauser, interview with Bromberg, 20, 21, and 23 May 2002, 26 (AIP). However, a copy of an original slide from Aspect's presentation shows that he acknowledged all the previous experiments, including Freedman and Clauser (1972), Holt and Pipkin (1973), Clauser (1976), and Fry and Thomson (1976); Aspect, note to Baggott, 24 November 2022.

46. Feynman to Aspect, 28 September 1984. We are grateful to Alain Aspect for providing a copy of this letter; Aspect, note to Baggott, 24 November 2022.

47. John Gribbin, note to Baggott, 28 December 2021.

48. Herbert, *Quantum reality* (1985), 118–19, 131, 149–52, 215–24; Bell's appreciation is from a blurb printed on the paperback edition of the book.

49. Bell, in Davies and Brown, *Ghost* (1986) 52.

50. Grangier, Roger, and Aspect, *Europhys. Lett.*, **1**:4 (1986), 179.

Adventures in Quantum Information

1. Vigier at al., in Hiley and Peat, *Implications* (1987), 201; Rauch, in Ellis and Amati, *Reflections* (2000), 61–3.

2. Zeilinger, interview with Freire, 30 June 2014, 15 ('most interesting').

3. Rauch, et al., *Phys. Lett. A*, **54**:6 (1975), 425–7.

4. Several of which are nicely illustrated in Rauch, in Ellis and Amati, *Reflections* (2000), 28–68.

5. Zeilinger, interview with Freire, 30 June 2014, 3 ('there is a conference', 'crucial').

6. Horne, interview with Bromberg, 12 September 2002, 2 ('Bell's theorem'), 3 ('we found out') (AIP).

7. Ibid., 2.

8. Zeilinger, interview with Freire, 30 June 2014, 4 ('talking that way').

9. Horne, interview with Bromberg, 12 September 2002, 2 ('weekends and summers'), 4 ('know someone').

10. Greenberger, in *Q(Un)S* I (2002), 285.

11. Greenberger et al., in Kafatos, *Bell's theorem* (1989), 73–6; cf. the scholastic article by Clifton, Redhead, and Butterfield, *Found. Phys.*, **21**:2 (1991), 149–84.

12. Greenberger, in *Q(Un)S* I (2002), 286.

13. Greenberger et al., *Am. J. Phys.*, **58**:12 (1990), 1131–43; Mermin, *Phys. Today*, **43**:6 (1990), 9–11.

14. Bouwmeester et al., *Phys. Rev. Lett.*, **82**:7 (1999), 1345–9.

15. Pan et al., *Nature*, **403** (3 February 2000), 515–9.

16. Weihs et al., *Phys. Rev. Lett.*, **81**:23 (1998), 5039–43.

17. Gisin and Gisin, *Phys. Lett. A*, **260**:5 (1999), 323–7.

18. Rowe et al., *Nature*, **409** (15 February 2001), 791–4.

19. Shiedl et al., *PNAS*, **107**:46 (2010), 19708–13.

20. Handsteiner et al., *Phys. Rev. Lett.*, **118**:6 (2017), 060401.

21. Rauch et al., *Phys. Rev. Lett.*, **121**:8 (2018), 080403; Kaiser, *Quantum legacies* (2020), 57–67.

22. Nakajima, in Namiki, *Foundations* (1987), 'Preface'; Gerry and Bruno, *Quantum divide* (2013), 151–62.

23. Mehlhop and Piccioni, in Namiki, *Proceedings* (1987), 72; Yang, in ibid., 183.

24. Leggett, in Namiki, *Proceedings* (1987), 287–97, and in Hiley and Peat, *Implications* (1987), 85 (all quotes).

25. Suppose that Bob chooses the prime numbers $p = 3$ and $q = 17$ (in practice, these numbers would be much, much larger). The product $n = p \times q = 51$. Bob first computes the 'Carmichael function', $\lambda(n)$, as the lowest common multiple of $p - 1 (= 2)$ and $q - 1 (= 16)$. Hence $\lambda(51) = 16$. He then chooses an encryption exponent e between 1 and $\lambda(n)$ that is coprime with $\lambda(n)$ i.e., a number for which the greatest common divisor of e and $\lambda(n)$ is 1. Suppose he chooses $e = 13$ (another prime number, as he then has only to ensure that $\lambda(n)$ is not evenly divisible by e). The decryption exponent d is the modular multiplicative inverse of e, such that $e \times d - 1 \equiv \bmod \lambda(n)$, meaning that $\lambda(n)$ must evenly divide $e \times d - 1$. He uses the Extended Euclidean Algorithm to calculate $d = 5$ (and Bob confirms that $13 \times 5 - 1 = 64$ is indeed evenly divisible by 16).

26. The cipher text (C) is calculated using the encryption equation $C = P^e \bmod n$, where P is the original message (the plaintext). Suppose Alice wants to send the number 7. The cipher text is then $C = 7^{13} \bmod 51 = 40$. The decryption equation is $P = C^d \bmod n$. So, in our example $P = 40^5 \bmod 51 = 7$.

27. https://en.wikipedia.org/wiki/Integer_factorization_records

28. Bennett and Brassard, *ICCSSP*, **1** (1984), 175–9.
29. Deutsch, *Proc. R. Soc. London A*, **400**:1818 (1985), 97–117; Ekert, *Phys. Rev. Lett.*, **67**:6 (1991), 661–3.
30. Deutsch and Jozsa, *Proc. R. Soc. London A*, **439**:1907 (1992), 553–8.
31. Shor, *Proc. 35th Annual Symposium on Foundations of Computer Science*, (1994), 124–134.
32. https://www.cnet.com/tech/computing/quantum-computers-could-crack-todays-encrypted-messages-thats-a-problem/
33. Gisin, interview with Freire, 7 December 2013 (AIP).
34. Gisin, *Quantum chance* (2014), 59.
35. Tittel et al., *Phys. Rev. Lett*, **81**:17 (1998), 3563–6.
36. Bouwmeester et al., *Nature*, **390** (1 December 1997), 575–9. See also Zeilinger, *Sci. Am.* **282** (April 2000), 50–9.
37. Yin et al., *Science*, **356** (16 June 2017), 1140–4.
38. Ren et al., *Nature*, **549** (9 August 2017), 70–3.
39. Bell, in Miller, *Years* (1990), 17–31, in *Speakable* (2004), 213–31, and *Phys. World*, **3**:8 (1990), 33–40. See also Baggott, *Phys. World*, **33**:12 (2020), 30–4.
40. Mermin, in *Q(Un)S* **1** (2002), 271 ('came close', 'music of his voice', 'wonderful conversations').
41. Whitaker, *Bell* (2016), 386.
42. Mott to Peierls, 7 May 1991, in Peierls, *Select. corr.* (2009), **2**, 1007; Gell-Mann, text of 1976, quoted by Stapp, in Hiley and Peat, *Implications* (1987), 257.
43. Cf. Peat, *Potential* (1973), 155, 159, 161; Pylkkhänen, *Prog. Biophys. Mol. Biol.*, **131** (31 August 2017), 174.
44. Peat, *Potential* (1973), 154–5, 159, 161, 167, 175–6; quotes on 132, 111, 149, resp.; Freire, *Bohm* (2019), 200.
45. Bohm, *Epistemologia*, **3** (1980), 53–74; Feyerabend, in Peat, *Potential* (1997), 188; Gross, in Hiley and Peat, *Implications* (1987), 46–9.
46. Bohm, in Bohm and Biedermann, *Corresp.* (1999), 26 March 1960, 4–5 (Kierkegaard), and 24 April 1960, 8, 11–14, 16 (new categories).
47. Ibid., 6 June and 1 August 1960, 29–34, 45, 48–9, and 2 February 1961, 93 (quote).
48. Ibid., 2 and 24 February, and 22 December 1961, 93, 95, 130–1.
49. Ibid., 23 March 1962, 201 ('unity'), 15 April 1962, 211 ('individual'), 216 (habit), 218 ('competitive society').
50. Ibid., 24 April 1962, 231.
51. Bohm and Peat, *Science, order* (2000), 26–7, 88 (Bohm and Kuhn's ideas), 98–100; Moody, *Collaboration* (2017), 259 (1st quote), 257 (2nd).

Where to Cut? Which Way to Go?

1. Bohm, interview with Wilkins, Session XI, 3 April 1987, 11 (AIP).
2. Van Kampen, *Physica A*, **153**:1 (1988), 99, quoted by Bell, *Phys. World*, **3**:8 (1990), 38, = *Speakable* (2004), 227.

3. Bohr, *Atomic physics* (1961), 71.
4. Bohm and Aharonov, *Phys. Rev.*, **108**:4 (1957), 1072.
5. Bohr, *Atomic physics* (1961), 88–9.
6. Heisenberg, unpublished manuscript, in *PWB*, **2**, 414. Trans. Bacciagaluppi and Crull, *Einstein's paradox* (in press), 263.
7. Bell, *Phys. World,* **3**:8 (1990), 36, = *Speakable* (2004), 220.
8. Bell, *Speakable* (1987), 117.
9. Zeh, *Found. Phys.*, **1**:1 (1970), 74.
10. Bell, *Phys. World,* **3**:8 (1990), 36, = *Speakable* (2004), 220.
11. Penrose, *Human mind* (2000), 82.
12. Penrose, *Gen. Rel. Grav.*, **28**:5 (1996), 584–5, 596–8, and *New mind* (1990), 475.
13. Leggett, *Prog. Theor. Phys. Suppl.*, **170** (May 2007), 103–4.
14. Leggett, *Les Prix Nobel. The Nobel Prizes 2003* (2004).
15. Leggett, interview with Baggott and Heilbron, 10 December 2021.
16. Leggett, *Les Prix Nobel. The Nobel Prizes 2003* (2004).
17. Leggett, *Second Order*, **1**:2 (1972), 80.
18. Leggett, interview with Baggott and Heilbron, 10 December 2021.
19. Leggett, *Prog. Theor. Phys. Suppl.*, **170** (May 2007), 104.
20. Ibid., 105 ('rather crude'), ('Schrödinger's cat').
21. Leggett and Garg, *Phys. Rev. Lett.*, **54**:9 (1985), 857–60.
22. Arndt et al., *Nature*, **401** (14 October 1999), 680–2.
23. Bohr, in Schilpp, *Einstein* (1949), 217–18.
24. Arndt et al., *Q(Un)S* **1** (2002), 333–50. See also Leggett, *Prog. Theor. Phys. Suppl.*, **170** (May 2007), 108–10.
25. Nairz et al., *Phys. Rev. A*, **65**:3 (2002), 032109.
26. Awschalom et al., *Phys. Rev. Lett.*, **68**:20 (1992), 3092–5; Gerlich et al., *Nature Comm.*, **2** (5 April 2011), 1–5; Marletto et al., *J. Phys. Commun.*, **2** (10 October 2018), 101001.
27. Leggett, *Prog. Theor. Phys. Suppl.*, **69** (March 1980), 93–9.
28. Clarke, *Sci. Am.*, **271** (August 1994), 49.
29. Freitas, in Freire, *Oxford companion* (2022), 504–7, 512–13.
30. Friedman et al., *Nature*, **406** (6 July 2000), 43–6; van der Wal et al., *Science*, **290** (27 October 2000), 773–7.
31. Leggett, *Prog. Theor. Phys. Suppl.*, **170** (May 2007), 100–18. Leggett later lowered these estimates to between 1 and 10 million; Leggett, note to Baggott, 4 August 2009.
32. Emary et al., *Rep. Prog. Phys.*, **77**:1 (2014), 016001.
33. Knee et al., *Nature Comm.*, **7** (4 November 2016), 1–5.
34. http://maqro-mission.org/
35. 'Assessment of quantum physics payload platform', CDF Study Report CDF-183(C), ESA, July 2018, 260.
36. Greenberger and Yasin, *Phys. Lett. A*, **128**:8 (1988), 391–4; Englert, *Phys. Rev. Lett.*, **77**:11 (1996), 2154. See also Jaeger et al., *Phys. Rev. A*, **51**:1 (1995), 54–67.
37. Scully and Drühl, *Phys. Rev. A*, **25**:4 (1982), 2208–13.

38. Herzog et al., *Phys. Rev. Lett.*, **75**:17 (1995), 3037.

39. Ma et al., *Rev. Mod. Phys.*, **88**:1 (2016), 015005.

40. Ma et al., *PNAS*, **110**:4 (2013), 1221–6.

41. Leggett, *Found. Phys.*, **33**:10 (2003), 1469–93.

42. Jarrett, *Noûs*, **18**:4 (1984), 569–89. We are grateful to Don Howard for drawing this work to our attention.

43. Howard, note to Baggott and Heilbron, received 14 December 2022.

44. Gröblacher, et al., *Nature*, **446** (19 April 2007), 875.

45. Colbeck and Renner, *Phys. Rev. Lett.*, **101** (2008) 050403.

46. Leggett, note to Baggott, 9 January 2023.

47. Pusey, Barrett, and Rudolph, *Nature Phys.*, **8** (2012), 475–8.

48. Heilbron, *Rev. Hist. Sci.*, **38**:3/4 (1985), 210, 47n.

Epilogue

1. Heilbron, *Dilemmas* (2000), 52.

2. Barilier, *Einstein* (2022), 159.

3. Nathansen, *Jude oder Europäer?* (1931), 39–42, 46, 50, 60, 78, 88, 94, as quoted in Heilbron, *Bohr* (2020), 5–8.

4. Zeilinger, *Dance* (2010), 208–36.

5. Gröblacher, et al., *Nature*, **446** (19 April 2007), 871–5.

6. Leggett, quoted in Roebke, *Seed*, **16** (June 2008), 58 (online version, 3).

7. Horgan, *End* (1996), 91.

8. Genesis. 11:6.

9. Cf. Tushna Commissariat and Hamish Johnston, https://physicsworld.com/a/and-the-winner-is-our-2015-nobel-prize-predictions/

10. The Royal Swedish Academy of Sciences, press release, 'The Nobel Prize in Physics 2022', 4 October 2022.

Sources

Unpublished Sources

Aspect, Alain. Interview with Baggott and Heilbron, 1 December 2021. Recording in JB's possession.

Bohm, David. Interview with Maurice Wilkins, Session III, IV, and XI, 7 July and 25 September 1986, and 3 April 1987 (AIP).

Bohr Scientific Correspondence, Niels Bohr Archive, Copenhagen (BSC).

Clauser, John. Interview with Joan Bromberg, 20, 21, and 23 May 2002 (AIP).

Clauser, John. Interview with Baggott and Heilbron, 24 November 2021. Recording in JB's possession.

Einstein Papers Project, California Institute of Technology: Einstein Papers Project at Caltech (EPP): https://www.einstein.caltech.edu/

Everett III, Hugh. Interview with Charles Misner, May 1977 (AIP).

Gisin, Nicolas. Interview with Olival Freire, Jr., 7 December 2013 (AIP).

Hiley, Basil. Interview with Olival Freire, Jr., 11 January 2008 (AIP).

Horne, Michael. Interview with Joan Bromberg, 12 September 2002 (AIP).

Kerr, Leslie W. 'Recollections of John Stewart Bell'. Unpublished manuscript, 23 January 1998. We are grateful to Andrew Whitaker for providing a copy of this document.

Leggett, Anthony. Interview with Baggott and Heilbron, 10 December 2021. Recording in JB's possession.

Margenau, Henry. Interview with Bruce Lindsay and W. James King, 6 May 1964 (AIP).

Peters, Bernard, ed. 'Notes on quantum mechanics: Physics 221, Oppenheimer, 1939'. Manuscript notes on Oppenheimer's lectures, Doe Library, University of California, Berkeley.

Phillips, Melba. Interview with Katherine Sopka, 5 December 1977 (AIP).

Phillips, Melba. Interview with Alexei Kojevnikov, 2 December 1997 (AIP).

Selleri, Franco. Interview with Olival Freire, Jr., Sessions I and II, 24 and 25 June 2003 (AIP).

Shikhovtsev, Eugene, and Kenneth W. Ford. 'Biographical Sketch of Hugh Everett, III'. space.mit.edu/home/tegmark/everett/everett.html

Shimony, Abner. Interview with Joan Bromberg, 9–10 September 2002 (AIP).

Werner, Frederick G., ed. 'Conference on the foundations of quantum mechanics. Cincinnati: Xavier University, Oct 1–5 1962'. Typescript in Hugh Everett III Papers, University of California, Irvine.

Zeilinger, Anton. Interview with Olival Freire, Jr., 30 June 2014. We are grateful to Olival Freire, Jr., for providing a copy of the interview transcript.

Zeilinger, Anton. Interview with Baggott, 17 December 2021. Recording in JB's possession.

Published Sources

Aaserud, Finn, and Helge Kragh, eds. *One hundred years of the Bohr atom*. Copenhagen: Royal Danish Academy of Sciences and Letters, 2015.

Aczel, Amir D. *Entanglement: The greatest mystery in physics*. Chichester: Wiley, 2002.

Aharonov, Yakir, and David Bohm. 'Significance of electromagnetic potentials in the quantum theory'. *Phys. Rev.*, **115**:3 (1959), 485–91.

Aharonov, Yakir, and David Bohm, and Lev Vaidman. 'About position measurements which do not show the Bohmian particle position'. In Cushing, et al., eds., *Bohmian mechanics* (1996), 141–54.

Alvarez, Luis W. *Adventures of a physicist*. New York: Basic Books, 1987.

Andrade E Silva, J. 'Une formulation causale de la théorie quantique de la mesure'. In d'Espagnat, *Foundations* (1971), 368–97.

Arndt, Markus, et al. 'Wave-particle duality of C_{60} molecules'. *Nature*, **401** (14 October 1999), 680–2.

Arndt, Markus, Olaf Nairz, and Anton Zeilinger. 'Interferometry with macromolecules: quantum paradigms tested in the mesoscopic world'. *Q(Un)S*, **1** (2002), 333–50.

Aspect, Alain. 'Proposed experiment to test the nonseparability of quantum mechanics'. *Phys. Rev. D.*, **14**:8 (1976), 1944–51, = *QT&M* (1983), 435–42.

Aspect, Alain. 'Foreword'. In Gisin, *Quantum chance* (2014), v–ix.

Aspect, Alain, Phillipe Grangier, and Gérard Roger. 'Experimental tests of realistic local theories via Bell's theorem'. *Phys. Rev. Lett.*, **47**:7 (1981), 460–3.

Aspect, Alain, Phillipe Grangier, and Gérard Roger. 'Experimental realization of Einstein-Podolsky-Rosen-Bohm *Gedankenexperiment*: A new violation of Bell's inequalities'. *Phys. Rev. Lett.*, **49**:2 (1982), 91–94.

Aspect, Alain, Phillipe Grangier, and Gérard Roger. 'Experimental test of Bell's inequalities using time-varying analyzers'. *Phys. Rev. Lett.*, **49**:25 (1982), 1804–7.

Awschalom, D. D., et al. 'Macroscopic quantum tunneling in magnetic proteins'. *Phys. Rev. Lett.*, **68**:20 (1992), 3092–5.

Bacciagaluppi, Guido. 'Did Bohr understand EPR?' In Aaserud and Kragh, *Hundred years* (2015), 374–96.

Bacciagaluppi, Guido, and Antony Valentini. *Quantum theory at the crossroads. Reconsidering the 1927 Solvay Conference*. Cambridge: Cambridge University Press, 2009.

Bacciagaluppi, Guido, and Elise Crull, eds. *Grete Hermann between physics and philosophy*. Dordrecht: Springer, 2017.

Bacciagaluppi, Guido, and Elise Crull. *The Einstein paradox. The debate on non-locality and incompleteness in 1935*. Cambridge: Cambridge University Press, in press.

Badino, Massimiliano, and Jaume Navarro, eds. *Research and pedagogy: A history of quantum physics through its textbooks*. Berlin: Max-Planck-Gesellschaft zur Förderung der Wissenschaften, 2013. Edition Open Access, 2017.

Baggott, Jim. *Perfect symmetry: The accidental discovery of buckminsterfullerene*. Oxford: Oxford University Press, 1994.

Baggott, Jim. *The quantum story: A history in 40 moments*. Oxford: Oxford University Press, 2011.

Baggott, Jim. *Higgs: The invention and discovery of the 'God particle'*. Oxford: Oxford University Press, 2012.

Baggott, Jim. *Quantum reality: The quest for the real meaning of quantum mechanics – a game of theories*. Oxford: Oxford University Press, 2020.

Baggott, Jim. 'Thirty years of "Against measurement"'. *Phys. World*, **33**:12 (2020), 30–4.

Ballentine, L.E. 'The statistical interpretation of quantum mechanics'. *Rev. Mod. Phys.*, **42**:4 (1970), 358–81.

Baracca, Angelo, Silvio Bergia, and Flavio del Santo. 'The origins of research on the foundations of quantum mechanics (and other critical activities) in Italy during the 1970s'. *SHPMP*, **57** (February 2017), 66–79.

Barilier, Etienne. Albert Einstein. *L'harmonie du monde*. Lausanne: Savoir Suisse, 2011.

Barut, Asim O., Alwyn van der Merwe, and Jean-Pierre Vigier, eds. *Quantum, space and time – the quest continues : Studies and essays in honour of Louis de Broglie, Paul Dirac and Eugene Wigner*. Cambridge : Cambridge University Press, 1984.

Bavink, Bernhard. 'Die Naturwissenschaften im Dritten Reich'. *Unsere Welt*, **25** (1933), 225–36.

Bavink, Bernhard. *Science and God*. London: Bell, 1933.

Bell, J.S. 'On the Einstein-Podolsky-Rosen Paradox'. *Physics Physique Fizika*, **1**:3 (1964), 195–200, = *Speakable* (1987), 14–21.

Bell, J.S. 'On the problem of hidden variables in quantum mechanics'. *Rev. Mod. Phys.*, **38**:3 (1966), 447–52, = *Speakable* (1987), 1–13.

Bell, J.S. 'Introduction to the hidden-variable question'. In d'Espagnat, *Foundations* (1971), 172–81.

Bell, J.S. 'Locality in Quantum Mechanics: Reply to Critics'. *Epistemological Lett.*, November 1975, 2–6, = *Speakable* (1987), 63–6.

Bell, J.S. 'Testing quantum mechanics'. *Prog. Sci. Culture*, **1**:4 (1977), 439–45.

Bell, J.S. 'Bertlmann's socks and the nature of reality'. *J. de Physique*, **42** Colloque C2, Suppl. 3 (1981), 41–61, = *Speakable* (1987), 139–58.

Bell, J.S. 'On the impossible pilot wave'. *Found. Phys.*, **12**:10 (1982), 989–99, = *Speakable* (1987), 159–68.

Bell, J.S. 'Are there quantum jumps?' In Kilmister, *Schrödinger* (1987), 41–52, = *Speakable* (1987), 201–12.

Bell, J.S. 'Beables for quantum field theory'. In Hiley and Peat, *Foundations* (1987), 227–34, = *Speakable* (1987), 173–80.

Bell, J.S. *Speakable and unspeakable in quantum mechanics*. Cambridge: Cambridge University Press, 1987. Cited as *Speakable* (1987); 2nd edn., 2004.

Bell, J.S. 'Against "measurement"'. In Arthur I. Miller, ed., *Sixty-two years of uncertainty*. New York: Plenum Press, 1990, 17–31. A slightly revised version was published in *Phys. World*, **3**:8 (1990), 33–40, = *Speakable* (2004), 213–31.

Bell, J.S. In M. Bell, K. Gottfried, and M. Veltman, eds., *John S. Bell on the foundations of quantum mechanics*. Singapore: World Scientific, 2001.

Bell, J.S., and Michael Nauenberg. 'The moral aspect of quantum mechanics'. In A. De-shalit, H. Feshbach, and L. Van Hove, eds., *Preludes in theoretical physics: In honor of V.F. Weisskopf*. Amsterdam: North-Holland, 1966, 279–86, = *Speakable* (1987), 22–28.

Bell, Mary. 'Some reminiscences'. In *Q(Un)S* 1 (2002), 1–5.

Beller, Mara. 'Pascual Jordan's influence on the discovery of Heisenberg's indeterminacy principle'. *Arch. Hist. Exact Sci*, **33**:4 (1985), 337–49.

Beller, Mara. *Quantum dialogue: The making of a revolution*. Chicago, IL: University of Chicago Press, 1999.

Belloni, Franco. 'Il Congresso di Como del 1927 visutto e descritto da un ospite dell'unione Sovietica'. *Museoscienza*, **20** (1981), 14–21.

Belousek, Darrin W. 'Einstein's unpublished hidden-variable theory'. *SHPMP*, **27**:4 (1996), 437–61.

Bennett, Charles, and Giles Brassard. 'Quantum cryptography: public key distribution and coin tossing'. *International Conference on Computers, Systems, and Signal Processing*, Bangalore, 9–12 December 1984. Vol. 1, 175–9. Cited as *ICCSSP*, **1** (1984), 175–9.

Bernhard, Carl Gustav, Elizabeth T. Crawford, and Per Sörbom, eds. *Science, technology and society in the time of Alfred Nobel*. Oxford: Pergamon, 1983.

Bernstein, Jeremy. *Quantum profiles*. Princeton, NJ: Princeton University Press, 1991.

Bernstein, Jeremy. 'Max Born and the quantum theory'. *Am. J. Phys.*, **73**:11 (2005), 999–1008.

Bertlmann, Reinhold A. 'Magic moments: A collaboration with John Bell'. In *Q(Un)S*, **1** (2002), 29–47.

Bertlmann, Reinhold A., and Anton Zeilinger, eds. *Quantum (un)speakables: From Bell to quantum information*. Berlin: Springer, 2002. Cited as *Q(Un)S* **1** (2002).

Bertlmann, Reinhold A., and Anton Zeilinger, eds. *Quantum (un)speakables II: Half a century of Bell's theorem*. Berlin: Springer, 2017. Cited as *Q(Un)S* **2** (2017).

Bialobrzeski, Czeslav. 'Sur l'interprétation concrète de la mécanique quantique'. *Revue de Métaphysique et de Morale*, **41**:1 (1934), 83–103.

Bird, Kai, and Martin Sherwin. *American Prometheus. The triumph and tragedy of J. Robert Oppenheimer*. New York: Vintage, 2006.

Black, T.D., et al., eds. *Foundations of quantum mechanics*. Singapore: World Scientific, 1992.

Bloch, Felix. 'Heisenberg and the early days of quantum mechanics'. *Phys. Today*, **29**:12 (1976), 23–27.

Blum, Alexander, et al. 'Translation and heuristics: Heisenberg's turn to matrix mechanics'. *SHPMP*, **60** (November 2017), 3–22.

Bohm, David. *Quantum theory*. Englewood Cliffs, NJ: Prentice Hall, 1951; New York: Dover, 1989.

Bohm, David. 'A suggested interpretation of the quantum theory in terms of "hidden variables"'. *Phys. Rev*, **85**:2 (1952), 166–79, 180–93.

Bohm, David. 'A discussion of certain remarks by Einstein on Born's probability interpretation of the ψ-function'. In [Edinburgh,] *Scientific papers* (1953), 13–19.

Bohm, David. *Causality and chance in modern physics*. [1957] New York: Harpers, 1961.

Bohm, David. 'Quantum theory as an indication of a new order in physics. Part A. The development of new orders as shown through the history of physics'. *Found. Phys.*, **1**:4 (1971), 359–81.

Bohm, David. 'Quantum physics as an indication of a new order in physics'. In d'Espagnat, *Foundations* (1971), 412–69.

Bohm, David. 'On insight and its significance'. *Epistemologia* [Genoa], **3** (1980), 53–74.

Bohm, David. 'Hidden variables and the implicate order'. *Zygon*, **20**:2 (1985), 111–24.

Bohm, David. 'Hidden variables and the implicate order'. In Hiley and Peat, *Implications* (1987), 33–45.

Bohm, David, and J.P. Vigier. 'Model of the causal interpretation of quantum theory in terms of a fluid with irregular fluctuations'. *Phys. Rev.*, **96**:1 (1954), 208–16.

Bohm, David, and Yakir Aharonov, 'Discussion of experimental proof for the paradox of Einstein, Rosen, and Podolsky'. *Phys. Rev.*, **108**:4 (1957), 1070–6.

Bohm, David, et al. *Quanta and reality. A symposium*. American Research Council, 1962.

Bohm, David, and Charles Biedermann. In Paavo Pykkänen, ed., *Bohm-Biederman correspondence*. London: Routledge, 1999.

Bohm, David, and F. David Peat. *Science, order, and creativity*. [1987] 2nd edn. London: Routledge, 2000.

Bohr, Niels. 'Atomic theory and mechanics'. *Nature*, **116** Suppl. (5 December 1925), 845–52, = *BCW*, **5**, 273–80.

Bohr, Niels, 'Can quantum mechanical description of physical reality be considered complete?' *Phys. Rev.*, **48**:8 (1935), 696–703, = *QT&M* (1983), 145–51.

Bohr, Niels. 'The quantum postulate and the recent development of atomic theory'. *Nature*, **121** Suppl. (14 April 1928), 580–90, = *Naturwiss.*, **16**:15 (1928), 245–57, = *BCW*, **6**, 148–58,

Bohr, Niels. 'Wirkungsquantum und Naturbeschreibung'. *Naturwiss.*, **17**:26 (1929), 483–6, = *BCW*, **6**, 203–6.

Bohr, Niels. 'Kausalität und Komplementarität'. *Erkenntnis*, **6**:1 (1936), 293–303, = 'Causality and complementarity', *Phil. Sci.*, **4**:3 (1937), 289–98, in *BCW*, **10**, 39–48.

Bohr, Niels. 'Natural philosophy and human culture'. *Nature*, **143** (18 February 1939), 268–72, = *BCW*, **10**, 237–49.

Bohr, Niels. 'The causality problem in modern physics'. In *New theories in physics*. Paris: IIIC, 1939, 11–30, = *BCW*, **7**, 303–22.

Bohr, Niels. 'Discussion with Einstein on epistemological problems in atomic physics'. In Schilpp, *Einstein* (1949), 201–41.

Bohr, Niels. *Atomic physics and human knowledge*. New York: Wiley, 1958. (Essays, 1933 to 1949.)

Bohr, Niels. *Atomic theory and the description of nature*. Cambridge: Cambridge University Press, 1961.

Bohr, Niels. *Collected works*. 12 vols. Amsterdam: North-Holland, 1972–2007. Cited as *BCW*.

Born, Max. 'Quantentheorie und Störungsrechnung'. *Naturwiss.*, **11**:27 (1923), 537–42.

Born, Max. 'Über Quantenmechanik'. *Z. Phys.*, **26**:1 (1924), 379–95.

Born, Max. *Vorlesungen über Atommechanik. Erster Band*. Berlin: Springer, 1925.

Born, Max. *Problems of atomic dynamics*. Cambridge: MIT Press, 1926.

Born, Max. 'Zur Wellenmechanik der Stossvorgänge'. Gesellschaft der Wissenschaft, Göttingen, *Nachrichten* (1926), 146–60.

Born, Max. 'Zur Quantenmechanik der Stossvorgänge'. *Z. Phys.*, **37**:12 (1926), 863–7. Trans. Wheeler and Zurek, in *QT&M* (1983), 52–5.

Born, Max. 'Quantenmechanik der Stossvorgänge'., *Z. Phys.*, **38**:11/12 (1926), 803–27.

Born, Max. *The mechanics of the atom*. London: Bell, 1927. (German original, 1925.)

Born, Max. *Natural philosophy of cause and chance*. Cambridge: Cambridge University Press, 1949.

Born, Max. 'Einstein's statistical theories'. In Schilpp, *Einstein* (1949), 163–77.

Born, Max. *Physics in my generation. A selection of papers*. New York: Pergamon, 1956.

Born, Max. 'The interpretation of quantum mechanics'. *Brit. J. Phil. Sci.*, **4**:14 (1953), 95–106.

Born, Max. *Atomic physics* [1935]. 6th edn. New York: Hafner; London: Blackie, 1959.

Born, Max. *Physik im Wandel meiner Zeit*. 4th edn. Braunschweig: Vieweg, 1966.

Born, Max. *My life and my views*. New York: Scribners, 1968.

Born, Max. *Physics in my generation*. 2nd edn. New York: Springer, 1969.

Born, Max, and Werner Heisenberg. 'La mécanique des quanta'. In Solvay V, 14–81.

Born, Max, and Pascual Jordan. *Elementare Quantenmechanik*. Berlin: Springer, 1930. (*Zweiter Band der Vorlesungen über Atommechanik*.)

Born, Max, and Pascual Jordan, and Albert Einstein. *The Born Einstein letters*. Trans. Irene Born, with commentary by Max Born. London: Macmillan, 1971; reprint, Diana Buchwald, et al., eds. Basingstoke: Macmillan, 2005.

Bouwmeester, Dik, et al., 'Experimental quantum teleportation'. *Nature*, **390** (1 December 1997), 575–9.

Bouwmeester, Dik et al., 'Observation of three-photon Greenberger-Horne-Zeilinger entanglement'. *Phys. Rev. Lett.*, **82**:7 (1999), 1345–9.

Bridgman, Percy W. *The logic of modern physics*. New York: Macmillan, 1927.

Bridgman, Percy W. 'Permanent elements in the flux of present-day physics'. *Science*, **21** (10 January 1930), 19–23.

Bridgman, Percy W. 'Einstein's theories and the operational point of view'. In Schilpp, *Einstein* (1949), 333–54.

Broad, C.D. 'The present relations of science and religion'. *Philosophy: J. Brit. Inst. Phil.*, **14** (April 1939), 131–54.

Broglie, Louis de. 'La mécanique ondulatoire et la structure atomique de la matière et du rayonnement'. *J. de Physique et le Radium*, **8**:5 (1927), 225–41.

Broglie, Louis de. 'Determinisme et causalité dans la physique contemporaine'. *Revue de Métaphysique et de Morale*, **36**:4 (1929), 433–43.

Broglie, Louis de. *An introduction to the study of wave mechanics*. Trans. H.T. Flint. London: Methuen, 1930.

Broglie, Louis de. 'Sur la complementarité des idées d'individu et de système'. *Dialectica*, **2**:3/4 (1948), 325–30.

Broglie, Louis de. 'A general theory of the scientific work of Albert Einstein'. In Schilpp, *Einstein* (1949), 109–27.

Broglie, Louis de. 'Will quantum mechanics remain indeterminist?' In de Broglie. *New perspectives in physics*. New York: Basic Books, 1962, 83–107.

Broglie, Louis de. 'The reinterpretation of wave mechanics'. *Found. Phys.*, **1**:1 (1970), 5–15.

Broglie, Louis de. 'L'interprétation de la mécanique ondulatoire par la théorie de la double solution'. In d'Espagnat, *Foundations* (1971), 346–67.

Bub, Jeffrey. 'Von Neumann's theory of quantum measurement'. In Rédei and Stöltzner, *Neumann* (2001), 63–74.

Bush, Steven. 'The chimerical cat: Philosophy of quantum mechanics in historical perspective'. *Soc. Stud. Sci.*, **10**:4 (1980), 393–447.

Byrne, Peter. *The many worlds of Hugh Everett III: Multiple universes, mutually assured destruction, and the meltdown of a nuclear family*. Oxford: Oxford University Press, 2009.

Byrne, Peter. 'Everett and Wheeler: The untold story'. In Saunders, et al., *Many worlds* (2010), 521–41.

Calvino, Italo. *Why read the classics?* New York: Pantheon, 1999.

Camilleri, Kristian. 'Heisenberg, Bohr and the divergent viewpoints of complementarity'. *SHPMP*, **38**:3 (2007), 514–28.

Camilleri, Kristian. 'Constructing the myth of the Copenhagen interpretation'. *Perspect. Sci.*, **17**:1 (2009), 26–57.

Camilleri, Kristian. *Heisenberg and the interpretation of quantum mechanics*. Cambridge: Cambridge University Press, 2009.

Capra, Fritjof. *The Tao of physics*. Boulder, CO: Shambhala, 1975.

Cassidy, David. *Uncertainty. The life and science of Werner Heisenberg*. New York: Freeman, 1992.

Cassidy, David and Martha Baker. *Werner Heisenberg. A bibliography of his writings*. Berkeley: Office for History of Science and Technology, University of California, 1984.

Chu, Steven 'A random walk in science'. Art and symmetry in experimental physics. AIP Conf. Proc., 596 (2001), 21–36.

Clarke, John. 'SQUIDs'. *Sci. Am.*, **271** (August 1994), 46–53.

Clauser, John F. 'Proposed experiment to test local hidden-variable theories'. *Bull. Am. Phys. Soc.*, **14** (1969), 578.

Clauser, John F. 'Experimental investigation of a polarization correlation anomaly'. *Phys. Rev. Lett.*, **36**:21 (1976), 1223–6.

Clauser, John F. 'Early history of Bell's theorem and experiment'. In Black, *Foundations* (1992), 168–74.

Clauser, John F. 'Review of David Kaiser, *How the hippies saved physics*'. *Phys. Perspect.*, **18**:4 (2016), 395–401.

Clauser, John F. 'Early history of Bell's theorem'. *Q(Un)S* **1** (2002), 61–98.

Clauser, John F. 'Bell's theorem, Bell's inequalities, and the "probability normalization loophole"'. *Q(Un)S* **2** (2017), 451–84.

Clauser, John F. 'Laboratory-space and configuration-space formulations of quantum mechanics, versus Bell-Clauser-Horne-Shimony local realism, versus Born's ambiguity'. In Jaeger, et al., eds., *Quantum arrangements* (2021), 35–91.

Clauser, John F. et al. 'Proposed experiment to test local hidden-variable theories'. *Phys. Rev. Lett.*, **23**:15 (1969), 880–4.

Clauser, John F. and Michael A. Horne. 'Experimental consequences of objective local theories'. *Phys. Rev. D*, **10**:2 (1974), 526–34.

Clauser, John F. and Abner Shimony. 'Bell's theorem: experimental tests and implications'. *Rep. Prog. Phys.*, **41**:12 (1978), 1881–927.

Clifton, Robert K., Michael L.G. Redhead, and Jeremy N. Butterfield. 'Generalization of the Greenberger-Horne-Zeilinger algebraic proof of non-locality'. *Found. Phys.*, **21**:2 (1991), 149–84.

Cohen-Tannoudji, Claude, Bernard Diu, and Frank Laloë. *Mécanique quantique*. Paris: Hermann, 1973.

Compton, Arthur H. 'The uncertainty principle and free will'. *Science*, **74** (14 August 1931), 172–3.

Compton, Arthur H. *Atomic quest: A personal narrative*. London: Oxford University Press, 1956.

Compton, Arthur H. *Scientific papers*. Robert S. Shankland, ed. Chicago, IL: University of Chicago Press, 1973.

Congrès international de physique, Paris. *Rapports*. 3 vols, Paris: Gauthier-Villars, 1900–01.

Cornu, Alfred. 'Discours de l'ouverture'. In *Congrès* (1900–01), **1**, 5–8.

Crawford, Elisabeth. *The Nobel population 1901–1950. A census of the nominators and nominees for the prizes in physics and chemistry*. Tokyo: Universal Academy Press, 2002.

Cross, Andrew. 'The crisis of physics. Dialectical materialism and the quantum theory'. *Soc. Stud. Sci.*, **21**:4 (1991), 735–59.

Cushing, James T. 'Bohm's theory: Common sense dismissed'. *SHPS*, **24**:5 (1993), 815–42.

Cushing, James T. *Quantum mechanics: Historical contingency and the Copenhagen hegemony*. Chicago, IL: University of Chicago Press, 1994.

Cushing, James T. 'Determinism versus indeterminism in quantum mechanics: A free choice'. In Russell, et al., *Quantum mechanics* (2001), 99–110.

Cushing, James T. and Ernan McMullin, eds. *Philosophical consequences of quantum theory. Reflections on Bell's theorem*. Notre Dame: University of Notre Dame Press, 1989.

Cushing, James T. et al., eds. *Bohmian mechanics and quantum theory: An appraisal*. Dordrecht: Kluwer, 1996.

Darwin, Charles Galton. 'The uncertainty principle'. *Science*, **73** (19 January 1931), 653–60.

Dass, N.D. Hari. 'The superposition principle in quantum mechanics – did the rock enter the foundation surreptitiously?' In Aaserud and Kragh, *Hundred years* (2015), 435–49.

Davidson, Martin. *Free will or determinism?* London: Watts, 1937.

Davies, P.C.W., and J.R. Brown. *The ghost in the atom*. Cambridge: Cambridge University Press, 1986.

Delft, Dirk van. 'Paul Ehrenfest's final years'. *Phys. Today*, **61**:1 (2014), 41–7.

Denman, Henry H. 'Boris Podolsky'. *Phys. Today*, **20**:3 (1967), 141.

Deutsch, David. 'Quantum theory, the Church-Turing principle and the universal quantum computer'. *Proc. R. Soc. London A*, **400**:1818 (1985), 97–117.

Deutsch, David. 'Apart from universes'. In Saunders et al., *Many worlds?* (2010), 242–52.

Deutsch, David and Richard Jozsa. 'Rapid solution of problems by quantum computation'. *Proc. R. Soc. London A*, **439**:1907 (1992), 553–8.

DeWitt, Bryce S. 'Quantum mechanics and reality'. *Phys. Today*, **23**:9 (1970), 30–5.

DeWitt, Bryce S. 'The many-universe interpretation of quantum mechanics'. In d'Espagnat, *Foundations* (1971), 211–62.

DeWitt, Bryce S. and Neill Graham. *The many worlds interpretation of quantum mechanics*. Princeton, NJ: Princeton University Press, 1973.

Dieks, D. 'Communication by EPR devices'. *Phys. Lett. A*, **92**:6 (1982), 271–2.

Dirac, P.A.M. *The principles of quantum mechanics*. Oxford: Oxford University Press, 1930.

Dirac, P.A.M. 'The evolution of the physicist's picture of nature. An account of how physical theory has developed in the past and how, in the light of this development, it can perhaps be expected to develop in the future'. *Sci. Am.*, **208** (May 1963), 292–302.

Dirac, P.A.M. *Directions in physics*. H. Hora and J.R. Shepanski, eds. New York: Wiley, 1978.

Dongen, Jeroen van. 'Communicating the Heisenberg uncertainty relations: Niels Bohr, complementarity and the Einstein-Rupp experiments'. In Aaserud and Kragh, *Hundred years* (2015), 310–43.

Dorling, Jon. 'Shrödinger's original interpretation of the Schrödinger equation: A rescue attempt'. In Kilmister, *Schrödinger* (1987), 16–40.

Dörries, Matthias, ed. *Michael Frayn's Copenhagen in debate. Historical essays and documents on the 1941 meeting between Niels Bohr and Werner Heisenberg*. Berkeley: Office for History of Science and Technology, University of California, 2005.

Douglas, A. Vibert. *The life of Arthur Stanley Eddington*. London: Nelson, 1956.

Drake, Stillman and C.D. O'Malley. eds. *The controversy on the comets of 1618*. Philadelphia: University of Pennsylvania Press, 1960.

Duncan, Anthony, and Michael Janssen. 'Pascual Jordan's resolution of the conundrum of the wave-particle duality of light'. *SHPMP*, **39**:3 (2008), 634–66.

Dyson, Freeman. 'Innovation in physics'. *Sci. Am.*, **199** (September 1958), 74–83.

Dyson, Freeman *Maker of pattens*. New York: Liveright, 2018.

Eddington, Arthur. *The nature of the physical world*. Cambridge: Cambridge University Press, 1928.

[Edinburgh University]. *Scientific papers presented to Max Born on his retirement from the Tait Chair of Natural Philosophy*. Edinburgh: Oliver and Boyd, 1953.

Ehrenfest, Paul. *Collected scientific papers*. M.J. Klein, ed. Amsterdam: North-Holland, 1959.

Einstein, Albert. 'Strahlungs-Emission und -Absorption nach der Quantentheorie'. *Verhandlungen der deutschen physikalischen Gesellschaft*, **18**:13/14 (1916), 318–23, = *ECP*, **6**, 364–9.

Einstein, Albert. 'Zur Quantentheorie der Strahlung'. *Phys. Zeit.*, **18** (1917), 121–28, = *ECP*, **6**, 382–97.

Einstein, Albert. 'Über den Äther'. *ECP*, **14**, 515–23 (text of October 1924).

Einstein, Albert. 'Bestimmt Schrödingers Wellenmechanik die Bewegung eines Systems vollständig oder nur im Sinne der Statistik?' *ECP*, **15**, 810–14 (text of May 1927, withdrawn before publication).

Einstein, Albert. 'Physics and reality'. *J. Franklin Inst.*, **221**:3 (1936), 349–82.

Einstein, Albert. 'Quantenmechanik und Wirklichkeit'. *Dialectica*, **2**:3/4 (1948), 320–4.

Einstein, Albert. 'Autobiographical notes'. In Schilpp, *Einstein* (1949), 3–95.

Einstein, Albert. *Out of my later years*. Scranton, PA: Philosophical Library, 1950.

Einstein, Albert. 'H.A. Lorentz, his creative genius and his personality'. In G.L. de Haas-Lorentz, ed., *H.A. Lorentz: Impressions of his life and work*. Amsterdam: North-Holland, 1957, 5–9.

Einstein, Albert. 'Elementare Überlegungen zur Interpretation der Grundlagen der Quanten-mechanik'. In Edinburgh, *Scientific papers* (1953), 33–40.

Einstein, Albert. *The collected papers*. John Stachel et al., eds., 16 vols to date. Princeton, NJ: Princeton University Press, 1987–2022. Cited as *ECP*.

Einstein, Albert, and Paul Ehrenfest. 'Quantentheoretische Bemerkung zum Experiment von Stern und Gerlach'. *Z. Phys.*, **11**:1 (1922), 31–34, = *ECP*, **13**, 441–4.

Einstein, Albert, Boris Podolsky, and Nathan Rosen. 'Can quantum-mechanical description of physical reality be considered complete?' *Phys. Rev.*, **47**:10 (1935), 777–80, = *QT&M* (1983), 138–41.

Ekert, Artur K. 'Quantum cryptography based on Bell's theorem'. *Phys. Rev. Lett.*, **67**:6 (1991), 661–3.

Ellis, John, and Daniele Amati, eds. *Quantum reflections*. Cambridge: Cambridge University Press, 2000.

Emary, Clive, Neill Lambert, and Franco Nori. 'Leggett-Garg inequalities'. *Rep. Prog. Phys.*, **77**:1 (2014), 016001.

Englert, Berthold-Georg. 'Complementarity'. In Black, *Foundations* (1992), 181–92.

Englert, Berthold-Georg. 'Fringe visibility and which-way information: an inequality'. *Phys. Rev. Lett.*, **77**:11 (1996), 2154–57.

Enz, Charles P. *No time to be brief. A scientific biography of Wolfgang Pauli*. Oxford: Oxford University Press, 2001.

Esfeld, Michael. 'Essay review: Wigner's view of physical reality'. *SHPMP*, **30**:1 (1999), 145–54.

Espagnat, Bernard d', ed. *Foundations of quantum mechanics. Proceedings of the International School of Physics 'Enrico Fermi', Course IL, 29 June–11 July 1970*. New York: Academic Press, 1971.

Espagnat, Bernard d'. 'Mesure et non séparabilité'. In d'Espagnat, *Foundations* (1971), 84–96.

Espagnat, Bernard d'. *Conceptual foundations of quantum mechanics*. 2nd ed. [1976] Reading, Mass: Addison-Wesley, 1989.

Espagnat, Bernard d'. 'The quantum theory and reality'. *Sci. Am.*, **241** (November 1979), 158–81.

Espagnat, Bernard d'. *In search of reality*. New York: Springer, 1983.

Espagnat, Bernard d'. 'Meaning and being in contemporary physics'. In Hiley and Peat, *Implications* (1987), 151–68.

Espagnat, Bernard d'. *Reality and the physicist*. Cambridge: Cambridge University Press, 1989.

Espagnat, Bernard d'. 'My interaction with John Bell'. *Q(Un)S* **1** (2002), 21–8.

Espagnat, Bernard d'. 'Consciousness and the Wigner's friend problem'. *Found. Phys.*, **35**:12 (2005), 1943–66.

Everett III, Hugh. 'The Theory of the Universal Wave Function'. In DeWitt and Graham, eds. *Many worlds* (1973). (Everett's 137-page draft dissertation of 1956.)

Faye, J., and Folse, F.J., eds. *Niels Bohr and contemporary philosophy*. Dordrecht: Kluwer, 1994.

Fedak, William A., and Jeffrey J. Prentis. 'The 1925 Born and Jordan paper "On quantum mechanics"'. *Am. J. Phys.*, **77**:2 (2009), 128–39.

Feynman, Richard. 'Simulating physics with computers'. *Int. J. Theor. Phys.*, **21**:6/7 (1982), 467–88.

Feynman, Richard. 'Quantum mechanical computers'. *Found. Phys.*, **16**:6 (1986), 507–31.

Fick, Dieter, and Horst Kant. 'Walter Bothe's contributions to the understanding of the wave-particle duality of light'. *SHPMP*, **40**:4 (2009), 395–405.

Filk, Thomas. 'Carl Friedrich von Weizsäcker's "Ortsbestimmung eines Elektrons" and its influence on Grete Hermann'. In Bacciagaluppi and Crull, *Hermann* (2017), 71–83.

Fine, Arthur. *The shaky game: Einstein realism and the quantum theory*, 2nd edn. Chicago, IL: University of Chicago Press, 1996.

Forman, Paul. 'Weimar culture, causality, and quantum theory, 1918-1927: Adaptation by German physicists and mathematicians to a hostile intellectual environment'. *HSPS*, **3** (1971), 1–115.

Frank, Lawrence K. 'Causation: An episode in the history of thought'. *J. Phil.*, **31**:16 (1934), 421–8.

Frank, Philipp. 'Philosophische Deutungen und Misdeutungen der Quantentheorie'. *Erkenntnis*, **6**:1 (1936), 303–17.

Franz, Marie-Louise. *Number and time: Reflections leading toward a unification of depth psychology and physics*. Evanston: Northwestern University Press, 1974.

Frappier, Mélanie. '"In the no-man's-land between physics and logic": On the dialectical role of the microscope experiment'. In Bacciagaluppi and Crull, *Hermann* (2017), 85–105.

Freire Jr., Olival. 'On the historical roots of "Foundations of quantum physics" as a field of research'. *Found. Phys.*, **34**:11 (2004), 1741–60.

Freire Jr., Olival. *The quantum dissidents. Rebuilding the foundations of quantum mechanics (1950–1990)*. Heidelberg: Springer, 2015.

Freire Jr., Olival. *David Bohm. A life devoted to understanding the quantum world*. Cham: Springer, 2019.

Freire Jr., Olival. 'Alain Aspect's experiments on Bell's theorem: A turning point in the history of the research on the foundations of quantum mechanics'. *The European Physics Journal D*, **76**:248 (2022), 1–8.

Freire Jr., Olival. ed. *The Oxford companion of the history of quantum interpretations*. Oxford: Oxford University Press, 2022.

Freitas, Fábio. 'Tony Leggett's challenge to quantum mechanics and its path to decoherence'. In Freire, *Oxford companion* (2022), 495–520.

French, A.P., and P.J. Kennedy, eds. *Niels Bohr. A centenary volume*. Cambridge: Harvard University Press, 1985.

Freedman, Stuart J., and John F. Clauser. 'Experimental test of local hidden-variable theories'. *Phys. Rev. Lett.*, **28**:14 (1972), 938–41.

Friedman, Jonathan R., et al. 'Quantum superposition of distinct macroscopic states'. *Nature*, **406** (6 July 2000), 43–6.

Fry, Edward S, and Shifong Li. 'Bell inequalities with nearly 100% efficient detectors'. In Black, *Foundations* (1992), 175–80.

Furry, Wendell H. 'Note on the quantum mechanical theory of measurement'. *Phys. Rev.*, **49**:5 (1936), 393–9, 476.

Gamow, George. *Thirty years that shook physics. The story of quantum theory*. New York: Doubleday, 1966; New York: Dover, 1985.

Gavroglu, Kostas. *Fritz London. A scientific biography*. Cambridge: Cambridge University Press, 1995.

George, André, ed. *Louis de Broglie physicien et penseur*. Paris: Albin Michel, 1953.

Gerlach, Walter. 'Der Solvay-Kongress 1930'. *Metallwirtschaft, Wissenschaft und Technik*, **9** (1930), 939–42, 965–67, 1003–6.

Gerlich, Stephen, et al., 'Quantum interference of large organic molecules'. *Nature Comm.*, **2** (5 April 2011), 1–5.

Gerry, Christopher C., and Kimberley M. Bruno. *The quantum divide: Why Schrödinger's cat is either dead or alive*. Oxford: Oxford University Press, 2013.

Gilder, Louisa. *The age of entanglement: When quantum physics was reborn*. New York: Vintage, 2009.

Gisin, Nicolas. *Quantum chance. Nonlocality, teleportation and other marvels*. Cham: Springer, 2014.

Gisin, Nicolas, and B. Gisin. 'A local hidden variable model of quantum correlation exploiting the detection loophole'. *Phys. Lett. A*, **260**:5 (1999), 323–7.

Giuliani, Domenico. 'Max Born's *Vorlesungen über Atommechanik, Erster Band*'. In Badino and Navarro, *Research and pedagogy* (2017), 203–26.

Giuntini, Roberto, and Federico Laudisa. 'The impossible causality: The no hidden variables theorem of John von Neumann'. In Rédei and Stöltzner, *Neumann* (2012), 173–88.

Gottfried, Kurt. 'Does quantum mechanics carry the seeds of its own destruction?' In Ellis and Amati, *Reflections* (2000), 165–85.

Gottfried, Kurt, and N. David Mermin. 'John Bell and the moral aspects of quantum mechanics'. In Ellis and Amati, *Reflections* (2000), 186–92.

Gowing, Margaret. *Britain and atomic energy, 1939–1949*. London: Macmillan, 1964.

Graham, Loren. 'Complementarity and Marxism-Leninism'. In French and Kennedy, *Bohr* (1985), 332–41.

Grangier, P., G. Roger, and A. Aspect. 'Experimental evidence for a photon anti-correlation effect on a beamsplitter: A new light on single-photon interferences'. *Europhys. Lett.*, **1**:4 (1986), 173–9.

Greenberger, Daniel M. 'The History of the GHZ Paper'. *Q(Un)S*, **1** (2002), 281–6.

Greenberger, Daniel M., and Allaine Yasin. 'A "haunted" version of the Einstein-Podolsky-Rosen experiment'. In Namiki, *Proceedings* (1987), 18–24.

Greenberger, Daniel M., and Allaine Yasin, 'Simultaneous wave and particle knowledge in a neutron interferometer'. *Phys. Lett. A*, **128**:8 (1988), 391–4.

Greenberger, Daniel M., Michael Horne, and Anton Zeilinger. 'Going Beyond Bell's Theorem'. In Kafatos, ed. *Bell's theorem* (1989), 73–6.

Greenberger, Daniel M., et al. 'Bell's theorem without inequalities'. *Am. J. Phys.*, **58**:12 (1990), 1131–43.

Gröblacher, Simon, et al. 'An experimental test of non-local realism'. *Nature*, **446** (19 April 2007), 871–5.

Gross, Eugene P. 'Collective variables in elementary quantum mechanics'. In Hiley and Peat, *Implications* (1987), 46–65.

Handsteiner, Johannes, et al. 'Cosmic Bell test: measurement settings from Milky Way stars'. *Phys. Rev. Lett.*, **118**:6 (2017), 060401.

Hansen-Schaberg, Inge. 'A biographical sketch of Prof. Dr Grete Henry-Hermann (1901–1984)'. In Bacciagaluppi and Crull, *Hermann* (2017), 3–16.

Hanson, Norwood Russell. 'Copenhagen interpretation of quantum physics'. *Am. J. Phys.*, **27**:1 (1959), 1–15.

Hartshorne, Charles. 'Contingency and the new era in metaphysics'. *J. Phil.*, **29**:16 (1932), 421–31, 457–69.

Hartz, Thiogo, and Olival Freire, Jr. 'Use and approximations of Niels Bohr's ideas about quantum field measurement'. In Aaserud and Kragh, eds. *Hundred years*, (2015), 397–418.

Harvey, Alex, ed. *On Einstein's path*. New York: Springer, 1999.

Haynes, John Earl, Harvey Klehr, and Alexander Vassiliev. *Spies: The rise and fall of the KGB in America*. New Haven: Yale University Press, 2009.

Hazelton, Roger. 'Human purpose and cosmic purpose'. *J. Phil.*, **36**:24 (1939), 656–66.

Heilbron, J.L. 'Robert Symmer and the two electricities'. *Isis*, **67**:1 (1976), 7–20.

Heilbron, J.L. 'Fin-de-siècle physics'. In Bernhard et al., *Science* (1982), 51–73.

Heilbron, J.L. 'The origins of the exclusion principle'. *HSPS*, **13**:2 (1983), 261–310.

Heilbron, J.L. 'The earliest missionaries of the Copenhagen spirit'. *Rev. Hist. Sci.*, **38**:3/4 (1985), 195–230.

Heilbron, J.L. *The dilemmas of an upright man. Max Planck as spokesman for German science*, 2nd edn. Cambridge: Harvard University Press, 2000.

Heilbron, J.L. 'Oppenheimer's guru'. In Cathryn Carson and David Hollinger, eds., *Reappraising Oppenheimer*. Berkeley: Office for History of Science and Technology, University of California, 2005, 275–91.

Heilbron, J.L. *Niels Bohr. A very short introduction*. Oxford: Oxford University Press, 2020.

Heims, Steve. *John von Neumann and Norbert Wiener*. Cambridge: MIT Press, 1980.

Heisenberg, Elisabeth. *Das politische Leben eines Unpolitischen. Erinnerungen an Werner Heisenberg*. Munich: Piper, 1980.

Heisenberg, Werner. 'Zur Quantentheorie der Multiplettstruktur und der anomalen Zeeman Effekte'. *Z. Phys.*, **32**:1 (1925), 841–60.

Heisenberg, Werner. 'Über quantentheoretische Umdeutung kinematischer und mechanischer Beziehungen'. *Z. Phys.*, **33**:1 (1925), 879–93.

Heisenberg, Werner. 'Über den anschaulichen Inhalt de quantentheoretischen Kinematik und Mechanik'. *Z. Phys.*, **43**:3/4 (1927), 172–98. English translation by Wheeler and Zurek in *QT&M* (1983), 62–86.

Heisenberg, Werner. *The physical principles of the quantum theory*. Chicago, IL: University of Chicago Press, 1930.

Heisenberg, Werner. 'Kausalgesetz und Quantenmechanik'. *Erkenntnis*, **2**:1 (1931), 172–82, = *HGW:C* I, 29–39.

Heisenberg, Werner. 'Prinzipien Fragen der modernen Physik'. In *Neuere Fortschritte in den exakten Naturwissenschaften*. Leipzig/Vienna: F. Deuticke, 1936, 91–102, = *HGW:C* I, 108–19.

Heisenberg, Werner. 'Die Begriffabgeschlossene Theorie in der modernen Naturwissenschaft'. *Dialectica*, **2**:3/4 (1948), 331–6.

Heisenberg, Werner. 'The development of the interpretation of the quantum theory'. In W. Pauli, L. Rosenfeld, and V. Weisskopf, eds., *Niels Bohr and the development of physics: Essays dedicated to Niels Bohr on the occasion of his seventieth birthday*. New York: McGraw-Hill; London: Pergamon, 1955, 12–29.

Heisenberg, Werner. *Physics and philosophy. The revolution in modern science*. New York: Harper & Row, 1958; 2nd edn, 1962.

Heisenberg, Werner. *Physics and beyond. Encounters and conversations*. New York: Harper & Row, 1971.

Heisenberg, Werner. *Gesammelte Werke*, Abt C. *Allgemeinverständliche Schriften. Physik und Erkenntnis*. 3 vols. Munich: Piper, 1984. Cited as *HGW:C*.

Heisenberg, Werner. *Liebe Eltern! Briefe aus kritischer Zeit 1918 bis 1945*. Munich: Langen Müller, 2003.

Hentschel, Klaus. 'Bernhard Bavink (1879–1947)'. *Sudhoffs Archiv*, **77** (1993), 1–32.

Herbert, Nick. *Quantum reality: Beyond the new physics*. Garden City, NY: Doubleday, 1985.

Hermann, Grete. 'Determinism and quantum mechanics'. (Manuscript of 1933). Trans. in Bacciagaluppi and Crull, *Hermann* (2017), 223–37.

Hermann, Grete. 'Die natur-philosophischen Grundlagen der Quantenmechanik [1935]'. Trans. in Bacciagaluppi and Crull, *Hermann* (2017), 239–78.

Hermann, Grete. 'Die naturphilosphischen Grundlagen der Quantenmechanik'. *Naturwiss*, **23**:42 (1935), 718–21, = Bacciagaluppi and Crull, *Hermann* (2017), 271–80.

Hermann, Grete. 'Zum Vortrag Schlicks'. *Erkenntnis*, **6**:1 (1936), 342–3.

Herzog, Thomas, et al. 'Complementarity and the quantum eraser'. *Phys. Rev. Lett.*, **75**:17 (1995), 3034–7.

Higgs, Peter W. 'Broken symmetries and the masses of gauge bosons'. *Phys. Rev. Lett.*, **13**:16 (1964), 508–9.

Hiley, Basil J., and F. David Peat, eds. *Quantum implications. Essays in honour of David Bohm*. London: Routledge & Kegan Paul, 1987.

Hnizdo, V. 'On Bohr's response to the clock-in-the-box thought experiment of Einstein'. *Eur. J. Phys.*, **23**:4 (2002), L9–L13.

Hoddeson, Lillian, et al., eds. *The rise of the standard model: Particle physics in the 1960s and 1970s*. Cambridge: Cambridge University Press, 1997.

Hofstadter, Richard. *Anti-intellectualism in American life*. New York: Vintage, 1966.

Holland, Peter. 'Jean-Pierre Vigier at seventy-five: La lutte continue'. *Found. Phys.*, **25**:1 (1995), 1–4.

Holland, Peter. 'What's wrong with Einstein's 1927 hidden-variable interpretation of quantum mechanics?', *Found. Phys.*, **35**:2 (2005), 177–96.

Horgan, John. *The end of science*. London: Little, Brown, 1996.

Hove, Léon van. 'Von Neumann's contributions to quantum theory'. *Bull. Am. Math. Soc.*, **65** (1958), 95–9.

Howard, Don. '"Nicht sein kann was nicht sein darf", or the prehistory of EPR, 1909–1935: Einstein's early worries about the quantum mechanics of compound systems'. In Miller, *Sixty-two years* (1990), 61–111.

Howard, Don. 'What makes a classical concept classical?' In Faye and Folse, eds. *Niels Bohr* (1994), 201–9.

Howard, Don. 'Who invented the Copenhagen interpretation? A study in mythology'. *Phil. Sci.*, **71**:5 (2004), 669–82.

Howard, Don. 'Revisiting the Einstein-Bohr dialogue'. *Iyyun: The Jerusalem Philosophical Quarterly*, **56** (January 2007), 57–90.

Howard, Don. 'Quantum mechanics in context: Pascual Jordan's 1936 Anschauliche Quantentheorie'. In Badino and Navarro, *Research* (2013), 265–83.

Howard, Don. 'The Copenhagen interpretation'. In Freire, Oxford companion (2022), 521–42.

HUAC. Committee on Un-American Activities, House of representatives, Eighty-Third Congress, First Session, February 25, 1953. *Stenographic record of hearing*. Washington, D.C.: Hart and Hopkins Shorthand and Stenographic reporting, 1953.

Im, Gyeong Soon. 'Experimental constraints on formal quantum mechanics: The emergence of Born's quantum theory of collision processes in Göttingen, 1924–1927'. *Arch. Hist. Exact Sci.*, **44**:1 (1996), 73–101.

Inge, William Ralph. *God and the astronomers*. London: Longmans Green, 1933.

Institut International de Physique Solvay. *Electrons et photons: Rapports et discussions du cinquième conseil de physique . . . 1927*. Paris: Gauthier-Villars, 1928. Cited as Solvay V.

Institut International de Physique Solvay. *Le magnétisme. Rapports et discussions du sixième conseil de physique . . . 1930*. Paris: Gauthier-Villars, 1931. Cited as Solvay VI.

Jacobsen, Anja Skaar. 'Léon Roenfeld's Marxist defense of complementarity'. *HSPS*, **37** Suppl. (2007), 3–34.

Jaeger, Gregg, Abner Shimony, and Lev Vaidman. 'Two interferometric complementarities'. *Phys. Rev. A*, **51**:1 (1995), 54–67.

Jaeger, Gregg, David Simon, Alexander V. Sergienko, Daniel Greenberger, and
Anton Zeilinger, eds. *Quantum arrangements: Contributions in honor of Michael
Horne*. Cham: Springer, 2021.

Jammer, Max. *The conceptual development of quantum mechanics*. New York: McGraw-
Hill, 1966.

Jammer, Max. *The philosophy of quantum mechanics*. New York: Wiley, 1974.

Jammer, Max. *Einstein and religion: Physics and theology*. Princeton, NJ: Princeton
University Press, 2002.

Jarrett, Jon P. 'On the physical significance of the locality conditions in the Bell
arguments'. *Noûs*, **18**:4 (1984), 569–89.

Jauch, J.M. 'Foundations of quantum mechanics'. In d'Espagnat, *Foundations* (1971),
20–55.

Jauch, J.M., and C. Piron. 'Can hidden variables be excluded in quantum
mechanics?' *Helv. Phys. Acta*, **36**:7 (1963), 470–85.

Jordan, Pascual. 'Philosophical foundations of quantum theory'. *Nature*, **119**
(16 April 1927), 566–9.

Jordan, Pascual. 'Quantentheoretische Bemerkungen zur Biologie und Psychologie'. *Erkenntnis*, **4**:1 (1934), 215–52.

Jordan, Pascual. 'Quantenmechanik und die Grundprobleme der Biologie und
Psychologie'. *Naturwiss.*, **20**:45 (1934), 815–21.

Jordan, Pascual. 'Über die positivistischen Begriff der Wirklichkeit'. *Naturwiss.*,
22:29 (1934), 485–90.

Jordan, Pascual. *Anschauliche Quantentheorie*. Berlin: Springer, 1936.

Jordan, Pascual. *Die Physik des 20. Jahrhunderts*. Braunschweig: Vieweg, 1936.

Jordan, Pascual. *Physics of the 20th century*. New York: Philosophical Library, 1944.

Jordan, Pascual. *Die Physik und das Geheim des organischen Lebens*. 4th edn.
Braunschweig: Vieweg, 1945.

Jordan, Pascual. *Verdrängen und Komplementarität*. Hamburg: Stromverlag, 1948.

Jordan, Pascual. *Science and the course of history. The influence of scientific research on
human events*. New Haven, CT: Yale University Press, 1955.

Jordan, Pascual. 'Die Anfangsjahre der Quantenmechanik—Erinnerungen'. *Phys.
Blätter*, **31**:3 (1975), 97–103.

Jung, Carl Gustav. *Psychology and alchemy*. In Jung, *Collected works*. 20 vols.
Princeton, NJ: Princeton University Press, 1953–1979. Vol. 12, 39–223.

Kafatos, Minas, ed. *Bell's theorem, quantum theory, and conceptions of the universe*.
Dordrecht: Kluwer, 1989.

Kaiser, David. 'Cold-War requisitions, scientific manpower, and the production of
American physicists after World War II'. *HSPS*, **33**:1 (2002), 131–59.

Kaiser, David. 'Turning physicists into quantum mechanics'. *Phys. World*, **20**:5 May
(2007), 28–33.

Kaiser, David. *How the hippies saved physics. Science, counterculture, and the quantum
revival*. New York: Norton, 2011.

Kaiser, David. *Quantum legacies. Dispatches from an uncertain world.* Chicago, IL: University of Chicago Press, 2020.

Kasday, Leonard Ralph. 'Experimental test of quantum predictions for widely separated photons'. In d'Espagnat, *Foundations* (1971), 195–210.

Kasday, Leonard Ralph. 'The distribution of Compton scattered annihilation photons and the Einstein-Podolsky-Rosen argument'. PhD thesis. Columbia University, 1971.

Kemble, Edwin C. 'The general principles of quantum mechanics, Part 1'. *Phys. Rev. Suppl.*, **1**:2 (1929), 1–241.

Kemble, Edwin C. 'Operational reasoning, reality, and quantum mechanics'. *J. Franklin Inst.*, **225**:3 (1938), 263–75.

Kilmister, C.W., ed. *Schrödinger: Centenary celebration of a polymath.* Cambridge: Cambridge University Press, 1987.

Kirsten, Christa, and Hans-Günther Körber, eds. *Wahlvorschläge zur Aufnahme von Physiker in der Berliner Akademie 1870 bis 1929.* Berlin: Akademie-Verlag, 1975.

Knee, George C., et al. 'A strict experimental test of macroscopic realism in a superconducting flux qubit'. *Nature Comm.*, **7** (4 November 2016), 1–5.

Kojevnikov, Alexei. *The Copenhagen network. The birth of quantum mechanics from a postdoctoral perspective.* Cham, Switzerland: Springer, 2020.

Komar, Arthur. 'The physical reality of the quantum wave function'. In Harvey, *Einstein's path* (1999), 275–81.

Konno, Hiroyuki. 'Kramers' negative dispersion: The virtual oscillator model, and the correspondence principle'. *Centaurus*, **36**:2 (1993), 117–66.

Körner, Stephen, ed. *Observation and interpretation in the philosophy of physics with special reference to quantum mechanics.* Proceedings of the ninth symposium of the Colston Research Society. London: Butterworth, 1957; New York: Dover, 1962.

Kožnjak, Boris. 'The missing history of Bohm's hidden variables theory: The ninth symposium of the Colston Research Society'. *SHPMP*, **62** (May 2018), 85–97.

Kragh, Helge. 'Erwin Schrödinger and the wave equation: The crucial phase'. *Centaurus*, **26**:2 (1982), 154–97.

Kragh, Helge. *Quantum generations: A history of physics in the twentieth century.* Princeton, NJ: Princeton University Press, 1999.

Kragh, Helge. *Neils Bohr and the quantum atom: The Bohr model of atomic structure 1913–1925.* Oxford: Oxford University Press, 2012.

Kragh, Helge. 'Paul Dirac and *The principles of quantum mechanics*'. In Badino and Navarro, *Research* (2017), 245–60.

Kramers, H.A. 'The quantum theory of dispersion'. *Nature*, **114** (30 August 1924), 310–11, Van der Waerden, *Sources* (1967), 199–201.

Kramers, H.A. *Quantum mechanics.* Trans. Dirk ter Haar. Amsterdam: North-Holland, 1958. This comprises the two parts of the German original, Leipzig: Akademische Verlagsgesellscaft, 1938: 'Die Grundlagung der Quantentheorie' (1933) and 'Quantentheorie des Elektrons und der Strahlung'.

Kuhn, Thomas S. *The structure of scientific revolutions.* Chicago: University of Chicago Press, 1962.

Kuhn, Thomas S. et al. *Sources for history of quantum physics. An inventory and report.* Philadelphia: American Philosophical Society, 1967.

Landau, L.D., and E.M. Lifschitz. *Quantum mechanics. Non-relativistic theory.* Trans. J.B. Sykes and J.S. Bell. London: Pergamon, 1958.

Landé, Alfred. 'Unity in quantum theory'. *Found. Phys.*, **1**:3 (1971), 191–202.

Lao, Tzu. *Tao te chung.* Trans. Victor M. Mair. New York: Quality Paperback, 1998.

Laurikainen, Kalervo V. *Beyond the atom: The philosophical thought of Wolfgang Pauli.* Berlin: Springer, 1988.

Leggett, A.J. 'Review: *Criticism and the growth of knowledge*'. *Second Order: An African Journal of Philosophy*, **1**:2 (1972), 80.

Leggett, A.J. 'Macroscopic quantum systems and the quantum theory of measurement'. *Prog. Theor. Phys. Suppl.*, **69** (March 1980), 80–100.

Leggett, A.J. 'Reflections on the quantum measurement problem'. In Hiley and Peat, *Implications* (1987), 85–104.

Leggett, A.J. 'The current status of quantum mechanics at the macroscopic level'. In Namiki, *Proceedings* (1987), 287–97.

Leggett, A.J. *The problems of physics.* Oxford: Oxford University Press [1987], 2000.

Leggett, A.J. 'Nonlocal hidden-variable theories and quantum mechanics: an incompatibility theorem'. *Found. Phys.*, **33**:10 (2003), 1469–93.

Leggett, A.J. Les Prix Nobel. *The Nobel Prizes 2003.* Tore Frängsmyr, ed. [Nobel Foundation], Stockholm, 2004.

Leggett, A.J. 'Probing quantum mechanics towards the everyday world: How far have we come?'. *Prog. Theor. Phys. Suppl.*, **170** (May 2007), 100–18.

Leggett, A.J. 'Realism and the physical world'. In Struppa and Tollaksen, *Quantum theory* (2014), 9–20.

Leggett, A.J. and Anupam Garg. 'Quantum mechanics versus macroscopic realism: Is the flux there when nobody looks?'. *Phys. Rev. Lett.*, **54**:9 (1985), 857–60.

Lenin, V.I. *Materialism and empirio-criticism. Critical comments on a reactionary philosophy.* In Lenin, *Collected works*, **14**. Moscow: Progress Publishers, 1968.

Lenzen, Victor F. 'Indeterminism and the concept of physical reality'. *J. Phil.*, **30**:11 (1933), 281–8.

Lenzen, Victor F. 'The interaction between subject and object in observation'. *Erkenntnis*, **6**:1 (1936), 326–45.

Lenzen, Victor F. 'Einstein's theory of knowledge'. In Schilpp, *Einstein* (1949), 357–84.

London, Fritz., and Edmond Bauer. *La théorie de l'observation en mécanique quantique.* Paris: Hermann, 1939. English translations (with a new paragraph by London) by Shimony, Wheeler and Zurek, and J. McGrath and S. McLean McGrath, are reconciled in *QT&M* (1983), 217–59.

Ludwig, G. 'The measuring process and the axiometric foundation of quantum mechanics'. In d'Espagnat, *Foundations* (1971), 287–315.

Ma, Xiao-song, et al. 'Quantum erasure with causally disconnected choice'. *PNAS*, **110**:4 (2013), 1221–26.

Ma, Xiao-song, Johannes Kofler, and Anton Zeilinger. 'Delayed-choice gedanken experiments and their realizations'. *Rev. Mod. Phys.*, **88**:1 (2016), 015005.

Maccone, Lorenzo. 'A simple proof of Bell's inequalities'. *Am. J. Phys.*, **81**:11 (2013), 854–9.

MacKinnon, Edward. 'Bohr on the foundations of quantum theory'. In French and Kennedy, *Bohr* (1985), 101–20.

Margenau, Henry. 'Probability and causality in quantum physics'. *The Monist*, **42**:2 (1932), 166–88.

Marletto, C., et al. 'Entanglement between living bacteria and quantized light witnessed by Rabi splitting'. *J. Phys. Commun.*, **2** (10 October 2018), 101001.

McCrea, William. 'Eamon de Valera, Erwin Schrödinger and the Dublin Institute'. In Kilmister, *Schrödinger* (1987), 119–35.

Means, Blanchard W. 'Freedom, determinacy, and value'. *J. Phil.*, **33**:4 (1936), 85–95.

Mehlhop, Werner, and Oreste Piccioni. 'The EPR experiment is a most interesting puzzle, now more than ever'. In Namiki, *Proceedings* (1987). New York: Springer, 72–78.

Mehra, Jagdish, and Kimball A. Milton. *Climbing the mountain. The scientific biography of Julian Schwinger*. Oxford: Oxford University Press, 2000.

Mermin, N. David. 'Is the moon there when nobody looks? Reality and the quantum theory'. *Phys. Today*, **38**:4 (1985), 38–47.

Mermin, N. David. 'Can you help your team tonight by watching on TV? More experimental metaphysics from Einstein, Podolsky, and Rosen'. In Cushing and McMullin, *Consequences* (1989), 38–59.

Mermin, N. David. 'What's wrong with this pillow?' *Phys. Today*, **42**:4 (1989), 9–11.

Mermin, N. David. 'What's wrong with these elements of reality?' *Phys. Today*, **43**:6 (1990), 9–11.

Mermin, N. David. 'Hidden variables and the two theorems of John Bell'. *Rev. Mod. Phys.*, **65**:3 (1993), 803–15.

Mermin, N. David. 'Could Feynman have said this?' *Phys. Today*, **57**:5 (2004), 10–11.

Mermin, N. David. 'Quantum mysteries revisited'. *Am. J. Phys.* **58**:8 (1998), 731–4.

Mermin, N. David. 'Whose knowledge?'. In *Q(Un)S* I (2002), 271–80.

Meyenn, Karl von. 'Pauli, Schrödinger und der Streit um die Deutung der Quantenmechanik'. *Gesnerus*, **44**:1/2 (1987), 99–123.

Miller, Arthur I., ed. *Sixty-two years of uncertainty*. New York: Plenum, 1990.

Moody, David Edmund. *An uncommon collaboration. David Bohm and J. Krishnamurti*. Ojai, CA: Alpha Centauri, 2017.

Moore, Walter. *Schrödinger: life and thought*. Cambridge: Cambridge University Press, 1989.

Motterlini, Matteo, ed. *For and against method: Imre Lakatos, Paul Feyerabend*. Chicago, IL: University of Chicago Press, 1999.

Nagel, Ernest, *On the logic of measurement*. New York: the author, 1930. (PhD Thesis, Columbia University.)

Nairz, Olaf, Markus Arndt, and Anton Zeilinger. 'Experimental verification of the Heisenberg uncertainty principle for fullerene molecules'. *Phys. Rev. A*, **65**:3, (2002), 032109.

Namiki, Mikio, et al., eds. *Proceedings of the 2nd International Symposium, Foundations of quantum mechanics in the light of new technology* [1986]. Tokyo: Physical Society of Japan, 1987.

Nathansen, Henri. *Jude oder Europäer? Porträt von Georg Brandes*. Frankfurt am Main: Rütten & Loening, 1931.

Newton, Isaac. *Mathematical principles of natural philosophy*. Trans. Andrew Motte (1729), ed. Florian Cajori. Berkeley, CA: University of California Press, 1934.

Newton, Isaac. *Correspondence*. 7 vols. Cambridge: Cambridge University Press, 1959–1977.

Olwell, Russell. 'Physical isolation and marginalization in physics. David Bohm's cold war exile'. *Isis*, **90**:4 (1999), 738–56.

Oppenheimer, J. Robert. *The open mind*. New York: Simon and Schuster, 1963.

Osnaghi, Stefano, Fábio Freitas, and Olival Freire Jr. 'The Origin of the Everettian Heresy', *SHPMP*, **40**:2 (2009), 97–123.

Pais, Abraham. *'Subtle is the Lord': The science and life of Albert Einstein*. Oxford: Oxford University Press, 1982.

Pais, Abraham. *Inward bound: Of matter and forces in the physical world*. Oxford: Oxford University Press, 1986.

Pais, Abraham. *Niels Bohr's times: In physics, philosophy, and polity*. Oxford: Oxford University Press, 1991.

Pais, Abraham. *J. Robert Oppenheimer: A life*. Oxford: Oxford University Press, 2006.

Pan, Jian-Wei, et al. 'Experimental test of quantum nonlocality in three-photon Greenberger-Horne-Zeilinger entanglement'. *Nature*, **403** (3 February 2000), 515–19.

Pauli, Wolfgang. 'Allgemeine Grundlagen der Quantentheorie des Atombaus [1929]'. In Pauli, *Coll. sci. papers*, **1**, 626–769. Reprinted from *Müller-Pouillets Lehrbuch*, **2**:2 (1929), 1709–842.

Pauli, Wolfgang. 'Die allgemeine Prinzipien der Wellenmechanik [1933]'. In Pauli, *Coll. sci. papers*, **1**, 771–938. Reprinted from *Handbuch der Physik*, S. Flügge, ed., **5**:1 (1958), 1–168.

Pauli, Wolfgang. 'Editorial'. *Dialectica*, **2**:3/4 (1948), 307–11.

Pauli, Wolfgang. 'Einstein's contributions to quantum theory'. In Schilpp, *Einstein* (1949), 149–60.

Pauli, Wolfgang. 'Remarques sur le problème des paramètres cachés dans la mécanique quantique et sur la théorie de l'onde pilote'. In George, *De Broglie* (1953), 39–42.

Pauli, Wolfgang. *Collected scientific papers*. Ralph Kronig and V.F. Weisskopf, eds., 2 vols. New York: Wiley, 1964.

Pauli, Wolfgang. *Physik und Erkenntnistheorie*. Karl von Meyenn, ed. Braunschweig: Vieweg, 1984.

Pauli, Wolfgang. *Wissenschaftlicher Briefwechsel*. 5 vols. Karl von Meyenn, et al., eds. Berlin: Springer, 1979–2005. Cited as *PWB*.

Pauli, Wolfgang. *Writings on physics and philosophy*. Charles P. Enz and Karl von Meyenn, eds. trans. Robert Schlapp. Berlin: Springer, 1994.

Peacock, George. *Life of Thomas Young*. London: J. Murray, 1855.

Peat, F. David. *Infinite potential. The life and times of David Bohm*. Reading, MA: Addison-Wesley, 1997.

Peierls, Rudolf. *Selected private and scientific correspondence*. Sabine Lee, ed., 2 vols. Singapore: World Scientific, 2009.

Penrose, Roger. *The emperor's new mind*. London: Vintage, 1990.

Penrose, Roger. 'On gravity's role in quantum state reduction'. *Gen. Rel. Grav.*, **28**:5 (1996), 581–600.

Penrose, Roger. *The large, the small and the human mind*. Cambridge: Cambridge University Press, 2000.

Peres, Asher. 'How the no-cloning theorem got its name'. *Fort. der Physik*, **51**:4/5 (2003), 458–61.

Petersen, Aage. *Quantum physics and the philosophical tradition*. Cambridge, MA: MIT Press, 1968.

Philippidis, C., C. Dewdney, and B.J. Hiley. 'Quantum interference and the quantum potential'. *Nuovo Cimento*, **52B**:1 (1979), 15–28.

Phillips, William D., and Jean Dalibard. 'Experimental tests of Bell's inequalities: A first-hand account by Alain Aspect'. *Eur. Phys. J. D* **77**:8 (2023), 1–14.

Pinch, Trevor. 'What does a proof do if it does not prove? A study of the social conditions and metaphysical divisions leading to David Bohm and John von Neumann failing to communicate in quantum physics'. In Everett Mendelsohn, et al., eds., *The social production of scientific knowledge*. Dordrecht: Reidel, 1977, 171–215.

Piron, C. 'Observables in general quantum theory'. In d'Espagnat, *Foundations* (1971), 374–86.

Planck, Max. *Wege zur physikalischen Erkenntnis*. Leipzig: Hirzel, 1933.

Planck, Max *Die Physik im Kampf um die Weltanschauung*. Leipzig: Barth, 1935.

Popper, Karl R. *Quantum theory and the schism in physics: Postscript to the logic of scientific discovery*. W.W. Bartley III, ed. London: Unwin Hyman, 1982.

Prosperi, G.M. 'Macroscopic physics and the problem of measurement in quantum mechanics'. In d'Espagnat, *Foundations* (1971), 97–126.

Przibram, Karl, ed. *Schrödinger, Planck, Einstein, Lorentz: Briefe zur Wellenmechanik*. Vienna: Springer, 1963.

Pusey, Matthew F., Jonathan Barrett, and Terry Rudolph, 'On the reality of the quantum state'. *Nature Physics*, **8** (2012), 475–8.

Pylkkänen, Paavo. 'The role of Eastern approaches in David Bohm's philosophical odysseia'. *Prog. Biophys. Mol. Bio.*, **131** (December 2017), 171–8.

Rauch, Dominik, et al. 'Cosmic Bell test using random measurement settings from high-redshift quasars'. *Phys. Rev. Lett.*, **121**:8 (2018), 080403.

Rauch, Helmut, et al. 'Verification of coherent spinor rotation of fermions'. *Phys. Lett. A*, **54**:6 (1975), 425–7.

Rauch, Helmut. 'Reality in neutron interference experiments'. In Ellis and Amati, *Reflections* (2000), 28–68.

Rayski, Jerzy. 'Problematic interpretation of quantum mechanics'. In Namiki, *Proceedings* (1987), 282–6.

Rédei, Miklós, and Michael Stöltzner, eds. *John von Neumann and the foundations of quantum physics*. Dordrecht: Kluwer, 2001.

Ren, Ji-Gang, et al. 'Ground-to-satellite quantum teleportation'. *Nature*, **549** (9 August 2017), 70–3.

Roebke, Joshua. 'The reality tests'. *Seed: Science is Culture*, **16** (June 2008), 50–9. Although the magazine ceased publication in 2012, this article is archived online at https://web.archive.org/web/20090324045345/http://seedmagazine. com/content/article/the_reality_tests/

Robertson, H.P. 'The uncertainty principle'. *Phys. Rev.*, **34**:1 (1929), 163–4.

Romano, Luigi. 'Franco Selleri and his contributions to the debate on particle physics, foundations of quantum mechanics and foundations of relativity theory'. PhD thesis, University of Bari, 2020.

Rosenfeld, Léon. 'L'évidence de la complémentarité'. In George, *De Broglie* (1953), 43–65.

Rosenfeld, Léon. 'Physics and metaphysics'. *Nature*, **181** (8 March 1958), 658.

Rosenfeld, Léon. 'Heisenberg, physics and philosophy'. *Nature*, **186** (11 June 1960), 830–1.

Rosenfeld, Léon. 'Niels Bohr's contributions to epistemology'. *Phys. Today*, **16**:10 (1963), 47–54.

Rosenfeld, Léon. 'The epistemological conflict between Bohr and Einstein'. *Z. Phys.*, **171**:1 (1963), 242–45.

Rosenfeld, Léon. 'Niels Bohr in the thirties. Consolidation and extension of the concept of complementarity'. In Rozental, *Niels Bohr* (1967), 114–36.

Rosenfeld, Léon 'The evolution of the idea of causality [1942]'. In Léon Rosenfeld. *Selected papers*. R.S. Cohen and J.J. Stachel, eds. Dordrecht: Reidel, 1979, 446–64.

Rowe, M.A., et al. 'Experimental violation of a Bell's inequality with efficient detection'. *Nature*, **409** (15 February 2001), 791–4.

Rozental, Stefan. 'The forties and fifties'. In Rozental, *Niels Bohr* (1967), 149–90.

Rozental, Stefan. *Niels Bohr. His life and work as seen by his friends and colleagues*. Amsterdam: North-Holland, 1967.

Russell, Robert John, et al., eds. *Quantum mechanics. Scientific perspectives on divine action*, Vol 5. Vatican City: Vatican Observatory, 2001.

Sachs, Mendel. *Einstein versus Bohr. The continuing controversies in physics.* La Salle, IL: Open Court, 1988.

Samuel, Herbert. 'New science and old philosophy'. *Phil.*, **11**:41 (1936), 3–15.

Saunders, Simon, et al., eds. *Many worlds? Everett, quantum theory, and reality.* Oxford University Press, 2010.

Sayen, Jamie. *Einstein in America.* New York: Crown, 1985.

Schiff, Leonard J. *Quantum mechanics.* New York: McGraw-Hill, 1949.

Schilpp, Paul Arthur, ed. *Albert Einstein—philosopher scientist.* Evanston, IL: Library of Living Philosophers, 1949.

Schlick, Moritz. 'Quantentheorie und Erkennbarkeit der Natur'. *Erkenntnis*, **6**:1 (1936), 317–26.

Schroer, Bert. 'Pascual Jordan, his contributions to quantum mechanics and his legacy in contemporary local quantum physics'. arXiv: hep-th/0303241v2 15 May 2003.

Schrödinger, Erwin. 'Discussion of probability relations between separated systems'. Cambridge Philosophical Society, *Proceedings*, **31**:4 (1935), 555–63.

Schrödinger, Erwin. 'Probability relations between separated systems'. Cambridge Philosophical Society, *Proceedings*, **32**:3 (1936), 446–52.

Schrödinger, Erwin. 'Die gegenwärtigen Situation in der Quantenmechanik'. *Naturwiss.*, **23**:48 (1935), 807–12; **23**:49 (1935), 823–8; **23**:50 (1935), 844–9. Trans. John D. Trimmer in *Proc. Am. Phil. Soc.*, **124**:5 (1980), 323–38, = *QT&M* (1983), 152–67.

Schrödinger, Erwin. 'Die Besonderheit des Weltbildes der Naturwissenschaften'. *Acta Phys. Austriaca*, **1** (1948), 201–45.

Schrödinger, Erwin. 'What is an elementary particle?' *Endeavour*, **9**:35 (1950), republished in Smithsonian Institution, Board of Regents, *Annual report*, 1950, 183–96.

Schrödinger, Erwin. 'Are there quantum jumps?' *Brit. J. Phil. Sci.*, Part I, **3**:10 (1952), 109–23; Part II, **3**:11 (1952), 233–42.

Schrödinger, Erwin. 'The meaning of wave mechanics'. In George, *De Broglie* (1953), 16–32.

Schrödinger, Erwin. *Collected papers on wave mechanics.* 3rd (augmented) English edn. Providence, RI: AMS Chelsea, 1982.

Schrödinger, Erwin. *The interpretation of quantum mechanics: Dublin seminars (1949–1955) and other unpublished essays.* Woodbridge, CT: Ox Bow Press, 1995.

Schücking, Engelbert L. 'Jordan, Pauli, politics, Brecht, and a variable gravitational constant'. *Phys. Today*, **52**:10 (1999), 26–31.

Schweber, Silvan S. *QED and the men who made it: Dyson, Feynman, Schwinger, and Tomonaga.* Princeton, NJ: Princeton University Press, 1994.

Schweber, Silvan S. *Einstein and Oppenheimer: The meaning of genius.* Cambridge: Harvard University Press, 2008.

Scully, Marlan O., and Kai Drühl. 'Quantum eraser: A proposed photon correlation experiment concerning observation and "delayed choice" in quantum mechanics'. *Phys. Rev. A*, **25**:4 (1982), 2208–13.

Seevinck, Michiel. 'Challenging the gospel: Grete Hermann on von Neumann's no-hidden variable proof'. In Bacciagaluppi and Crull, *Hermann* (2017), 107–17.

Selleri, Franco. 'Realism and the wave-function of quantum physics'. In d'Espagnat, *Foundations* (1971), 398–406.

Selleri, Franco. 'Sull'ideologia nella fisica contemporanea'. *Critica Marxista. Quaderni*, **6** (1972), 120–50.

Selleri, Franco. *Quantum paradoxes and physical reality*. Alwyn van der Merwe, ed. Dordrecht: Kluwer, 1990.

Serber, Robert. 'The early years'. *Phys. Today*, **20**:10 (1967), 35–39.

Seth, Suman. '*Zweideutigkeit* about "Zweideutigkeit": Sommerfeld, Pauli, and the methodological origins of quantum mechanics'. *SHPMP*, **40**:4 (2009), 303–15.

Seth, Suman. *Crafting the quantum. Arnold Sommerfeld and the practice of theory, 1890–1926*. Cambridge: MIT Press, 2010.

Shankland, R.S. 'Conversations with Einstein'. *Am. J. Phys.*, **31**:1 (1963), 47–57.

Shiedl, Thomas, et al. 'Violation of local realism with freedom of choice'. *PNAS*, **107**:46 (2010), 19708–13.

Shimony, Abner. 'Role of observer in quantum mechanics'. *Am. J. Phys.*, **31**:10 (1963), 755–73.

Shimony, Abner 'Experimental test of local hidden-variable theories'. In d'Espagnat, *Foundations* (1971), 192–4.

Shimony, Abner 'Philosophical comments on quantum mechanics'. In d'Espagnat, *Foundations* (1971), 470–80.

Shimony, Abner 'Wigner on foundations of quantum mechanics'. In Eugene Wigner. *Collected works*, Vol. A:3. *Particles and fields, foundations of quantum mechanics*. Berlin: Springer, 1997, 401–14.

Shimony, Abner. 'The reality of the quantum world'. In Russell et al., *Quantum mechanics* (2001), 3–16.

Shimony, Abner. 'John S. Bell: Some reminiscences and reflections'. *Q(Un)S* **1** (2002), 51–60.

Shor, P.W. 'Algorithms for quantum computation: Discrete logarithms and factoring'. *Proc. 35th Annual Symposium on Foundations of Computer Science*. Shafi Goldwasser, ed. IEEE Computer Society Press, 1994, 124–34.

Silva, Indiana. 'Chien-Shiung Wu's contributions to experimental philosophy'. In Freire, *Oxford companion* (2022), 735–54.

Singh, Rajinder, and Falk Riess. 'Belated Nobel Prize for Max Born'. *Indian J. Hist. Sci.*, **48**:1 (2013), 79–104.

Slater, John C. 'Quantum physics in America between the wars'. *Phys. Today*, **21**:1 (1968), 43–51.

Soler, Léna. 'The convergence of transcendental philosophy and quantum physics: Grete Henry-Hermann's 1935 pioneering proposal'. In Bacciagaluppi and Crull, *Hermann* (2017), 55–69.

Solovey, Mark. *Social science for what? Battles over public funding for the 'other sciences' at the National Science Foundation*. Cambridge: MIT Press, 2021.

Sommerfeld, Arnold. *Atombau und Spektrallinien* [1919], 2nd edn. Braunschweig: Vieweg, 1921.

Stapp, Henry Pierce. 'The Copenhagen interpretation'. *Am. J. Phys.*, **40**:8 (1972), 1098–116.

Stapp, Henry Pierce. 'Bell's theorem and world process'. *Nuovo Cimento B*, **29**:2 (1975), 270–6.

Stapp, Henry Pierce. 'Light as a foundation of being'. In Hiley and Peat, *Foundations* (1987), 255–66.

Steyer, Daniel F., et al. 'Nine formulations of quantum mechanics'. *Am. J. Phys.*, **70**:3 (2002), 288–97.

Stoeger, William. 'Epistemological and ontological issues arising from quantum theory'. In Russell, et al., *Quantum mechanics* (2001), 81–98.

Struppa, Daniele, and Jeffrey M. Tollaksen, eds. *Quantum theory: A two-time success story. Yakir Aharonov Festschrift*. Milan: Springer, 2014.

Szilard, Leo. 'Über die Entropieverminderung in einem thermodynamischen System bei Eingriffen intelligenter Wesen', *Z. Phys.*, **53**:11/12 (1929), 840–56; 'On entropy reduction in a thermodynamic system by interference by intelligent beings'. NASA: Technical Translation F-16,723 (1976).

Talbot, Chris. *David Bohm: Causality and chance, letters to three women*. Springer International, 2017.

Taylor, E.F. 'The anatomy of a collaboration'. In John R. Klauder, ed. *Magic without magic: John Archibald Wheeler*. San Francisco: Freeman, 1972, 475–85.

Teller, Paul. 'The projection postulate and Bohr's interpretation of quantum mechanics'. *Proc. Biennial Meeting Phil. Sci. Assoc.*, **2** (1980), 201–23.

Thomas, Kelly Devine. 'T.S. Eliot at the Institute for Advanced Study'. *The Institute Letter*, Spring 2007.

Thomas, Kelly Devine. 'The Advent and Fallout of EPR'. *The Institute Letter*, Fall 2013.

Tittel, W., et al. 'Violations of Bell inequalities by photons more than 10 km apart'. *Phys. Rev. Lett.*, **81**:17 (1998), 3563–6.

Toqueville, Alexis de. *Democracy in America*. 2 vols. Trans. Henry Reeve. New edn. London: Longmans, 1875.

Torre, A.C. de la, A. Daleo, and I. Garcia-Mata. 'The photon-box Bohr-Einstein debate demythologized'. *Eur. J. Phys.*, **21**:3 (2000), 253–60.

Tyrrell, G.N.M. 'Physics and the ontological program'. *Phil.*, **7**:28 (1932), 404–12.

Van der Wal, Caspar H., et al. 'Quantum superposition of macroscopic persistent-current states'. *Science*, **290** (27 October 2000), 773–7.

Van Kampen, Nicolaas Godfried. 'Ten theorems about quantum mechanical measurements'. *Physica A*, **153**:1 (1988), 97–113.

Vigier, J.P., et al. 'Causal particle trajectories and the interpretation of quantum mechanics'. In Hiley and Peat, *Implications* (1987), 169–204.

Vogel, Heinrich. *Zum philosophischen Wirken Max Plancks*. Berlin: Akademie Verlag, 1961.

Von Neumann, John von. *Mathematical foundations of quantum mechanics* [1932]. Trans. Robert T. Beyer. Princeton, NJ: Princeton University Press, 1955.

Waerden, Bartel L. van der, ed. *Sources of quantum mechanics*. New York: Dover, 1967.

Wallace, David. 'Decoherence and ontology'. In Saunders, *Many worlds?* (2010), 53–72.

Washer, Charles T., and Glenn A. Slack. 'Who named the -ON's?' *Am. J. Phys.*, **38**:12 (1970), 1380–9.

Weihs, Gregor, et al. 'Violation of Bell's inequality under strict Einstein locality conditions'. *Phys. Rev. Lett.*, **81**:23 (1998), 5039–43.

Weinberg, Steven. *Dreams of a final theory*. London: Vintage, 1993.

Weiner, Charles, ed. *History of twentieth century physics*. New York: Academic Press, 1977. (Proceedings of the International School of Physics 'Enrico Fermi', Course LVII, 1972.)

Weiner, Charles. 'The foundations of quantum mechanics'. *Phys. Today*, **17**:1 (1964), 53–60. (Account of the conference at Xavier University.)

Weizsäcker, Carl Friedrich von. 'Ortsbestimmung eines Elektrons durch ein Mikroscop'. *Z. Phys.*, **70**:1/2 (1931), 114–30.

Weizsäcker, Carl Friedrich von. 'C.F. von Weizsäcker über sein Studium in Leipzig'. *NTM*, **1** (1993), 3–18.

Wheeler, J.A. 'Polyelectrons'. New York Academy of Sciences, *Annals*, **48**:3 (1946), 219–38.

Wheeler, J.A. 'Physics in Copenhagen in 1934 and 1935'. In Aaserud and Kragh, *Hundred years* (2015), 221–6.

Wheeler, J.A. *At home in the universe*. Woodbury, CT: AIP Press, 1994.

Wheeler, J.A., and K.W. Ford. *Geons, black holes, and quantum foam: A life in physics*. New York: Norton. 1998.

Wheeler, J.A., and Wojciech H. Zurek. *Quantum theory and measurement*. Princeton, NJ: Princeton University Press, 1983. Cited as *QT&M*.

Whitaker, Andrew. 'John Bell in Belfast: Early years and education'. In *Q(Un)S* **1** (2002), 7–20.

Whitaker, Andrew. *John Stewart Bell and twentieth-century physics: Vision and integrity*. Oxford: Oxford University Press, 2016.

Wigner, Eugene Paul. 'The limits of science'. American Philosophical Society, *Proceedings*, **94**:5 (1950), 422–7.

Wigner, Eugene Paul. 'Die Messung quantummechanischer Operatoren'. *Z. Phys.*, **133**:1/2 (1952), 101–08.

Wigner, Eugene Paul. 'Remarks on the Mind–Body Question'. In I. J. Good, ed. *The scientist speculates: an anthology of partly-baked ideas*. London: Heinemann, 1961, 284–302, = *Symmetries* (1970), 171–84, = *QT&M* (1983), 168–81.

Wigner, Eugene Paul. 'The problem of measurement'. *Am. J. Phys.*, **31**:1 (1963), 6–15, = *Symmetries* (1970), 153–70, = *QT&M* (1983), 324–41.

Wigner, Eugene Paul. 'Two kinds of reality'. *The Monist*, **48**:2 (1964), 248–64.

Wigner, Eugene Paul. *Symmetries and reflections. Scientific essays* [1967]. Cambridge: MIT Press, 1970.

Wigner, Eugene Paul. 'On hidden variables and quantum mechanical probabilities'. *Am. J. Phys.*, **38**:8 (1970), 1005–8.

Wigner, Eugene Paul. 'Physics and the explanation of life'. *Found. Phys.*, **1**:1 (1970), 35–45.

Wigner, Eugene Paul 'The subject of our discussion'. In d'Espagnat, *Foundations* (1971), 1–19.

Wigner, Eugene Paul. *Recollections as told to Andrew Szanton*. New York: Plenum, 1992.

Wise, M. Norton. 'Pascual Jordan: Quantum mechanics, psychology, National Socialism'. In Monika Renneberg and Mark Walker, eds., *Science, technology, and National Socialism*. Cambridge: Cambridge University Press, 1994, 224–54.

Wootters, W. K., and W. H. Zurek. 'A single quantum cannot be cloned'. *Nature*, **299** (28 October 1982), 802–3.

Wu, C.S., and I. Shaknov. 'The angular correlation of scattered annihilation radiation'. *Phys. Rev.*, **77**:1 (1950), 136.

Yang, Chen Ning. 'Square root of minus one, complex phases and Erwin Schrödinger'. In Kilmister, *Schrödinger* (1987), 53–64.

Yang, Chen Ning. 'Complex phases in quantum mechanics'. In Namiki, *Proceedings* (1987), 181–4.

Yin, Juan, et al. 'Satellite-based entanglement distribution over 1200 kilometers'. *Science*, **356** (16 June 2017), 1140–4.

Young, Thomas. *A course of lectures on natural philosophy and the mechanical arts* [1807]. Philip Kelland, ed., 2 vols. London: Taylor and Walton, 1845.

Zeh, H. Dieter. 'On the interpretation of measurement in quantum theory'. *Found. Phys.*, **1**:1 (1970), 69–76, = *QT&M* (1983), 342–9.

Zeh, H. Dieter. 'On the irreversibility of time and observation in quantum theory'. In d'Espagnat, *Foundations* (1971), 263–73.

Zeilinger, Anton. 'Quantum teleportation'. *Sci. Am.*, **282** (April 2000), 50–9.

Zeilinger, Anton. *Dance of the photons. From Einstein to quantum teleportation*. New York: Farrar, Strauss and Giroux, 2010.

Zinkernagel, Henrik. 'Are we living in a quantum world? Bohr and quantum fundamentalism'. In Aaserud and Kragh, *Hundred years* (2015), 419–34.

Zukav, Gary. *The dancing Wu Li masters: An overview of the new physics*. New York: Morrow, 1979.

Zurek, Wojciech. 'Quantum jumps, Born's rule, and objective reality'. In Saunders, *Many worlds?* (2010), 409–32.

Index

For the benefit of digital users, indexed terms that span two pages (e.g., 52–53) may, on occasion, appear on only one of those pages.

Note: references to figures are indicated by an italic *f* following the page number.